Applying the Classification of Finite Simple Groups

A User's Guide

Mathematical
Surveys
and
Monographs

Volume 230

Applying the Classification of Finite Simple Groups

A User's Guide

Stephen D. Smith

American Mathematical Society
Providence, Rhode Island

2010 *Mathematics Subject Classification.* Primary 20-02, 20D05, 20Bxx, 20Cxx, 20Exx, 20Gxx, 20Jxx.

For additional information and updates on this book, visit
www.ams.org/bookpages/surv-230

Library of Congress Cataloging-in-Publication Data

Names: Smith, Stephen D., 1948– author.
Title: Applying the classification of finite simple groups: A user's guide / Stephen D. Smith.
Other titles: Classification of finite simple groups
Description: Providence, Rhode Island: American Mathematical Society, [2018] | Series: Mathematical surveys and monographs; volume 230 | Includes bibliographical references and index.
Identifiers: LCCN 2017044767 | ISBN 9781470442910 (alk. paper)
Subjects: LCSH: Finite simple groups. | Representations of groups. | AMS: Group theory and generalizations – Research exposition (monographs, survey articles). msc | Group theory and generalizations – Abstract finite groups – Finite simple groups and their classification. msc | Group theory and generalizations – Permutation groups – Permutation groups. msc | Group theory and generalizations – Representation theory of groups – Representation theory of groups. msc | Group theory and generalizations – Structure and classification of infinite or finite groups – Structure and classification of infinite or finite groups. msc | Group theory and generalizations – Linear algebraic groups and related topics – Linear algebraic groups and related topics. msc | Group theory and generalizations – Connections with homological algebra and category theory – Connections with homological algebra and category theory. msc
Classification: LCC QA177 .S645 2018 | DDC 512/.2–dc23
LC record available at https://lccn.loc.gov/2017044767

Printed in the United States of America.

∞ The paper used in this book is acid-free and falls within the guidelines
established to ensure permanence and durability.
Visit the AMS home page at `http://www.ams.org/`

10 9 8 7 6 5 4 3 2 1 23 22 21 20 19 18

To Judy—for exemplary patience

Contents

Preface

Most mathematicians are at least aware of the Classification of Finite Simple Groups (CFSG), a major project involving work by hundreds of researchers. The work was largely completed by about 1983, though final publication of the "quasithin" part was delayed until 2004. Since the 1980s, the result has had a huge influence on work in finite group theory and in many adjacent fields of mathematics. This book attempts to survey and sample a number of topics from the very large and increasingly active research area of *applications* of the CFSG.

The book cannot hope to systematically cover all applications. Indeed, the particular applications chosen were mainly provided by contacting colleagues who are experts in many related areas. The resulting collection of material might seem somewhat scattered; however, I have tried to present the choices within contexts of wider areas of applications, and I am grateful to the referee for helpful suggestions in that direction.

Origin of the book and structure of the chapters

This book began life as a series of 10 two-hour lectures that I was invited to give during the September 2015 Venice Summer School on Finite Groups.

The material in the book has been adapted only fairly lightly from that lecture format, mainly in order to try to preserve the introductory, and comparatively informal, tone of the original. The primary difference in the book form has been to try to bring the cross-references and the literature citations up to the more formal level expected of a published reference book.

I will mention briefly one particular feature of the chapters in the book, arising from the lectures. Each original lecture was divided into two parts: typically, the first part introduced some basic theory from a particular aspect of the CFSG, and the second ("applications") part then demonstrated ways in which that theory can be put to use. The 10 lectures have now become the 10 chapters in this book format. The reader will notice, even from the Table of Contents, that the later sections of each chapter usually reflect the applications focus of the second part of the original lecture.

Some notes on using the book as a course text

The material is intended to be accessible to an audience with basic mathematical training: for example, to beginning graduate students with at least an undergraduate course in abstract algebra. Preferably, the background should also include first-year graduate algebra, especially some experience with examples of the most familiar types of groups and, ideally, at least the basics of Lie algebras. It will also be helpful, at times, to have some exposure to various adjacent areas of mathematics: for example, the fundamentals of algebraic topology and homological algebra, and possibly also a little combinatorics.

Of course, the original audience for the lectures varied from early graduate students through postdoctoral researchers; this variation led to the inclusion, during the lectures, of some additional explicit background material and references at various levels.

The style of presentation is deliberately fairly informal: for example, statements of some results (and even some definitions) are given with a warning such as "roughly". The purpose is to communicate mainly just the overall flavor of the original. These approximations are normally accompanied by fuller references to the precise statements, since one main motivation for the survey is to get the reader interested enough, for at least some of the topics, to pursue the details—and maybe even to look for research problems.

The book has retained the exercises from the original lectures, typically without providing solutions. Very often, these exercises come directly after a more fully worked-out example; the exercise is then to mimic that work for some other groups so that the example in effect provides a "hint" for the exercise. Furthermore, some of the exercises also provide explicit hints, including reference to sources where similar material is worked out. Finally, some exercises are detailed further in Appendix B.

The Appendix also provides some supplementary material to the text. Much of this was generated during the lectures, in response to student questions, and was originally provided to the students via pdf files on the Web, as the lectures progressed.

The Index is intended to be substantial enough to help indicate where the main ideas (and relevant papers) are used in practice. In particular, the most substantial applications of many entries are indicated in **boldface**.

Boldface is also often used in the Index to indicate definitions, but there is some variation in the level of these definitions: Fairly standard background concepts, likely to be familiar to most readers, may be indicated in passing within the text or recalled via a footnote. Definitions which are fairly brief may be indicated in a LaTeX equation environment, while longer definitions appear in a more formal definition environment. Finally, as noted earlier, definitions which are only roughly approximated in the text should also be accompanied by a reference to an appropriate source for the full details.

Acknowledgments

I am deeply indebted to the organizers of the Summer School: Mario Mainardis, Clara Franchi, John van Bon, and Rebecca Waldecker. In particular, they made many helpful suggestions during the preparation and delivery of the lectures, as well as further urging me to publish the material afterwards. I emphasize that this post-course-updated version of the material also owes very much to the contributions of the students during the course.

During the preparation of the original lectures, I consulted many colleagues with deep expertise in applications of the CFSG (since by contrast I had personally been more active in the classification itself, rather than applications). They were most generous in directing me to their favorite applications and in improving my outsider's description of their work. I particularly mention Michael Aschbacher, Russell Blyth, Persi Diaconis, David Green, Jesper Grodal, Bob Guralnick, Jon Hall, Derek Holt, Bill Kantor, Bob Oliver, Cheryl Praeger, Gary Seitz, Ron Solomon, Gernot Stroth, and P. H. Tiep, though many others also provided recommendations and assistance.

In an area as broad as applications of the CFSG, I had to select a comparatively small number of topics. I apologize to readers in advance if I have omitted their favorite applications.

Finally, I am grateful to the referees for very constructive recommendations.

A note on certain references. As a primary reference on the CFSG, I am mainly using the recent "CFSG outline" book:

Aschbacher–Lyons–Smith–Solomon [**ALSS11**].

For convenience of reference, within the lecture-course format, I tended to refer to this central source for many results, rather than to the corresponding original papers in the literature. (If the reader in fact requires those original references, they can in turn be found in the corresponding areas of [**ALSS11**].)

While many of the applications described in this book were suggested to me by experts in other areas, I also chose some from areas more well known to me; in particular, it was convenient for me to refer at various times to some of my other publications, notably:

- the quasithin classification: Aschbacher-Smith [**AS04a, AS04b**];
- the Quillen Conjecture: Aschbacher-Smith [**AS93**];
- a survey of subgroup complexes: Smith [**Smi11**].

I am grateful to Sergei Gelfand of the AMS for allowing me to provide the students with temporary online access to these during the 2015 Venice Summer School.

Stephen D. Smith

CHAPTER 1

Background: Simple groups and their properties

Of course the long-term goal of the CFSG is to be used in subsequent *applications*—in the many problems which reduce to simple groups.

For that purpose, we will first need the *list* of the simple groups in the CFSG; but equally, we will need various *properties* of the groups in that list—and indeed these features provide the main avenue for making those applications. So this initial chapter gives the list of simple groups, and describes their basic properties.

Our first remark is perhaps obvious, but is important throughout this book:

REMARK 1.0.1 (Use of the *groups* in the CFSG-list—regarded as an application). (1) The CFSG proof uses induction on group order: namely for G a counterexample to the CFSG of minimal order, and a proper subgroup $H < G$, we may assume by induction that the simple composition factors of H are "known"—that is, that they are among the conclusion groups in the CFSG-list (see Theorem 1.0.2 below). Such a group H, with known composition factors, is called a *$\mathcal{K}$-group* in the literature; where the $\mathcal{K}$ refers to that list of known groups.[1]

Thus we see that intermediate results within the CFSG proof itself can be regarded as "applications", at least of the CFSG-*list*; and so are analogous with applications of the proved-CFSG.

(2) And indeed more generally: we will also typically regard as an "application" any work which references the groups in the CFSG-list and their properties; even though that work might not necessary invoke the final statement of the CFSG itself—namely that the list is *complete*. $\diamondsuit$

That said, let's get started—with the fundamental CFSG-list:

Introduction: Statement of the CFSG—the list of simple groups

Here is the usual summary-form [**ALSS11**, Thm 0.1.1] of the CFSG:

THEOREM 1.0.2 (CFSG). *A nonabelian*[2] *finite simple group G is one of:*
i) An alternating group $A_n (n \geq 5)$.
ii) A group of Lie type. (*"Most" simple groups are of Lie type.*)
iii) One of 26 sporadic *groups.*

[1]In much of the literature, $\mathcal{K}$ stands for the known *simple* groups. But in the GLS "revisionism" series, $\mathcal{K}$ also includes their *quasi*simple covers (cf. [**GLS94**, 21.2]). To prevent confusion, here we will avoid using the abbreviation "$\mathcal{K}$", and instead write "the CFSG-list 1.0.2" for the known simple groups.

[2]Warning: Often when I write "simple", I implicitly mean "nonabelian simple"—of course, the abelian simple groups are just those of prime order. Ideally this nonabelian-intention will always be clear from the context.

Note that the three classes in this summary don't even include the names of the groups. So in the next three corresponding sections, we expand on the above summary—not just with their names, but also with a description of some of their most basic properties.

As a standard source for properties of the simple groups (especially in this chapter), I use "GLS3"—Gorenstein-Lyons-Solomon [**GLS98**]. But also Wilson's excellent book [**Wil09**] will be frequently cited. For elementary group theory, I often use Aschbacher's book [**Asc00**]; as well as GLS2—Gorenstein-Lyons-Solomon [**GLS96**].

What are some kinds of basic properties of a simple group G that we will be interested in? To mention a few:

- subgroups (maximal; p-local; elementary abelian; p-rank $m_p(-)$; etc)
- extensions; outer automorphisms $\mathrm{Out}(G)$; Schur multipliers; etc)
- representations (permutation; linear—characteristics 0, and p; etc.)

This chapter will introduce some of those properties—typically with few details, but indicating further references. Later chapters will develop various properties more deeply, as needed.

So now, let's turn to exploring the various types of simple groups in the CFSG Theorem 1.0.2.

1.1. Alternating groups

Recall that A_n consists of the even permutations of n points, and that it is a normal subgroup of index 2 in the full symmetric group S_n of all permutations.

Most mathematicians encounter these groups early on, typically in Galois theory; and many probably regard them as fairly familiar, and easy to work with. So we will frequently use A_n to provide comparatively elementary examples of various concepts.

More typically, it is often convenient to instead use the full symmetric group S_n, which is in fact *almost-simple*, in the sense of later Definition 1.4.7—in effect, this means a simple group extended only by some outer automorphisms. It is usually easy enough to adjust observations from larger S_n back to the simple subgroup A_n.

REMARK 1.1.1 (Some properties of A_n and S_n). For fuller details, see sources such as [**GLS98**, Sec 5.2] or [**Wil09**, Ch 5]. Here we extract here just a few fairly standard observations;

(1) multiple transitivity—cf. Definition (1.6.1): S_n is n-transitive, and A_n is $(n-2)$-transitive, on the n permuted points. Indeed S_n is transitive on partitions of n of any fixed type. See later Section 1.6 for more on multiple transitivity.

(2) p-rank: See [**GLS98**, 5.2.10.a]. For odd p, $m_p(S_n) = m_p(A_n) = \lfloor \frac{n}{p} \rfloor$. This can be seen by partitioning n into parts of size p, as far as possible.

(3) Maximal subgroups: These are more subtle—see later Section 6.1. Indeed we'll give an informal preview of the analysis at the beginning of Chapter 6: The "obvious" maximals arise as stabilizers of various *structures* on the n points; for example, a $(k, n-k)$-partition is stabilized by a subgroup $S_k \times S_{n-k}$. Also, p-local subgroups can sometimes be maximal. But beyond these, simple groups can also arise as maximals—in an "unpredictable" way. The maximal subgroups are in fact described in the O'Nan-Scott Theorem, which we discuss later as Theorem 6.1.1. $\diamondsuit$

EXERCISE 1.1.2 (2-rank). How does $m_p(-)$ in 1.1.1(2) become more complicated for $p = 2$? Hint: Consult e.g. [**GLS98**, 5.2.10.b] and its proof, if needed. ♢

EXERCISE 1.1.3 (other structures?). As in 1.1.1(3), find (when possible!) some other structures and stabilizers, in small examples—such as S_5, S_6, S_7, S_8, etc.

Hint: There are no others for the prime degress $n = 5, 7$. But for $n = 6$, a partition $(2, 2, 2)$ is stabilized by the *wreath product*[3] $S_2 \wr S_3$. This equal-sizes partition provides a system of *blocks of imprimivity*, in the standard language of later Definition 6.0.1. An exhaustive list of the possible structures can of course be derived using the O'Nan-Scott Theorem 6.1.1. ♢

1.2. Sporadic groups

In the usual ordering of the simple groups in the CFSG-list 1.0.2, the Lie-type groups would come next. But we had informally remarked, in that statement, that "most" simple groups have Lie type; and we will wish to devote a fair amount of time in this chapter to the very rich structure available in that important model-case of Lie-type groups. So first, in this section, we will present our much briefer discussion of the sporadic groups.

The term *sporadic* here means just: not fitting into any natural *infinite* family (such as the alternating groups; or the linear groups; or the orthogonal groups; etc). That is, we combine the 26 sporadics as one "case" in the CFSG; but in effect, they give 26 separate exceptional cases.

Hence for reasons of space, we won't attempt to provide any substantial details on their individual properties. For fuller reference, we point the reader to standard sources: we have mainly used [**GLS98**, Sec 5.3]; but cf. also Wilson [**Wil09**, Ch 5], and the Atlas [**CCN$^+$85**], and Griess [**Gri98**], and Aschbacher [**Asc94**].

Here we will instead just select a few sample properties, to try to give some of the flavor of sporadic groups.

REMARK 1.2.1 (Naming conventions). We mention first of all that these groups are usually just named after their discoverers, and denoted by an abbreviation of that name; with a subscript in the case of an author discovering several groups: for example, Co_2 for the second Conway group (see below).

But the literature also contains a number of variant-notations; typically given by adding the name of the researcher who first gave an explicit construction of the group: For example, the group discovered by Harada is most frequently denoted by HN—reflecting in addition the construction of the group by Norton.

The convention of this book is usually to use the shorter name—such as J_2 for the second Janko group, rather than HaJ indicating the construction by M. Hall. The reader moving on to the primary literature can consult basic sources such as the Atlas [**CCN$^+$85**] or [**GLS94**, Table I, pp 8–9] for further variant-notations. ♢

[3]Recall for B a permutation group of degree d, the wreath product $A \wr B$ means a direct product of d copies of A—which are then permuted naturally by B.

In spite of their individuality, there are connections among various sporadic groups. In fact, some of them can be viewed as belonging to "families":

The Mathieu groups: $M_{11} < M_{12}$ and $M_{22} < M_{23} < M_{24}$. These arise as permutation groups, of the degree indicated by the subscript. In the context of multiple transitivity (1.6.1), they are exceptional in that their action is 3-, 4-, or 5-transitive. But they can also be instructively viewed in their actions on *Steiner systems* $\mathcal{S}(5, 6, 12)$ and $\mathcal{S}(5, 8, 24)$; and on the (extended) perfect ternary and binary *Golay codes*. For these terms and other details, see the discussion of these groups in e.g. the Atlas [**CCN**$^+$**85**].

We mention that the Mathieu groups were discovered in the 1860s; and no more sporadic groups were found for about a century. All the others, described below, were discovered in an intense burst of activity during the 1960s and 1970s.

The Conway groups: $\{Co_3, Co_2\} < Co_1$. These arise from subgroups of the automorphism group of the 24-dimensional *Leech lattice*; for this lattice and details, again see the discussion of these groups in the Atlas [**CCN**$^+$**85**, p 180]. The subgroups Co_3 and Co_2 in fact arise as stabilizers of vectors in the lattice with (suitably-normalized) lengths 3,2. Indeed sub-stabilizers inside them give one possible approach to some further sporadic groups indicated below, namely HS and McL. Furthermore Co_1 also involves as sections[4] some more of the sporadic groups, namely M_{12}, M_{24}, J_2, and Suz.

The Fischer groups: Fi_{22}; $Fi_{23} < Fi_{24}'$. Here the subscript n in F_n indicates that F_n contains a 2-local subgroup of structure[5] $V_n : M_n$—where M_n indicates the Mathieu group M_{22}, M_{23}, or M_{24}; acting on an elementary-abelian 2-subgroup V_n, which arises as a suitable section of the *cocode* module. This module is the quotient of 24-space over $\mathbb{F}_2$ by the 12-dimensional subspace given by the Golay code, which we indicate in our discussion leading up to later Example 3.3.14.

These groups are exceptional in being the only almost-simple groups G, other than the symmetric groups and some classical matrix groups, to be generated by a conjugacy class C of:

(1.2.2) 3-*transpositions*: for $x, y \in C$, xy has order 1, 2, or 3.

Some other sporadic groups also arose via related transposition-like properties.

The Janko groups: J_1, J_2, J_3, J_4. These were all discovered by Janko—but in different contexts, so they are not really inter-related. However, the latter three are among the fairly many sporadic groups with a *large extraspecial subgroup*—see later Definition 8.1.3 and the subsequent discussion, as well as that of the treatment of branch (3) of the Trichotomy Theorem 2.2.8.

---

[4]Recall *section* means a quotient of a subgroup.

[5]The colon in an expression $A : B$ indicates a *semidirect* product: having a normal subgroup A, with $A \cap B = 1$; but B may act nontrivially on A.

The Monster series: In the Atlas [**CCN+85**, p 231], these groups are denoted by F_5, F_3, F_2, F_1. They arise from automorphisms of the 196884-dimensional Griess algebra; for this algebra see the discussion of the Monster F_1 in e.g. the Atlas [**CCN+85**, p 228]. The subscript n in F_n indicates that F_n is involved in the centralizer in F_1 of a suitable element of order n. In fact the earlier, and still more customary, names and notation for the groups are: the Harada-Norton group HN, the Thompson group Th, the Baby Monster BM discovered by Fischer, and the Monster M discovered by Fischer and Griess. It turns out that 20 of the 26 sporadic groups are involved as sections of the Monster.

The others: There are 7 more sporadic groups, mostly independent of the rest: the Held group He; the Higman-Sims group HS; the Lyons group Ly; the McLaughlin group McL; the O'Nan group $O'N$, the Rudvalis group Ru; and the Suzuki sporadic group Suz. There are only occasional inter-relations among these, e.g.: Some arise as rank-3 permutation groups—see later Definition 4.3.1 and the subsequent discussion; some appear in the Leech-lattice context indicated above; etc.

As promised, we now turn to a more extensive discussion of the "model case" for simple groups:

1.3. Groups of Lie type

We had commented, in stating the CFSG-list 1.0.2, that "most" simple groups are of Lie type. For example, we'll see in Remark 1.3.18 that this class contains 16 infinite families of groups. Indeed, since the development of the theory of Lie-type groups in the 1950s, their role in finite group theory has become more and more central; certainly students beginning in group theory will want to build up a comfortable familiarity with these groups. Hence in this section, we will try to present a somewhat fuller introduction to the Lie-type groups, than we did for the alternating and sporadic classes in the previous two sections.

The material below can be found in many standard sources; indeed because of space limitations, we'll often just give terminology and statements, without fully detailed definitions—so the reader may wish, for fuller reference, to continue to the relevant literature.

For example: I often tend to refer to Carter's book [**Car89**],[6] which approaches the material via the Chevalley construction of the groups; we will sketch the construction later in this section. Some prefer Carter's later book [**Car93**], which approaches the groups via fixed points of automorphisms of infinite algebraic groups—a viewpoint we will touch on more briefly, toward the end of the section. A further traditional source is given by the lecture notes of Steinberg [**Ste68**]; and a great deal of useful information is conveniently collected in Chapters 1–4 of the more recent [**GLS98**].

Prior to sketching the fully general Lie-theoretic context, we first explore various features, in the more familiar explicit examples given by:

[6](Note to students:) mainly because it's the one I learned from—as a postdoc, in the Caltech group theory seminar around 1974.

The classical matrix groups. These are the usual linear, unitary, symplectic, and orthogonal groups—defined over *finite* fields.

Here are some additional sources for classical groups: a traditional reference is Artin [**Art88b**]; but see also the more recent treatments of Taylor [**Tay92**] or Wilson [**Wil09**, Ch 3,4].

The linear groups. We had commented before Remark 1.1.1 that the symmetric group S_n, consisting of *all* permutations of n points, is perhaps the most accessible example of a permutation group—though this almost-simple group is slightly larger than simple A_n.

Analogously, the full linear group $GL_n(q)$ of dimension n, consisting of *all* invertible matrices, is (though again it's not quite simple) the most natural and accessible example among the classical groups—and indeed among all the Lie-type groups. In particular, using it allows us to demonstrate many concepts, just by drawing pictures of square matrices. We collect some basic features of the group:

REMARK 1.3.1 (The linear group $GL_n(q)$—as Standard Example). We recall that the general linear group $GL_n(q)$, for a power $q = p^a$ of a prime p, is the group of all $n \times n$ invertible matrices over the field $\mathbb{F}_q$. We refer to the charactistic p—of the field—also as the characteristic *of the group* $GL_n(q)$.

Largely in analogy with our earlier comparison of S_n to A_n: The general linear group $GL_n(q)$—when considered modulo the central subgroup of scalar matrices, as $PGL_n(q)$—becomes almost-simple as in Definition 1.4.7. In fact, its commutator subgroup is given by the special linear group $SL_n(q)$, namely the determinant-1 matrices—which usually[7] is just quasisimple.[8] And then the quotient modulo the central scalars, namely the projective special linear group $L_n(q) := PSL_n(q)$, is the corresponding simple group. Nonetheless, again much like S_n, using $GL_n(q)$ is "close enough" for most expository purposes—since it is typically easy enough to transfer properties of interest over to its simple section $L_n(q)$.

Usually we realize $GL_n(q)$ via its natural action: by multiplication from the right, on row-vectors of an n-dimensional vector space over $\mathbb{F}_q$—which we typically denote by V; we call V the *natural* module for the group. $\diamondsuit$

As we proceed, we will encounter many more analogies between S_n and $GL_n(q)$. One reason for the relationship is that S_n is the Lie-theoretic *Weyl group* of $GL_n(q)$—cf. later 1.3.20.

In introducing properties of the linear group, we start with the vector-space analogue of the multiple- and partition-transitivity of S_n in 1.1.1(1):

REMARK 1.3.2 (Flag-transitivity of $GL_n(q)$). If we fix a choice of dimensions (adding up to n) for a set of subspaces V_i of V, then $GL_n(q)$ is transitive on the set of direct-sum decompositions of the form $V = V_1 \oplus V_2 \oplus \cdots$. Similarly the group is transitive on *flags* $V_1 < V_2 < \cdots$ in V, again with a fixed choice of dimensions for the V_i. This latter property is called *flag-transitivity*. These transitivity statements are consequences of the usual linear-algebra statement about existence of a linear self-map transforming one basis to any other; in more group-theoretic language, that statement just expresses the *conjugacy* of bases under $GL_n(q)$.

[7]That is, away from some small dimensions, and small fields—see later 1.3.17.

[8]Recall L *quasisimple* means that $L/Z(L)$ is simple, and L is perfect ($L = [L, L]$).

Later at 1.3.5 and 1.3.20(7), we will get analogous notions of flag-transitivity for other classical groups, and indeed for all Lie-type groups; suitable analogues even also hold for certain sporadic groups. This property of flag-transitivity is very heavily used in applications. $\diamondsuit$

There are also analogies between certain subgroups of S_n and $GL_n(q)$, including some maximal subgroups. For example, for S_n consider the partition stabilizers, including the maximal subgroups indicated in 1.1.1(3). These correspond with certain subgroups of $GL_n(q)$, which we now explore in a fairly concrete setting:

Later in the section at 1.3.20, we will be describing these subgroups more abstractly as *parabolic* subgroups. But before working in that abstract setting, we will first approach these subgroups from the more elementary, but equally important, viewpoint of *p-local subgroups*: that is, as the normalizers of non-trivial p-subgroups. Here for the prime p, we are focusing just on the characteristic prime p of $GL_n(p^a)$. As promised earlier, we can begin to use matrix-pictures to illustrate the concepts:

EXAMPLE 1.3.3 (The Sylow normalizer etc of $GL_3(q)$). For convenience, we use the small dimension $n = 3$; larger dimensions yield similar pictures. Consider the subgroup B of lower-triangular matrices. We can express B as a semi-direct product $U : H$, where:

$$U := \begin{pmatrix} 1 & & \\ * & 1 & \\ * & * & 1 \end{pmatrix}, \; H := \begin{pmatrix} * & & \\ & * & \\ & & * \end{pmatrix}$$

Here the lower-unitriangular subgroup U is a p-group—it is even a Sylow p-subgroup of $GL_n(q)$. So is the transpose U^- of U, the upper-unitriangular group; and one finds that $\langle U, U^- \rangle = SL_n(q)$—that is, they generate most of $GL_n(q)$.

We also find that B is the Sylow normalizer: namely $N_G(U) = UH = B$, using the above subgroup H of diagonal matrices. Furthermore the *monomial* matrices, corresponding to linear transformations which just permute the members of the standard basis, are given by HW, where W is the subgroup of permutation matrices. Indeed usually $HW = N_G(H)$—except when $q = 2$ so that $H = 1$. (For a general Lie-type group, the monomial subgroup "N" with $N/H \cong W$ arises as an apartment stabilizer, in the language around later (7.2.7).)

The experienced reader will of course notice that our notation here is chosen for consistency with the conventions of general Lie-type groups, later in the section—notably Remark 1.3.20. $\diamondsuit$

We note that B above is the stabilizer of a maximal (or *full*) flag $V_1 < V_2$, consisting of 1- and 2- dimensional subspaces of V.

Similarly for a full decomposition $V = X_1 \oplus X_2 \oplus X_3$ into 1-dimensional subspaces X_i: we see first that H above is the stabilizer of the individual subspaces; while its three diagonal positions are in turn permuted by the permutation matrices $W \cong S_3$. This decomposition is the analogue in the 3-dimensional vector space V of the trivial $(1, 1, 1)$-partition of the 3 points for the symmetric group; the latter has trivial blockwise stabilizer, and blocks permuted by the full S_3. Thus the subgroup B of the linear group corresponds with a partition stabilizer in the symmetric group—in this case, given by the full S_3.

Now we may as well turn from the small case $n = 3$ back to general n. Then corresponding to the stabilizer in S_n of a k-set—or equivalently of the stabilizer of the $(k, n-k)$-partition in 1.1.1(3)—we have:

EXAMPLE 1.3.4 (The k-subspace stabilizer in $GL_n(q)$). We examine the subgroup of $GL_n(q)$ stabilizing the k-subspace V_k—generated by the first k basis vectors. In the literature, this is sometimes denoted just by P_k; but for consistency with our later notation for parabolic subgroups in Remark 1.3.20(4) (and its implementation for the present group in Example 1.3.21), we instead write $P_{\hat{k}}$. Then we find that $P_{\hat{k}}$ is a product $P_{\hat{k}} = U_{\hat{k}} : L_{\hat{k}}$, where:

$$U_{\hat{k}} := \left(\begin{array}{c|c} I_k & 0 \\ \hline * & I_{n-k} \end{array}\right), \quad L_{\hat{k}} := \left(\begin{array}{c|c} GL_k(q) & 0 \\ \hline 0 & GL_{n-k}(q) \end{array}\right)$$

Here $U_{\hat{k}}$ is also a p-group; with $N_G(U_{\hat{k}}) = U_{\hat{k}} L_{\hat{k}} = P_{\hat{k}}$. Further $N_G(P_{\hat{k}}) = P_{\hat{k}}$; indeed $P_{\hat{k}}$ is a maximal subgroup of $GL_n(q)$.

In this example, note that $U_{\hat{k}}$ acts trivially on V_k, and on the quotient V/V_k; while $L_{\hat{k}}$ is the product of the natural GL_k on V_k, with the GL_{n-k} on V/V_k. ♢

The classical matrix groups of forms—unitary, symplectic, orthogonal. These are the subgroups of $GL_n(q)$ preserving suitable *forms* (e.g. bilinear) on V: a symmetric form, giving the orthogonal group $\Omega_n(q)$; an anti-symmetric form, giving the symplectic group $Sp_{2n}(q)$; or a conjugate-symmetric form, giving the unitary group $U_n(q)$.

Again see the recommended sources, for example [**Car89**, Ch 1], for full definitions and details. Some examples in small dimensions are developed in Section 2.1 of [**Smi11**]; these may be helpful in some of the Exercises below. Here, we will just briefly mention a number of analogies of these groups with the linear groups.

First, they similarly provide good examples—since we can still draw pictures of square matrices.

Next, we get analogues of the flag-transitivity properties in 1.3.2:

REMARK 1.3.5 (Witt's Lemma and flag-transitivity for forms). One property crucial for our purposes is *Witt's Lemma*. See e.g. [**GLS98**, 2.7.1] for a detailed statement; very roughly: This gives the vector-space analogue of earlier 1.3.2, which described $GL_n(q)$ on decompositions; namely transitivity of the group on *suitable* decompositions of the natural module V. Here, the summand subspaces W should be appropriately natural for the form: typically *nondegenerate*—that is, with zero radical, namely $W \cap W^{\perp} = 0$; or contrastingly, *totally isotropic*—with $W \leq W^{\perp}$.

Similarly the Lemma gives transitivity on *flags* of isotropic subspaces; and this is the appropriate notion of flag-transitivity for the classical groups of forms. ♢

Finally: subject to a choice of basis appropriate to the form, we get analogues of the structural results on p-local subgroups in Examples 1.3.3 and 1.3.4. For example: the unitriangularity of a Sylow p-subgroup, as well as the product form $B = UH$ for its normalizer; along with the product form $U_{\hat{k}} L_{\hat{k}}$, as well as maximality, for the isotropic k-space stabilizers $P_{\hat{k}}$.

EXERCISE 1.3.6 (Practice with forms). In several examples of small-dimensional groups of forms, verify the above remarks about analogies with the linear case.

Hint: I have in mind examples like those in [**Smi11**, 2.1.18—2.1.25]. In particular: It is customary to decompose the space V via a basis consisting of mutually orthogonal *hyperbolic pairs*; namely isotropic vectors a, b with $(a, b) = 1$. In most sources, these pairs are taken adjacent, in an ordered basis. However, if you split the pairs, spacing them at opposite ends of the ordering (that is, hyperbolic pairs v_1, v_n and v_2, v_{n-1} etc), then matrices preserving the form will have some symmetry about the "anti"-diagonal (/). For example, a p-Sylow U will be lower-unitriangular. ◊

This completes our introductory overview, of the more concrete classical matrix groups. So we now turn to the more abstract context, which unifies all the groups of Lie type:

A high-speed sketch of the theory of Lie-type groups. The theory of Lie-type groups was developed around the 1950s: as a single unified context for many types of simple groups—including the classical matrix groups above. In quick summary, the idea is to imitate parts of the Lie theory in characteristic 0: in which simple complex Lie groups arise via connections with simple complex Lie algebras.

There are in fact two main approaches to the finite groups of Lie type: We'll mainly follow the somewhat more concrete Chevalley construction, which produces a suitable version of classic matrix exponentiation; but we'll close the section with some brief remarks on the other approach, via fixed points in infinite algebraic groups.

Both of these approaches naturally require some background from Lie algebras; so before discussing exponentiation, we first present a few pages of corresponding preliminary material:

Some features of simple Lie algebras over $\mathbb{C}$. One standard reference for this material is the book of Humphreys [**Hum78**]; and I also like Varadarajan [**Var84**]—for the detailed relationship of the Lie algebras with Lie groups over $\mathbb{C}$. As before, below we will essentially just indicate terminology and results—with occasional references given for full definitions and details.

The simple Lie algebras $\mathcal{G}$ defined over the complex numbers $\mathbb{C}$ were classified by Cartan and Killing in the 1890s; see e.g. [**Hum78**, 11.4]. In the lengthy Remark 1.3.7 below, we collect together many features of the standard setup used for such $\mathcal{G}$; and we will give the classification at subsequent Remark 1.3.11. The reader not yet familiar with Lie algebras may wish to first skip down to the example given as 1.3.10—and read it concurrently with the onslaught of notation now approaching:

REMARK 1.3.7 (Lie algebras and their root systems). The simple Lie algebra $\mathcal{G}$ has a self-normalizing nilpotent[9] subalgebra $\mathcal{H}$, called a *Cartan subalgebra*. It is abelian: we have $[\mathcal{H}, \mathcal{H}] = 0$ in the Lie multiplication $[-, -]$; and its dimension n is called the *Lie rank* of $\mathcal{G}$. The action of $g \in \mathcal{G}$ by right multiplication $[-, g]$ on $\mathcal{G}$ itself determines the *adjoint* representation of $\mathcal{G}$. And the representation theory[10] of Lie algebras on general modules, when applied to the special case of the adjoint

[9] Recall this means products $[u_1, [u_2, \cdots, [u_{k-1}, u_k] \cdots]$, are 0 for $k \geq$ some fixed M.

[10] Cf. aspects of the analogous representation theory for groups, at Remark 5.2.3.

module $\mathcal{G}$ and especially to the action of $\mathcal{H}$ on it, leads to the very special properties of the algebra's *root system*, which we now sketch:

First the action of $\mathcal{H}$ on $\mathcal{G}$ is *completely reducible*—into irreducible subspaces, which are in fact 1-dimensional since $\mathcal{H}$ is abelian. The eigenvalues of $\mathcal{H}$, on any such invariant subspace X of dimension 1, define a linear functional $\alpha : \mathcal{H} \to X \cong \mathbb{C}$; so we may naturally regard α as a member of the dual space $\mathcal{H}^* = \mathrm{Hom}_{\mathbb{C}}(\mathcal{H}, \mathbb{C})$. On a general $\mathcal{G}$-module, α is called a *weight*, and such an X a *weight space*; but on the special case of the adjoint module $\mathcal{G}$, α is called a *root* (when non-zero) and X a *root space*. Using a natural inner product on $\mathcal{H}$ called the *Killing form*, duality gives a natural isomorphism $\mathcal{H}^* \cong \mathcal{H}$; so we may also regard weights and roots as vectors in $\mathcal{H}$. Indeed it turns out that the image of the weights in fact falls into a Euclidean subspace $\mathbb{R}^n \subseteq \mathbb{C}^n \cong \mathcal{H}$. Thus we can regard the weights (including the roots) as vectors in a real space—with corresponding lengths and angles. And we get many strong restrictions on the configuration of these root vectors:

We begin with a *Cartan decomposition* of the algebra, which has form:

$$\mathcal{G} = \mathcal{U}^+ \oplus \mathcal{H} \oplus \mathcal{U}^-. \tag{1.3.8}$$

Recall that the Cartan subalgebra $\mathcal{H}$ is abelian; so its action determines the zero-functional—which we do not consider a root. The subalgebra $\mathcal{U}^+$ is nilpotent; and the same holds for the negative analogue $\mathcal{U}^-$. We write Φ^+ for the roots occurring on $\mathcal{U}^+$, and call them the *positive roots*; and similarly Φ^- for those occurring on $\mathcal{U}^-$, and call them the *negative roots*. And it turns out that, in our view of the roots as vectors in $\mathbb{R}^n$, we do in fact have $\Phi^- = -\Phi^+$. We set $\Phi := \Phi^+ \cup \Phi^-$, and call this union the *root system* for $\mathcal{G}$. For each root $\alpha \in \Phi$, it turns out that the full subspace $\mathcal{U}_\alpha$ affording the functional α is exactly 1-dimensional (on a general module, that full weight space might conceivably have been larger); $\mathcal{U}_\alpha$ is called the *root subspace* for α. Thus the dimension of $\mathcal{U}^+$ is given by the size $|\Phi^+|$, with similar remarks for $\mathcal{U}^-$; this important dimension-value is often abbreviated by N.

We mention an important property of weight and root spaces: If V_λ is the weight space for λ under $\mathcal{H}$ on a module V, then for a root space $\mathcal{U}_\alpha$ we have:

$$[V_\lambda, \mathcal{U}_\alpha] \subseteq V_{\lambda+\alpha}. \tag{1.3.9}$$

For example, nilpotence of $\mathcal{U}^+$ follows, using this along with finiteness of Φ^+.

The terminology of positive and negative roots is natural in an even stronger sense: The set Φ^+ contains a *simple* system Π of size n, called the *simple* roots: These not only give a basis for $\mathbb{R}^n$, but in fact span the root system Φ, using integer coefficients: indeed the positive roots Φ^+ are determined via nonnegative integer coefficients; and hence the negative roots $\Phi^- = -\Phi^+$ are determined via nonpositive integer coefficients.

Also there are only a few possibilities for angles between roots; and there at most two lengths for roots—if both occur, they are called *long* and *short* roots.

Finally there is an associated finite group W of permutations of Φ, called the *Weyl* group. The root system Φ, and correspondingly the Weyl group W, can be axiomatized by means of a *Dynkin diagram*: a graph in which the vertices are the simple roots; and the number of edges or "bonds" between any pair encodes the angle between those simple roots. In particular, the absence of any bonds at all (a disconnected pair) indicates perpendicularity. ♢

We will list the possible Dynkin diagrams and root systems below, in Remark 1.3.11. But first we explore some of these concepts in a small example:

EXAMPLE 1.3.10 (Some features of the Lie algebra $sl_3(\mathbb{C})$). We mention that the 3×3 matrix pictures in the earlier group-Example 1.3.3 will still be suggestive, in the following algebra-situation:

We consider $\mathcal{G} := sl_3(\mathbb{C})$, namely the 3×3 trace-0 matrices—with the Lie-algebra multiplication ("bracket") from the additive commutator $[x, y] := xy - yx$; that is, by anti-symmetrizing the usual associative matrix multiplication xy. A standard linear-algebra fact about the trace, namely $\mathrm{Tr}(xy) = \mathrm{Tr}(yx)$, shows that the trace-0 condition is preserved by this Lie bracket.

Here one finds that a Cartan subalgebra $\mathcal{H}$ is generated by diagonal matrices $h_\alpha := \mathrm{diag}(1, -1, 0)$ and $h_\beta = \mathrm{diag}(0, 1, -1)$. Since $\mathcal{H}$ has dimension 2, the algebra $\mathcal{G}$ has Lie rank 2. Below we summarize some features, resulting from calculating the action $[-, \mathcal{H}]$ on $\mathcal{G}$:

The 8-dimensional algebra $\mathcal{G}$ has Cartan decomposition as in (1.3.8), where $\mathcal{U}^\pm$ are the 3-dimensional nilpotent subalgebras given by strictly lower-triangular, and stricly upper-triangular, matrices.

Now $\mathcal{H}$ acts on the 1-subspaces from the obvious bases for these subalgebras: Namely let $e_{i,j}$ denote the "matrix unit"—with 1 in the (i, j) position and 0 elsewhere. We get $\mathcal{U}^+$ from the cases with $i > j$, and $\mathcal{U}^-$ from $i < j$. The action of $\mathcal{H}$ on the the three 1-spaces for $i > j$ determines three corresponding linear functionals $\mathcal{H} \to \mathbb{C}$; and we find they have the linear-dependence relation $\alpha, \beta, \alpha + \beta$; where α is exhibited on the root space generated by $u_\alpha := e_{2,1}$, and β on $u_\beta := e_{3,2}$. Furthermore we check that $[u_\beta, u_\alpha] = e_{3,1}$, and we call this latter element $u_{\alpha+\beta}$.

Thus α and β are positive roots, arising on the immediately-subdiagonal positions, and they give a simple system Π; with $N = 3$ roots in the positive system $\Phi^+ = \{\alpha, \beta, \alpha + \beta\}$, whose root spaces generate the positive nilpotent subalgebra $\mathcal{U}^+$. (These matrices look much like the group U in 1.3.3—but with the diagonal set to 0, since they are nilpotent.) And the negatives of these functionals arise on the super-diagonal transposes of these matrices; giving the negative roots Φ^- and nilpotent subalgebra $\mathcal{U}^-$.

When we work in the $\mathbb{R}$-span of Φ inside $\mathcal{H}^*$, the functionals α and β are at an angle of $\frac{2\pi}{3}$; and this angle corresponds to a single bond in the conventions for Dynkin diagrams, so that the resulting diagram is of the type called A_2 ($\circ - \circ$) in the context of Remark 1.3.11 below. All six roots of Φ have the same length.

Finally the Weyl group $W \cong S_3$ corresponds to the 3×3 permutation matrices; its natural permutation-module action, on the 3-dimensional diagonal subspace in the larger algebra $gl_3(\mathbb{C})$ of all 3×3 matrices, induces a "reflection" action on the subspace of dimension 2 given by our Cartan subalgebra $\mathcal{H}$.

See e.g. [**Car89**, Sec 3.6(i)] for a more formal description of the A_2 root system. And we mention that some other small examples of root systems (beyond type A_2 here) will be featured in later Exercise 1.3.22, and are explored fairly explicitly in the corresponding appendix Remark B.1.1.

For the above, the actual calculations with 3×3 matrices of the underlying action $[-, \mathcal{H}]$ are routine—if tedious; you can easily check them in your favorite computer-algebra package. But I'm not aware of any source that does such an explicit example, fully demonstrating such calculations. Ideally a Lie-algebras course might do so. However, I believe that you *should* do such a calculation yourself, at least once in your life;[11] in order to get some more concrete feeling for what is actually going on in the abstract theory. ♢

In fact the Cartan-Killing classification of the algebras $\mathcal{G}$ proceeds by first classifying the corresponding root systems Φ—via their geometry: Roughly: there are very strong restrictions on the possible angles between roots (considered as vectors in $\mathbb{R}^n$); leading to just a few configurations. The final result is usually presented via the associated Dynkin diagrams, in the following picture—which has become fundamental across many areas of mathematics:

REMARK 1.3.11 (The Dynkin diagrams and root systems). A simple Lie algebra $\mathcal{G}$ over $\mathbb{C}$ must have a root system Φ corresponding to one of the Dynkin diagrams in the table below—often called the Lie *types*. The notation $>$ or $<$ in the diagram-column indicates which simple roots are long or short.

For later purposes, we have added a final column also listing certain "twisted types": In these, a Dynkin diagram is folded on itself, by identifying images under a diagram automorphism; and then for the twisted type, the order of the automorphism appears as a superscript on the left of the original type. In most cases, this automorphism is a left-right reflection, and in particular, has order 2; and the resulting twisted diagram is reasonably clear. (But see [**Car89**, 13.3.8] for details.) However, the D_4 diagram also admits a rotation of order 3, called *triality*; here the three outer nodes are identified, with a triple bond to the inner node, to produce the twisted type ${}^3D_4$. Below there are 7 twisted types, beyond the 9 Lie types—giving a total of 16 families:

Lie type	Dynkin diagram	twisted types?
(classical:) A_n	$\circ - \circ - \circ - \cdots - \circ - \circ$	${}^2A_n$
B_n	$\circ - \circ - \circ - \cdots - \circ \overset{>}{\equiv} \circ$	${}^2B_2$
C_n	$\circ - \circ - \circ - \cdots - \circ \overset{<}{\equiv} \circ$	$({}^2C_2 \simeq {}^2B_2)$
D_n	$\circ - \circ - \circ - \cdots - \circ <^{\circ}_{\circ}$	${}^2D_n, {}^3D_4$
(exceptional:) E_6	$\circ - \circ - \overset{\overset{\circ}{\vert}}{\circ} - \circ - \circ$	${}^2E_6$
E_7	$\circ - \circ - \overset{\overset{\circ}{\vert}}{\circ} - \circ - \circ - \circ$	
E_8	$\circ - \circ - \overset{\overset{\circ}{\vert}}{\circ} - \circ - \circ - \circ - \circ$	
F_4	$\circ - \circ = \circ - \circ$	${}^2F_4$
G_2	$\circ \equiv \circ$	${}^2G_2$

The root systems Φ, corresponding to the above Dynkin diagrams, are fully described in various sources; see e.g. [**Car89**, Sec 3.6] for the view in terms of sets of points in a suitable Euclidean space. We will use these descriptions at various later points. For example, root systems are used in the construction of parabolic subgroups—see Remark 1.3.20(4); and correspondingly some such uses are then explored in Example 1.3.21 and Exercise 1.3.22. ♢

[11](Note to students:) I did—indeed, in teaching a Lie-algebras course...long ago.

The twisted Dynkin diagrams will lead to twisted *groups*, in a later development at Remark 1.3.15; that will be a variation on the more fundamental procedure which we describe next:

From algebras to groups: the Chevalley construction. In classical analysis, one proceeds from a Lie algebra (for us, over $\mathbb{C}$) to a Lie group, by exponentiating matrices of the algebra. This makes sense over $\mathbb{C}$, which contains the factorial denominators in the infinite-sum definition of the exponential.

In fact it's not even necessary to worry about the *convergence* of the limit in that sum: For one can exponentiate elements just from the nilpotent algebras $\mathcal{U}^{\pm}$ in the Cartan decomposition (1.3.8) above—and then only finite sums arise in the exponential. These generate our desired group. (We *would* get infinite sums, if we exponentiated from the term $\mathcal{H}$ in the Cartan decomposition.)

In overview: This results in groups $U^{\pm}$, which are characteristic-0 analogues of the characteristic-p groups U, U^- in Example 1.3.3. The elements of these groups are in fact *unipotent*;[12] meaning that on subtracting the identity, they become nilpotent matrices. And again much as in 1.3.3, these unipotent groups in turn generate as $\langle U^+, U^- \rangle$ the corresponding *Lie group* G over $\mathbb{C}$.

In the background, we still have the question of denominators in the sums—if we wish to make a meaningful analogue of this exponentiation for *finite* groups in characteristic p. But we postpone that question, to first consider some explicit examples of classical characteristic-0 exponentiation:

EXAMPLE 1.3.12 (Exponentiating $\mathrm{sl}_3(\mathbb{C})$ to $SL_3(\mathbb{C})$). We continue with the setup and properties in Example 1.3.10.

First naively: we consider $\mathcal{G} = \mathrm{sl}_3(\mathbb{C})$ acting on its *natural* module V, given by the vectors of $\mathbb{C}^3$. One can directly exponentiate a nilpotent 3×3 matrix like $u_\alpha = e_{2,1}$ in that Example. Indeed since the matrix $e_{2,1}$ squares to 0, the result is just the identity plus $e_{2,1}$—which is indeed an element of the group $SL_3(\mathbb{C})$. It is natural to denote by U_α the *root group*[13] defined by the matrices $I + c \cdot e_{2,1}$ for the various scalars $c \in \mathbb{C}$. Doing this over all of $\mathcal{U}^+$ produces matrices generating the full unipotent group U^+, namely the lower-untriangular matrices. (As mentioned earlier, we had described the analogous group U, defined over a finite field, via a matrix picture in Example 1.3.3.) Similarly negative roots lead to U^-; and U^+, U^- then generate $SL_3(\mathbb{C})$.

Now less naively: we consider $\mathcal{G}$ in its *adjoint*-module action; that is, on $\mathcal{G}$ itself. We saw above that $\mathcal{U}^+$ is nilpotent, when regarded just as a 3-dimensional subalgebra, inside $\mathcal{G}$. But we observe now that it is even nilpotent, in its action on the full algebra 8-dimensional $\mathcal{G}$: This is because the representation-theory property (1.3.9), applied in computing the exponential, tells us that for any root γ:

(1.3.13) $$[u_\gamma, u_\alpha] \text{ involves higher } i\gamma + j\alpha \text{ with } i, j > 0.$$

Thus we get nilpotence of U^+, since there are only a finite number of choices for such higher roots in Φ^+. Viewed in terms of matrix pictures, this says that the action of u_α in effect pulls strictly "southwest"—that is, down and to-the-left. Consequently the 8×8 matrix M for this action of u_α is strictly lower-triangular. Now M

[12] Of course the analogous U, U^- over $\mathbb{F}_q$ in Example 1.3.3 are *nilpotent* as groups; in the somewhat different terminology of group theory.

[13] The root groups are examples of "1-parameter subgroups" in classical analysis.

does not square to 0, as was the case for u_α regarded as a 3×3 matrix above; but certainly $M^8 = 0$. And exponentiation of M and its analogues for other roots will produce a copy of $SL_3(\mathbb{C})$, indeed embedded in the larger group $GL_3(\mathbb{C})$—but all are now represented as 8×8 matrices inside $GL_8(\mathbb{C})$. As before with Example 1.3.3, the details of such computations are straightforward-if-tedious to implement. $\diamondsuit$

We now return to the crucial point mentioned earlier: If we want to produce groups over a field of characteristic p, such as $\mathbb{F}_{p^a}$, then factorial denominators present a problem for defining exponentiation.

See e.g. [**Car89**, Ch 4] for precise details on the following material:

We now sketch the *Chevalley construction*, which finessed this problem. That is, we can proceed from a simple Lie algebra $\mathcal{G}$ over $\mathbb{C}$ to a corresponding simple *Chevalley group* $G(q)$, defined over the field $\mathbb{F}_q$.

The basic idea is roughly: First, introduce an "integral" $\mathbb{Z}$-form $\mathcal{G}_\mathbb{Z}$ of the algebra. This gives us a notion of non-zero characteristic—for if k is a field say of characteristic p, then the tensor product $\mathcal{G}(k) := (\mathcal{G}_\mathbb{Z} \otimes_{\mathbb{F}_p} k)$ gives a k-form of the algebra. Second, and more difficult: choose this form in such a way that exponentiation to a group $G(k)$ makes sense; that is, so that the necessary denominators are already "built in" to the $\mathbb{Z}$-form.

Here is a little background on how that process was actually accomplished:

Chevalley around 1955 made the basic construction: He defined a special *Chevalley basis* for $\mathcal{G}_\mathbb{Z}$; and via some fairly intricate calculations, which in effect provided the necessary denominators, he could exponentiate the nilpotent elements of that basis of $\mathcal{G}(k)$, to the positive and negative unipotent elements generating the *Chevalley group* $G(k)$. To get the finite Chevalley groups, we take k to be a finite field $\mathbb{F}_q$.

Many standard references, such as [**Car89**, Ch 4], follow that original Chevalley construction. However, we also now briefly indicate a different viewpoint, which may be a little more intuitive, at least in our broad overview: One limitation of the original construction is that it describes the group $G(k)$ only via the *adjoint* module—essentially originating from the action of $\mathcal{G}$ on itself, via the Lie multiplication. An important generalization (see e.g. [**Cur71**, Sec 4] or [**Hum78**, Sec 27]) was provided by Kostant,[14] who established the existence of a $\mathbb{Z}$-form on *any* $\mathcal{G}$-module. This led to the full *representation theory* of $G(k)$ on such modules. Of course the restriction of this *Kostant $\mathbb{Z}$-form* back to the particular adjoint module $\mathcal{G}$ in effect reproduces the original Chevalley construction. But the Kostant viewpoint does not have an "elementwise" focus on Chevalley basis elements—this is primarily because the denominator-calculations are basically built in to the associated *universal enveloping algebra* $U(\mathcal{G})$, whose Kostant $\mathbb{Z}$-form then leads to those of the various modules. So when we apply the $\mathbb{Z}$-form to the Cartan decomposition (1.3.8) of $\mathcal{G}$, the nilpotent subalgebras $\mathcal{U}_\mathbb{Z}^\pm$, when tensored with k, exponentiate more or less automatically to full unipotent groups $U^\pm(k)$—which in turn generate $G(k)$.

EXAMPLE 1.3.14 (The Chevalley construction for $SL_3(q)$). We build on elements of the $\mathbb{C}$-case considered in Example 1.3.12 above.

[14]And independently by Cartier.

In the naive viewpoint of the natural module, the fact that the nilpotent matrices $e_{2,1}$ etc there square to 0 means that there are no denominator problems in exponentiation. So the integer-matrix result there, now tensored with $\mathbb{F}_q$, gives us $U^{\pm}$ and $SL_3(q)$. Thus the Lie-algebra context has now reproduced the groups U, U^- that we had previous just introduced directly, in earlier Example 1.3.3.

The wonderful accident of squaring to 0 is not available in the less-naive viewpoint of the adjoint module. But here is a quick sample of the kind of adjustment that would be needed: Since we saw that $[u_\beta, u_\alpha] = u_{\alpha+\beta}$, just naively exponentiating the action of u_α would put an entry of $\frac{1}{2}$ in the matrix for the action of u_α—in the row recording its effect on u_β; this would be undesirable for characteristic 2.

To avoid this denominator, we might try using say $\sqrt{2}u_\alpha$ and $\sqrt{2}u_\beta$ in our basis—so that their commutator would be $2u_{\alpha+\beta}$, which would multiply with the previously-problematic $\frac{1}{2}$ to give 1, in the square-power term of the exponential.

Of course, there is actually a lot more to take care of, in defining a full Chevalley basis. And again, I know no source giving full computational details, in an explicit example like this one. $\diamondsuit$

The Chevalley groups provided a unified approach to both the classical matrix groups, and the *exceptional* groups—that is, groups for the algebras of the exceptional types G_2, F_4, E_6, E_7, and E_8. And for example, groups which we can now recognize as being of type G_2 had in fact been known to Dickson since around 1901—but they had not been so explicitly understood via a common theory with the classical cases.

In a similar vein, the Suzuki groups $Sz(2^{2n+1})$ and Ree groups $\mathrm{Ree}(3^a)$, discovered around 1960, were directly constructed as matrix groups; and they were not explicitly covered by the original Chevalley construction. But soon, these and some further groups were also understood in a uniform fashion, in a variation on that construction:

REMARK 1.3.15 (twisted groups). Steinberg saw how to exploit the situation of a Dynkin diagram D with an automorphism τ—as in the final column of the table in 1.3.11; he obtained a *twisted* group corresponding to the quotient diagram D/τ. For details see e.g. [**Car89**, Ch 14]; as usual we only summarize rapidly:

Instead of any process of exponentiation based on D/τ, he worked inside the "already constructed" Chevalley group of type D: taking certain subgroups $U_\tau^{\pm}$ of its unipotent groups $U^{\pm}$, corresponding to products in *orbits* under τ on roots. Very roughly: our twisted diagram doesn't exactly have a root system; but it has root groups, which are based on root-orbits. These subgroups in turn generate the desired twisted group for the quotient diagram D/τ.

The twisted types are the 7 cases which we had listed in earlier 1.3.11; namely types ${}^2A_n$, ${}^2C_2$, ${}^2E_6$, ${}^2F_4$, ${}^2G_2$, ${}^2D_n$, ${}^3D_4$. The construction in fact covered certain classical groups, including the unitary groups, and the orthogonal groups of minus type; as well as recognizing the Suzuki and Ree groups via types ${}^2C_2$ and ${}^2G_2$. $\diamondsuit$

We examine the twisted group constructed from our usual example of SL_3:

EXAMPLE 1.3.16 ($U_3(q)$ as twisted group). The diagram-reflection for type A_2 just switches the two nodes, which correspond to the simple roots α and β in Example 1.3.10; the quotient diagram is a single node—so we expect just one positive root group. Correspondingly, from $SL_3(q^2)$ as in Example 1.3.14, we can take a diagonal element from the product $U_\alpha U_\beta$, suitably balancing it under the order-2 field automorphism in $\mathrm{Gal}(\mathbb{F}_{q^2}/\mathbb{F}_q)$; such elements and their negative-root analogues will generate a subgroup $SU_3(q)$ inside $SL_3(q^2)$. ♢

We now have the class of simple groups giving the topic of this section:

DEFINITION 1.3.17 (The Lie-type groups). Taken together, the Chevalley groups and the twisted groups give the finite *groups of Lie type*, or *Lie-type groups*. Most of the groups are simple—away from a few small dimensions, and small primes (see e.g. [**GLS98**, 2.2.8]).

The *Lie type* of such a group $G(q)$ is that of its diagram—including the types for twisted diagrams, as in 1.3.11. The *Lie rank* of $G(q)$ is also that of the diagram: for a Chevalley group, this is just the number of nodes in a standard Dynkin diagram; but for a twisted group, note that the quotient diagram D/τ has only about half as many nodes as the Dynkin diagram D of the overlying Chevalley group.

Chevalley groups are sometimes called *un*twisted Lie-type groups. ♢

Next we provide a table, which relates the viewpoint of the Lie types with earlier standard terminology and notation—e.g. for classical groups:

REMARK 1.3.18 (The Lie types—related to other naming conventions). In the table below, for brevity we usually omit any indication of the particular field, writing just A_n for $A_n(q)$ (etc)—except when only certain fields can arise. And the right-hand column contains just the most common notation: that is, $L_{n+1}, U_{n+1}, \cdots$, rather than more classical notation such as $PSL_{n+1}, PSU_{n+1}, \cdots$

Lie type X	usual term for $X(q)$	usual notation
(classical matrix types—including two twisted types:)		
A_n	linear	L_{n+1}
^2A_n	unitary	U_{n+1}
B_n	orthogonal	Ω_{2n+1}
C_n	symplectic	Sp_{2n}
D_n	orthogonal plus-type	Ω^+_{2n}
^2D_n	orthogonal minus-type	Ω^-_{2n}
(exceptional types:)		
G_2, F_4, E_6, E_7, E_8	(same: $G_2, \cdots$)	(same)
(the (non-classical) twisted types—some say also "exceptional":)		
$^2C_2(2^{2n+1})$	Suzuki groups	$Sz(2^{2n+1})$
$^3D_4$	triality D_4	(same)
$^2E_6$	twisted E_6	(same)
$^2F_4(2^{2n+1})$	twisted F_4	(same)
$^2G_2(3^a)$	Ree groups	$Ree(3^a)$

Some sources prefer $^2B_2$ to $^2C_2$ for the Suzuki groups. In practice, we use whichever notation seems best for the current context. ♢

REMARK 1.3.19 (Forms of the CFSG-list). We noted in stating our CFSG-list in Theorem 1.0.2 that it was in fact just a summary of the types of simple groups. But now in effect we have given a longer form of that list: Beyond the alternating groups in Section 1.1, in Section 1.2 we indicated the names of the 26 sporadic groups; and in Remark 1.3.18 just above, we have indicated the individual families of Lie-type groups.

During the remainder of the book, in describing works which examine the various groups in the CFSG-list, we will continue to make reference usually just to the summary-form in 1.0.2; but the reader should understand that often such an examination in fact requires the longer form. ♢

Properties of the Lie-type groups. Just as in the previous sections, for later applications we wish to collect together various properties of this class of simple groups. Many structural properties of the finite $G(q)$ are of course analogous to those of the corresponding Lie group $G_{\mathbb{C}}$ over $\mathbb{C}$. In particular, notation used in these properties will mirror the notation we used in our earlier more explicit classical-group Examples 1.3.3 and 1.3.4.

REMARK 1.3.20 (Some structural features of Lie-type groups). See for example [**Car89**, Sec 8.3] for fuller details on this material. Our statements below are sometimes made just for the easier case given by the Chevalley groups $G := G(q)$; and then the statements might typically need to be adjusted somewhat, for the more complicated case of twisted groups.

(1) Sylow structure: The root subspaces $\mathcal{U}_\alpha$ of Remark 1.3.7 exponentiate to *root subgroups* U_α. Those for positive roots generate a *full unipotent subgroup* U; it is of order q^N, where $N = |\Phi^+|$, and is Sylow in G. Further U and its negative-root analogue U^- generate the usually-simple group G.

(2) Cartan subgroup: The pair $U_{\pm\alpha}$ generates a copy of $SL_2(q)$ (in the untwisted case); inside which, from $[U_\alpha, U_{-\alpha}]$ we can extract a "diagonal" group H_α. The H_α as α varies generate a *Cartan* subgroup H—also sometimes called a diagonal group; it is an abelian p'-group of order roughly $(q-1)^n$, where n is the Lie rank of G. Elements of p'-order are said to be *semisimple*. In the algebraic-group viewpoint, H is called a *split torus*; for non-split tori see later Example 5.2.2.

(3) Sylow normalizer etc: We have $N_G(U) = UH =: B$; this defines the conjugacy class of *Borel subgroups* of G. Furthermore we have a *monomial* subgroup N, with $N/H \cong W$—where W is the Weyl group of the Dynkin diagram for G; and generically $N = N_G(H)$—except when $H = 1$, as mentioned in Example 1.3.3. Suitable properties of B and N can be used to define the axioms of a BN-*pair*; which provide an alternative approach to the Chevalley construction of G.

(4) Parabolic subgroups: Subgroups of G containing a Borel subgroup such as B are called *parabolic subgroups*; these are important p-local subgroups of G. A parabolic is determined by a subset $J \subseteq \Pi$ of the simple roots; and the parabolic then has a corresponding *Levi decomposition* of the form $P_J = U_J L_J$, where: the *unipotent radical* $U_J = O_p(P_J)$ is generated by the root subgroups U_α for positive roots which are *not* a linear combination from J; and the *Levi complement* L_J is generated by H and the $U_{\pm\alpha}$ for the positive α which *are* linear combinations from J.[15] We have $N_G(U_J) = P_J$; and further $N_G(P_J) = P_J$. The parabolics range between $P_\emptyset = B$ and $P_\Pi = G$; for a case with intermediate J, see 1.3.21 below.

[15]So our notational convention has the advantage that the subdiagram for J describes the structure of L_J. But it also has a disadvantage, already seen in Example 1.3.4: in the later language of simplex "types" in Remark 7.1.4, a simplex of type J has stabilizer $P_{\hat{J}}$ indexed by the complement $\hat{J}$. For this reason, some of the literature uses the convention opposite to ours: namely writing the complement $\hat{J} = \Pi \setminus J$ wherever we write J.

(5) The fundamental *Borel-Tits Theorem* (see e.g. [**GLS98**, 3.1.3]) states that any p-local subgroup of G lies in some parabolic subgroup.

(6) Maximal parabolics: The parabolics determined by the maximal subsets J of Π, namely those of size $n-1$, are in fact *maximal* subgroups of G. Notice in particular: The number of maximal p-local subgroups over a fixed Borel subgroup $B = N_G(U)$, and hence over a fixed p-Sylow U, is equal to "n"—namely the Lie rank; this is $|J| = |\Pi|$ = the number of nodes in the Dynkin diagram. Thus for G of Lie rank 1, using $J = \emptyset$, the Sylow group U lies in a *unique* maximal subgroup, namely $N_G(U) = B = P_\emptyset$. Similarly: when $|\Pi| > 1$, the singleton-subsets of J determine the *minimal* parabolics; here, the meaning is "minimal, subject to properly containing a Borel subgroup".

(7) Flag-transitivity: For a fixed J, G is transitive on the set of all parabolics of that type J—this is the relevant notion of flag-transitivity for Lie-type groups. $\diamondsuit$

EXAMPLE 1.3.21 (The k-space stabilizer revisited). We now express the k-space stabilizer in earlier Example 1.3.4, as a parabolic P_J in the language of (4) above—indeed a maximal parabolic as in (6); this will in particular explain the notation $P_{\hat{k}}$ of that earlier Example.

We can essentially mimic structures seen in the small case $SL_3(q)$, arising from the Lie algebra sl_3 in Example 1.3.10: recall there the two simple roots corresponded to the two positions just below the diagonal.

Indeed we expand this to a description of the general root system of type A_{n-1}, for $SL_n(q)$; for rerefence see e.g. [**Car89**, Sec 3.6(i)]: The corresponding Dynkin diagram has the form $\circ - \circ - \cdots - \circ$, with all roots of the same length. The simple system Π has simple roots denoted by $\alpha_1, \cdots, \alpha_{n-1}$. And the positive roots Φ^+ can be characterized as sums of simple roots which are adjacent in the ordering on Π: namely of form $\alpha_i + \alpha_{i+1} + \cdots + \alpha_{j-1} + \alpha_j$.

In Example 1.3.4, we take J to be the *complement* of the k-th node: so for the simple root α_k for the k-th subdiagonal position, P_J is the k-space stabilizer in 1.3.4. Indeed if we write $\hat{k}$ for the complement $J = \Pi \setminus \{k\}$ of $\{k\}$, then we get the notation $P_{\hat{k}}$; and we have finally explained why we wrote that form in Example 1.3.4: For in our present convention, the simpler notation P_k in fact refers to the k-th *minimal* parabolic; and the $\hat{k}$ in $P_{\hat{k}}$ emphasizes that root groups for all the *other* $\pm\alpha_j$ generate the typically-large Levi complement $L_{\hat{k}}$, as we will see below:

We now follow 1.3.20(4): For the unipotent radical $U_{\hat{k}}$, we need the root groups for the positive roots that are not combinations from J—namely those involving α_k. Using the characterization of roots above, these are sums of adjacent simple roots including α_k—and as matrices, they turn out to be those to the left of, and/or below, the subdiagonal position of $u_{\alpha_k} = e_{k+1,k}$. For the Levi complement $L_{\hat{k}}$, we need root subgroups which are combinations from $(\pm)$ J—namely those not involving α_k. These come from combinations of simple roots either before α_k (namely α_j for $j < k$), or after α_k (with $j > k$). As matrices, the corresponding root groups for $\pm\alpha_j$ generate first the SL_k on V_k, and then the SL_{n-k} on the quotient V/V_k—the factors of the Levi complement $L_{\hat{k}}$, which we had seen in Example 1.3.4. $\diamondsuit$

EXERCISE 1.3.22 (More practice with root systems and parabolics). Exhibit the above parabolics $P_{\hat{k}}$ in some small cases such as $L_4(2)$ and $Sp_4(2)$. Hint: Some details are provided in appendix Remark B.1.1. $\diamondsuit$

Some remarks on the approach via algebraic groups. We conclude the section with just a brief glimpse of the alternative approach to Lie-type groups, via fixed points of automorphisms of algebraic groups.

Here the story does not start with a root system Φ; instead Φ emerges only later, from the process of classifying the semisimple algebraic groups.

That approach is in the context of algebraic geometry. In order to use tools such as the Hilbert Nullstellensatz, it is necessary to work over an algebraically closed field k. (For us, k will be the algebraic closure $\overline{\mathbb{F}_p}$ of $\mathbb{F}_p$.) In particular, the groups of initial interest will all be infinite.

Here an *algebraic group* over k is defined "intrinsically"—as a group which is in fact a *variety*: namely the zero set of a collection of polynomials over k in suitable coordinates. For example, note that $SL_n(k)$ is the zero set of $(\det - 1)$—where we recall that the determinant is indeed a polynomial in the matrix entries. Furthermore the group operations of multiplication and inversion are also expressed via polynomials.

The classification of semisimple[16] algebraic groups during the 1950s used techniques of algebraic geometry, to directly produce "abstract" root subgroups (without exponentiating from some Lie algebra). The interrelations of the abstract root subgroups could then be shown to satisfy the restrictions corresponding to a root system Φ; so that the Cartan-Killing classification of root systems could be invoked, to give essentially the same set of answers for algebraic groups as for Lie algebras.

Now taking k to be $\overline{\mathbb{F}_p}$, we let F denote the field automorphism $x \mapsto x^p$; of course the fixed points in k under F^a are just the finite field $\mathbb{F}_{p^a}$. And so with the algebraic-group theory and classification in place, it is then relatively easy to obtain the finite Lie-type groups:

REMARK 1.3.23 (Finite Lie-type groups via algebraic groups). So the fixed points of F^a in the algebraic group G over $\overline{\mathbb{F}_p}$ give the finite Chevalley group $G(p^a)$. The twisted groups $G(p^a)$ arise as the fixed points under the product of F^a with a suitable "graph" automorphism—cf. the discussion after later 1.5.4. ◊

We mention, roughly conversely, that the Chevalley construction, applied to the algebraic closure $\overline{\mathbb{F}_p}$ rather than to a finite field, reproduces as $G(\overline{\mathbb{F}_p})$ the algebraic group—but in a non-intrinsic way, from the viewpoint of algebraic geometry.

Some easy applications of the CFSG-list

The main theme of this book is applications of the CFSG; in practice, this means: my personal selection (with much help from more knowledgeable colleagues) of a number of sample applications. So before starting those, let me try to provide at least a glimpse of the more general context of applications:

First I'll recommend some extremely valuable general surveys of CFSG applications, by major experts in the indicated areas:

- permutation groups: Cameron [**Cam81**];
- representation theory: Tiep [**Tie14**];
- maximal subgroups: Kleidman-Liebeck [**KL88**];
- various areas: Kantor [**Kan85**]; Guralnick [**Gur17**].

Of course I have used parts of these, in preparing the material for this book.

[16] Here semisimple also has an intrinsic definition; but it ends up being product-of-simple.

Now very roughly: The applications in the first five chapters of the book remain fairly close to aspects of group structure; while in the last five chapters, we consider applications in various additional areas. And correspondingly, we'll provide some further context on those broader areas of application, in later Section 6.4.

The particular theme of the remaining sections of the present chapter is to provide some "initial" applications: which can ideally be deduced fairly easily from the list of the simple groups in the CFSG 1.0.2, along with the basic properties of the three classes of groups recorded in the previous sections.

So we might begin with some fairly naive question like, what does knowing the CFSG-list immediately buy us?

1.4. Structure of $\mathcal{K}$-groups: Via components in $F^*(G)$

One place to start might be the inductive situation of the CFSG-proof itself; we review the application-situation of Remark 1.0.1(1):

In a minimal counterexample—some "unknown" simple G—we may apply the CFSG-list 1.0.2, by induction, to any proper subgroup $H < G$. What kind of description of H does the list give?

The most elementary answer might be to consider a composition series:

$$1 = H_0 \lhd H_1 \lhd \cdots \lhd H_{n-1} \lhd H_n = H$$

where each successive quotient H_{i+1}/H_i is simple. So by induction, each such quotient is either of prime order, or a simple group on the CFSG-list. That is, H is a $\mathcal{K}$-group, in the standard terminology introduced in 1.0.1(1).

This description of H via composition factors can be sufficient for approaching some problems. However, it is not really very informative; e.g. it does not deal with possibly-complicated extension problems among the sections.

Instead, much analysis in finite group theory of such subgroups H has proceeded by focusing on the generalized Fitting subgroup $F^*(H)$. So we wish to discuss how the CFSG-list can be applied in that viewpoint.

But first, we'll quickly review the basic theory of $F^*(H)$; the experienced reader can of course skip over this background.

The generalized Fitting subgroup and its properties. The motivating idea is roughly to analyze a finite group X by focusing on some important subgroup—which is suitably "crucial", as opposed to say peripheral, for the overall structure of X.

The Fitting subgroup. One notion of crucial arose classically, in the theory of solvable groups. For any group X, we define the *Fitting subgroup*:

(1.4.1) $$F(X) := \text{ product of all normal nilpotent subgroups of } X.$$

(In this definition, we could replace normality by its transitive extension, namely subnormality.) As an easy example of the group: $F(S_4 \times S_3) \cong (\mathbb{Z}_2 \times \mathbb{Z}_2) \times \mathbb{Z}_3$.

The "crucial" aspect of $F(X)$ is exhibited by the self-centralizing property (for example 31.10 in [**Asc00**, 31.10]):

THEOREM 1.4.2 (Fitting's Theorem). *If X is a finite solvable group, then we have $C_X\big(F(X)\big) \leq F(X)$.*

For notice that we get the following corollaries:

REMARK 1.4.3 (Consequences of the self-centralizing property). Note that if we had $F(X) = 1$ for solvable X, then we would obtain from Theorem 1.4.2 that $C_X(1) \leq 1$; hence $X = C_X(1)$ must be trivial. So since in practice we work with $X > 1$, we can always expect that also $F(X) > 1$.

Furthermore it follows from Theorem 1.4.2 that for solvable X:

$$C_X\big(F(X)\big) = Z\big(F(X)\big).$$

Hence $X/Z\big(F(X)\big)$ must act *faithfully*, as automorphisms of $F(X)$; indeed:

$X/F(X)$ must induce *outer* automorphisms of $F(X)$.

So, roughly: the rest of X can't just ignore $F(X)$; indeed when $\mathrm{Out}\big(F(X)\big)$ is small, we see that $F(X)$ even constitutes "most of" X. This gives a reasonable notion of "crucial".

These same consequences will in fact hold for general X, when we instead use $F^*(X)$ below in place of $F(X)$—in view of the analogous self-centralizing property (1.4.6). ♢

The generalized Fitting subgroup. Bender, building on ideas of Gorenstein and Walter, saw how to extend these notions to an arbitrary finite group X. He defined the *generalized Fitting subgroup*:

$$F^*(X) := E(X)F(X) \tag{1.4.4}$$

where the new term $E(X)$ is analogous to $F(X)$—but with respect to nonabelian simple groups:

$$E(X) := \text{ product of all subnormal quasisimple subgroups of } X. \tag{1.4.5}$$

We will say more about $E(X)$ in a moment. The main result is that we get the self-centralizing property for any X; see e.g. [**Asc00**, 31.13]:

THEOREM 1.4.6. $C_X\big(F^*(X)\big) \leq F^*(X)$.

And so $F^*(X)$ also has the "crucial" consequences we discussed in Remark 1.4.3.

We mention one standard situation where these properties are used:

DEFINITION 1.4.7 (almost-simple). A group X with $F^*(X)$ simple (say S) is called *almost-simple*. Notice then by 1.4.6 and the discussion in 1.4.3, that X is an extension of S, by some subgroup of $\mathrm{Out}(S)$. ♢

We still need to say a few more things about $E(X)$:

Quasisimple components in $E(X)$. First recall that L is quasisimple if $L/Z(L)$ is simple, and L is perfect ($L = [L, L]$). The perfect requirement is essentially a kind of irreducibility—e.g. to avoid multiple factors, or "inessential" elements in $Z(L)$.

Assuming that L is quasisimple, with central quotient S, we might ask: how much larger can L be than S? Or equivalently, how large can $Z(L)$ be? As noted just above, $Z(L)$ is limited by the restriction that L is perfect. The answer to this question is given by the *Schur multiplier* of S; and for given S, there are standard techniques for computing the multiplier. For Schur's classical theory of the multiplier, see e.g. [**Asc00**, Sec 33].

We next recall that a quasisimple group L, which is subnormal in X, is called a *component* of X. Thus $E(X)$ is generated by the components of X.

REMARK 1.4.8 (Some properties of components). We recall e.g. from 31.5 in [**Asc00**] that distinct components commute; so that $E(X)$ is in fact a central[17] product of the components.

In particular, no component could be diagonally embedded in a product of two or more components. It follows that elements of X not normalizing components must in fact *permute* them intact—rather than conjugating them to "mixed" products. (This is very different from the situation inside nilpotent $F(X)$—where a direct product of n copies of $\mathbb{Z}_p$ might conceivably admit the "mixing" action of any subgroup of $GL_n(p)$.) $\diamondsuit$

With these structures in hand, we can proceed to:

Applying the CFSG to $F^*(H)$. We return to our earlier inductive setup in the CFSG proof where $H < G$; and we indicate some effects that the CFSG-list has on the structure of $F^*(H)$.

First, we have in effect an extension of the CFSG-list 1.0.2 from simple to quasisimple groups:

REMARK 1.4.9 (Quasisimple CFSG-list). The possible quasisimple covers L of simple groups in the CFSG-list 1.0.2 (here we implicitly mean the longer-form list of Remark 1.3.19) are known. So the possible components L in $E(H)$ are known. $\diamondsuit$

For of course, the simple quotient $L/Z(L) =: S$ is known by induction, and appears in the CFSG-list. And then the possibilities for L are limited by the Schur multiplier of S.

And those Schur multipliers are also known: Indeed as the various simple groups (especially sporadics) were discovered, typically the discoverer would also determine the Schur multiplier. In fact the multipliers are mostly quite small. A number of other authors also contributed to this process; finally Griess in [**Gri72**] completed the remaining problems. Hence his list served for a long time as a standard reference. Cf. also more recent discussions, such as [**GLS98**, 5.1,6.1].

Hence the possible extensions L of S can also be (and indeed have been) constructed; so that they are also known.

Second, we observe just informally that:

[17] Recall this means that commuting components L, M satsify $L \cap M \leq Z(E(X))$.

The action of $H/F^*(H)$ on $E(H)$ is very restricted:
For we already saw in Remark 1.4.8 that elements of H not normalizing components can only permute them (indeed, in sets of isomorphic components). This reduces much of the corresponding analysis to questions about subgroups of the symmetric group S_n for suitable n.

So we turn to elements of H normalizing a component L. Since L is known via the quasisimple CFSG-list 1.4.9 above, inner automorphisms of L are known. Any other nontrivial action must induce an outer automorphism of L, and hence of the simple quotient S. And for $\mathrm{Out}(S)$, the situation is similar to that for Schur multipliers in the discussion following 1.4.9: namely the outer automorphism groups of the simple S in the CFSG-list had already been determined. Indeed we'll return to outer automorphism groups in the next section, and see that they are solvable—and indeed small compared to the size of S.

This more explicit view of H via $F^*(H)$ is typically more useful than just a composition series.

1.5. Outer automorphisms of simple groups

It had been observed early on that the outer automorphism groups of the simple groups are rather small, and of uncomplicated structure. Schreier in 1926 conjectured that they must be solvable.

And as we observed in the previous section, when new simple groups were discovered, their discoverers usually also determined their automorphism groups—in particular, verifying the conjecture in the new cases.

Indeed researchers became increasingly convinced of the truth of the conjecture; and sometimes assumed it, to deduce significant consequences. Furthermore it was used inductively, in certain parts of the CFSG itself.

Thus when the CFSG was finally proved, the work on the automorphisms had already been done—so that the Schreier Conjecture became an "instant" theorem. As did the various known consequences mentioned in the previous paragraph. So as one important application of the CFSG, we get:

THEOREM 1.5.1 (Schreier Conjecture). *For simple S, $Out(S)$ is solvable.*

For our present purposes of presenting applications of the CFSG-list, it will actually be comparatively easy to outline the verification of the Conjecture for the three usual classes of simple groups:

Outer automorphisms of alternating groups. It is a classical result that automorphisms of the alternating group mostly come from the symmetric group—see e.g. [**GLS98**, 5.2.1] for:

$$\mathrm{Out}(A_n) \cong \mathbb{Z}_2\text{—except } \mathbb{Z}_2 \times \mathbb{Z}_2 \text{ for } n = 6. \tag{1.5.2}$$

These groups are certainly solvable; even elementary abelian 2-groups.

Outer automorphisms of sporadic groups. The results, from various authors, are tabulated in e.g. [**GLS98**, Table 5.3.a–z]; we summarize them as:

$$\text{For } G \text{ sporadic, } |\mathrm{Out}(G)| = 2 \text{ in 12 of the 26 cases; } = 1 \text{ otherwise.} \tag{1.5.3}$$

Again the groups are solvable, and indeed elementary abelian 2-groups. But this doesn't hold for all simple groups; indeed:

Outer automorphisms of Lie-type groups. This final case has somewhat more interesting structure; the results arise naturally from the Lie context that we sketched in earlier Section 1.3. See e.g. [**GLS98**, 2.5.12,1.15.7] for the result which we summarize as:

THEOREM 1.5.4 (The "diagonal-field-graph" theorem). *For simple G of Lie type, $Out(G)$ has a normal subgroup D, with quotient $F \times \Gamma$; where:*
D induces diagonal *automorphisms;*
F induces field *automorphisms; and*
Γ induces graph *automorphisms.*
Each of the groups D, F, Γ is abelian; so $Out(G)$ is solvable.

We next quickly indicate some features of these automorphisms; they demonstrate what intuitively *could* be an automorphism of a matrix group. We mention that as before, we'll give statements for the easier case of Chevalley groups; adjustments are sometimes required for the case of twisted groups.

The *diagonal* automorphisms D roughly extend, outside simple G, the Cartan subgroup H in Remark 1.3.20(2). For example: consider any non-scalar diagonal matrices from $GL_n(q) \setminus SL_n(q)$. The order $|D|$ is at most that of the *fundamental group* (cf. [**Car89**, p 99]) of G—this value is at most 4, except for a cyclic group of order n in the case of linear $L_n(q)$ (or unitary $U_n(q)$).

The *field* automorphisms F for $G(p^a)$ are determined by the Galois group of $\mathbb{F}_{p^a}$ over $\mathbb{F}_p$; so this group is cyclic of order a. It is induced by applying the field automorphism to the matrix entries of the groups $G(p^a)$ from the Chevalley construction in the discussion prior to Example 1.3.14.

The *graph automorphisms* Γ are determined by the Dynkin-diagram automorphisms which we indicated in Remark 1.3.11; this group is cyclic of order at most 3 (or S_3, for type D_4). For example, the inverse-transpose automorphism of the linear group arises from a graph automorphism.

EXERCISE 1.5.5 (Practice with automorphisms). Explore the various automorphism types above—in some small-dimensional linear and other classical cases.

Hint: some of these can be compared and contrasted, using certain standard isomorphisms among small simple groups—see for example [**ALSS11**, p 261] and [**Wil09**, 3.11,3.12] and [**GLS98**, 2.2.10] and [**GLS94**, Table II, p 10]. Here are some small cases to explore:

(1) Transpositions in S_5 exhibit $\mathrm{Out}(A_5)$.
But A_5 can also be regarded as the Lie-type group $L_2(4)$, or $L_2(5)$.
When viewed as $L_2(4)$, the outer automorphism has field type.
Whereas when viewed as $L_2(5)$, it has diagonal type. (Explain...)
Similarly transpositions in S_8 exhibit $\mathrm{Out}(A_8)$.
But for A_8 viewed as Lie-type $L_4(2)$, the automorphism has graph type.

(2) Transpositions in S_6 give one automorphism in $\mathrm{Out}(A_6)$.
For A_6 viewed as M'_{10},[18] from $M_{10} \setminus M'_{10}$ we get more outer automorphisms.
But A_6 can viewed as Lie-type $L_2(9)$; or as $Sp_4(2)'$, proper in Lie-type $Sp_4(2)$.
When viewed as $L_2(9)$, outer automorphisms have diagonal and field types.
When viewed as $Sp_4(2)'$, they come from graph type and $Sp_4(2)$. (Explain.) ♢

Quite a few applications of the CFSG in fact only quote it in the "mild" form of the Schreier Conjecture 1.5.1. Indeed we'd mentioned that various such applications were made before the CFSG, on the assumption that the classification would eventually be available. So we close the chapter with a quick mention of some of these—as well as problems that involved closer consideration of the explicit groups in the CFSG-list.

1.6. Further CFSG-consequences: e.g. doubly-transitive groups

Various early applications of the CFSG arose in the area of permutation groups; we had mentioned Cameron [**Cam81**] as a standard reference for these. We will next extract from Cameron's discussion a very rapid overview of the solution of one major classical problem:

The classification of doubly-transitive groups. First we review some standard background; as usual, the experienced reader can skip ahead.

Multiple transitivity. Recall we say a subgroup of S_n is:

(1.6.1) *k-transitive*, if it is transitive on the set of ordered k-subsets.

For example, it is standard (1.1.1(1)) that S_n is n-transitive, and the simple subgroup A_n is still $(n-2)$-transitive.

Of course single transitivity is ubiquitous, since a group is transitive on the set of cosets of any subgroup. But double transitivity is already comparatively rare; so, long ago, permutation-group theorists proposed:

PROBLEM 1.6.2. *Classify doubly-transitive groups.*

There are a fair number of examples: e.g. rank-1 Lie-type groups are doubly (sometimes triply) transitive on the set of their Borel subgroups (recall 1.3.20(3)). But the condition of 4-transitivity and higher is quite rare: Aside from the standard symmetric and alternating groups as above, only certain sporadic examples were known, namely those we had mentioned in Section 1.2: 4-transitivity for M_{11} and M_{23}, along with 5-transitivity for M_{12} and M_{24}. So a solution of Problem 1.6.2 above would include a solution of the sub-problem:

[18]Recall M_{10} is the point stabilizer in M_{11}—cf. [**CCN⁺85**, p 4].

PROBLEM 1.6.3. *Classify multiply-transitive groups. (In particular, having* 6*-transitivity should imply action only of* S_n *or* A_n*.)*

In fact, pre-CFSG work of Wielandt, Nagao, and O'Nan had solved the sub-problem 1.6.3—modulo assuming the Schreier Conjecture 1.5.1. However, the main Problem would require the CFSG-list in fuller detail.

The 2-transitive classification. For the main Problem 1.6.2, around 1905 Burnside had given a basic reduction [**Cam81**, 5.2]:

THEOREM 1.6.4. *A 2-transitive group H has a unique minimal normal subgroup N—which must be either an elementary abelian p-group, or a nonabelian simple group.*

Of course the latter case of simple N provided one strong motivation for the CFSG.

But first we will summarize the treatment of the former case, in which N is an elementary abelian p-group—say of rank r: Here N must act regularly (in particular, transitively) on the points. Now an elementary argument (see later 3.0.1), going back at least to Frattini in 1885, shows that a transitive subgroup is supplemented by a point stabilizer; thus we have $H = NH_\alpha$, for the stabilizer H_α of a point α. The case of solvable H_α was handled by Huppert. Hering then showed that nonsolvable H_α (which must be a subgroup of $GL_r(p)$) has a unique nonabelian simple composition factor S. So here also the CFSG-list (again in the longer-form of 1.3.19) could be applied, to examine each S—and this was done by Hering and others; see for example [**Lie87a**, Appendix] for details.

The case of simple N was handled—of course using the CFSG-list to examine each N—by work of Maillet, Howlett, and Curtis-Kantor-Seitz.

And that completed the solution of the 2-transitive Problem 1.6.2; which of course also covered the multiply-transitive sub-problem 1.6.3.

We'll now give a rough summary of which groups can arise in the case above of simple N. Here N is called the *socle* of H, in the language of later Remark 6.0.4. For the detailed statement of the result, see [**Cam81**, 5.3].

THEOREM 1.6.5 (simple-socle 2-transitive groups). *For* 2*-transitive H, the possibilities for a simple socle N are as follows; several groups have two* distinct *doubly-transitive representations:*

(alternating:)

- A_n, *on the points of the natural permutation representation;*
- A_7, *two representations on* 15 *points (arising from $A_7 < A_8 \simeq L_4(2)$);*

(Lie type:)

- $L_n(V)$, *two representations of degree $\frac{q^n-1}{q-1}$: on points, or hyperplanes, of the projective space* 7.0.1 *of V;*
- $G(q)$ *of Lie rank* 1,[19] *on Borel subgroups (points of the projective line);*
- $Sp_{2n}(2)$, *two representations on orthogonal forms in 2d-space (of plus-type, or minus-type);*
- $L_2(11)$, *two representations on* 11 *points (i.e. not the usual* 12 *points of its projective line).*

(sporadic:)

- M_{11}, M_{12}, M_{22}, M_{23}, M_{24}; Co_3, *on* 276 *points; HS, two on* 176 *points.*

[19] These rank-1 groups are $L_2(q)$, $U_3(q)$, $Sz(q)$, and $Ree(q)$.

REMARK 1.6.6 (The successive-lists aspect). In addition to the list of the CFSG appearing as input to Theorem 1.6.5, the reader will have observed that the output of the Theorem is also a longish list—indeed, a refined sub-list of the input. And of course, applications of the Theorem could result in still-further lists. This aspect of successive lists might seem unexpected to the reader new to the area; but it is a very standard practice in finite group theory. ◇

Some other consequences of the Schreier Conjecture. A further instant consequence of the CFSG was the Hall-Higman reduction for the *restricted Burnside problem*—since the work of [**HH56**] had assumed the Schreier Conjecture 1.5.1.

The Hall-Higman reduction (and hence the CFSG via the Schreier Conjecture) was in turn used by Zelmanov [**Zel91**], in his celebrated solution of the restricted Burnside problem—work which was soon recognized by the award of a Fields Medal.

CHAPTER 2

Outline of the proof of the CFSG: Some main ideas

The first half of this chapter is a selection of some of the material appearing in Chapters 0–2 of Aschbacher-Lyons-Smith-Solomon [**ALSS11**]—outlining the main case divisions of the CFSG, and the eventual treatment of those cases.

Before starting, we provide some context on that particular outline:

Some history of the CFSG outline. Why did the authors of the outline in [**ALSS11**] decide to give such an extensive exposition?

Now that the participants in the great CFSG project are starting to retire, or at least to age (indeed a number have died), one important motivation was to make the CFSG proof more accessible for future generations—while our memories are still available.

For example, the 2005 *AMS Notices* article [**Dav05**] of Brian Davies, on the general topic of long proofs in mathematics, had suggested this problem of aging memory; in particular, he quoted Aschbacher to the effect that there was at that time "no published outline" of the CFSG proof.

An outline had in fact been begun by Gorenstein in 1982: he published an introduction [**Gor82**], and an extensive overview of the Odd Case [**Gor83**]); but he couldn't really finish his outline—since the final treatment of quasithin groups in the Even Case didn't appear until later [**AS04b**].

Thus the authors of [**ALSS11**] were in particular bringing Gorenstein's earlier outline project to completion; indeed the later Chapters 3–8 of that work provide an extensive overview of the treatment of the Even Case (now including of course the quasithin groups).

However, Chapters 0–2 of [**ALSS11**] provide a much briefer, introductory outline of the CFSG proof. And this chapter of the present work similarly has the goal of giving an expository overview of the main *case divisions* of the CFSG; along with a few important concepts in the eventual treatment of those cases.

We mention that the outline about to be presented refers to the "original" proof of the CFSG. Readers may be aware that there are more recent approaches being developed, towards improving and innovating the proof. We'll say a little more about these, in our Afterword in Section 2.3.

2.0. A start: Proving the Odd/Even Dichotomy Theorem

To begin on our outline, we provide some historical background—from which the initial case division, of Odd vs Even Cases, will begin to emerge:

Prelude: Approaching the CFSG via involution centralizers. Our mention of Odd and Even Cases has implicitly suggested a special role for $p = 2$.

This distinction was already clear, long before the CFSG. Indeed perhaps the most famous result from the "pre-history" of the CFSG is given by the Odd Order Theorem [**ALSS11**, 1.2.1] of Feit and Thompson:

(2.0.1) A group of odd order is solvable.

Thus a nonabelian simple group G contains some t of order 2—an *involution*.

Hence a natural subgroup to consider is the involution centralizer $C_G(t)$. This group is not simple, because it has t in its center; so as G is simple, it is a proper subgroup: $C_G(t) < G$.

Since we would expect to work with G a minimal counterexample to the CFSG, we might then wish to apply induction to $C_G(t)$. This suggests:

REMARK 2.0.2 (A naive centralizer-approach to the CFSG). We could hope to proceed along the following lines:

- determine, using induction, all possible H with a central involution t;
- then for each H, determine all simple G with $C_G(t) \cong H$. $\diamondsuit$

Indeed there is a classical result showing that this approach might be feasible, namely the Brauer-Fowler Theorem [**BF55**]—which shows that given such a group H with t central, only *finitely* many simple G can have $C_G(t) \cong H$.

And in fact the simple-minded approach in Remark 2.0.2, admittedly rendered less naive by substantial adjustments, is basically how the CFSG was eventually proved. As we will soon be seeing.

The fundamental odd/even distinction will be exhibited even more dramatically, when we next explore centralizers $C_G(t)$ in some typical examples of simple groups G.

Motivation: some examples of the notions of Odd and Even Cases. What should "typical" G mean? We had commented, in the statement of the CFSG-list 1.0.2, that most simple groups are of Lie type. Indeed for our present expository purposes, it will even suffice to explore inside the group $GL_n(q)$, which is just almost-simple modulo its center; and within $GL_n(q)$, we will be able to draw suggestive matrix pictures.

EXAMPLE 2.0.3 (Odd Case Example). Take $G := GL_n(p^a)$. And assume here that p is *odd*. For a typical involution, take t diagonal, with say k entries of -1 (we have $-1 \neq 1$ since p is odd!):

$$t := \left(\begin{array}{c|c} -I_k & 0 \\ \hline 0 & I_{n-k} \end{array}\right)$$

By elementary linear algebra, matrices commuting with t must preserve its (± 1)-eigenspaces. So in the centralizer we get block-partitioned matrices, as in the following picture:

$$C_G(t) = \left(\begin{array}{c|c} GL_k(q) & 0 \\ \hline 0 & GL_{n-k}(q) \end{array}\right)$$

Now the SL_r in each GL_r is usually[1] quasisimple; so since these SL_r are also normal in $C_G(t)$, they are in fact *components* of $C_G(t)$ (in the language of 1.4.8). $\diamondsuit$

This last feature of our Odd Case Example suggests how to abstractly define our general *Odd Case*:

DEFINITION 2.0.4 (component type). We say X is of *component type*, if for some involution $t \in X$, the quotient $C_X(t)/O_{2'}\big(C_X(t)\big)$ has a component. Or, we can use the alternative terminology that $C_X(t)$ has a 2-*component*. $\diamondsuit$

We mention that the "core" above, namely the largest normal $2'$-subgroup denoted by $O_{2'}\big(C_X(t)\big)$, is typically trivial—or at worst central in $C_X(t)$. And indeed the reason for working here modulo $O_{2'}\big(C_X(t)\big)$ will emerge later, in the proof of the Dichotomy Theorem 2.0.9; where we have a situation in which we can even prove that this core must be trivial.

We get this same feature of component type also in the other Lie-type groups over fields of odd characteristic; and indeed in (large-enough) alternating groups, as well as in some sporadic groups.

EXERCISE 2.0.5 (More examples of component type). Find some similar components of involution centralizers in some smallish (but large-enough) groups G of the other types mentioned just above. For example, the centralizer in A_9 of the involution $(1,2)(3,4)$ has a component A_5. $\diamondsuit$

The pictures will look very different from the Odd Case Example 2.0.3, when we instead take $p = 2$:

EXAMPLE 2.0.6 (Even Case Example). Now take $p = 2$; so that $G := GL_n(2^a)$. In our field of characteristic 2, the element -1 is now the same as 1. So our diagonal t from the Odd Case Example 2.0.3 would here be just the identity—rather than an involution.

Instead, we will now obtain an involution t via the smallest Jordan-form matrix of order 2: it has all its Jordan blocks, which are for the eigenvalue 1, of size 1×1—except for a single 2×2 block. In fact to get a more symmetric form for some later matrices, we can conjugate in G so as to move the subdiagonal entry from that 2×2 block down to the lower-left position of the full matrix; so that our t becomes:

$$t := \left(\begin{array}{c|c|c} 1 & 0 & 0 \\ \hline 0 & I_{n-2} & 0 \\ \hline 1 & 0 & 1 \end{array}\right)$$

[1]usually: Away from some small dimensions, and small fields—recall 1.3.17.

Then a little more linear-algebraic computation determines that:

$$C_t := C_G(t) = \left(\begin{array}{c|c|c} * & 0 & 0 \\ \hline * & GL_{n-2}(2^a) & 0 \\ \hline * & * & * \end{array}\right)$$

In fact C_t is a semi-direct product UL,[2] where:

$$U := \left(\begin{array}{c|c|c} 1 & 0 & 0 \\ \hline * & I_{n-2} & 0 \\ \hline * & * & 1 \end{array}\right), \quad L := \left(\begin{array}{c|c|c} * & 0 & 0 \\ \hline 0 & GL_{n-2}(2^a) & 0 \\ \hline 0 & 0 & * \end{array}\right)$$

Here U is a 2-group—in fact the largest normal 2-subgroup $O_2(C_t)$. And again we have a nearly-simple subgroup GL_{n-2}; but, in strong contrast to the situation in the Odd Case Example 2.0.3, this subgroup is *not* normal in C_t: Just note that conjugating L by elements of U results in nonzero entries in the sub-diagonal blocks—hence elements not in L. Indeed here the 2-subgroup U is roughly the "only" normal subgroup: For we can check that C_t does not have any subnormal quasisimple subgroups; or even any normal odd-order subgroups—work if necessary in the quotient modulo any central scalar matrices. This property is usually expressed in the language (1.4.4) of the generalized Fitting subgroup, as: $F^*(C_t) = O_2(C_t)$. ♢

This suggests how to abstractly define our *Even Case*:

DEFINITION 2.0.7 (Characteristic 2 type). We say a group X is of *characteristic 2 type*, if $F^*(N) = O_2(N)$ for all 2-local subgroups N. Indeed it suffices (e.g. [**ALSS11**, B.1.6]) to have $F^* = O_2$ for all involution centralizers $C_X(t)$. ♢

The property of characteristic 2 type can be verified in the other Lie-type groups over a field of characteristic 2; as well as in a few sporadic groups.

EXERCISE 2.0.8 (More examples of characteristic 2 type). Check this feature, in some Lie-type groups in characteristic 2, other than linear groups; say classical groups—symplectic, unitary, orthogonal. (E.g. in $Sp_4(2)$—which can also be regarded as S_6, as in Exercise 1.5.5(2).) ♢

We've now seen that our notions of Odd and Even Cases cover the actual examples of simple groups in the CFSG-list 1.0.2. But how might we *abstractly* show that some "unknown" simple G should be either of component type, or of characteristic 2 type? We now move on to the actual mathematics of this process:

[2] In the Lie-type language of Remark 1.3.20(4), this is the Levi decomposition for the parabolic subgroup C_t.

Result: The Odd/Even Dichotomy Theorem. We need one further piece of historical background:

In the earliest days of the CFSG, it was natural for researchers to begin with classes of groups which could be considered "small" by some measure.

One such measure is the 2-rank $m_2(G)$, the largest rank of an elementary abelian 2-subgroup of G. And the simple groups in the small case $m_2(G) \leq 2$ had been classified by about 1972—in work of Alperin-Brauer-Gorenstein, Lyons, and Walter; see e.g. [**ALSS11**, 1.4.6].

In fact the case $m_2(G) \leq 2$ had an independent importance: for the "generic" case, in which we have $m_2(G) \geq 3$, is required in order to use *signalizer functor methods*. We will be briefly indicating these, in outlining the proof (for fuller details, see 0.3.10 or B.3.5 in [**ALSS11**]) of the following result—which establishes the main case division that we have been leading up to:

THEOREM 2.0.9 ((Odd/Even) Dichotomy Theorem). *Assume that G is simple, with $m_2(G) \geq 3$. Then G is of component type, or G is of characteristic 2 type.*

The proof is not difficult, and is comparatively short, for finite group theory—around 4 pages,[3] in the treatment presented in the latter part of Section 0.3 of [**ALSS11**]. Furthermore, a number of the ideas involved in the proof were re-used repeatedly, in many other places in the CFSG. So in the remainder of the Section, we will present a roughly summarized form of that treatment in [**ALSS11**].

Sketch: Proof of the Dichotomy Theorem. We begin by assuming:

(2.0.10) G does *not* have component type.

Hence: we must end by showing that G has characteristic 2 type.

Now the initial assumption (2.0.10)—the denial of component type—means, using Definition 2.0.4, that for each involution t of G, there are no 2-components in $C_t := C_G(t)$. That is, when we pass to the quotient $\overline{C}_t := C_t/O_{2'}(C_t)$, there are no components; so that $E(\overline{C}_t) = 1$. On the other hand, we have $O_{2'}(\overline{C}_t) = 1$, since we have already quotiented out any odd-order normal subgroups. In particular, there are no odd-order normal subgroups in the Fitting group $F(\overline{C}_t)$; so we have:

(2.0.11) $$F^*(\overline{C}_t) = E(\overline{C}_t)F(\overline{C}_t) = O_2(\overline{C}_t) \text{ for all involutions } t.$$

We saw above that we must end up with characteristic 2-type; so:

(2.0.12) It will suffice to show that $O_{2'}(C_t) = 1$ for all t.

For then $\overline{C}_t = C_t$, so that (2.0.11) gives $F^*(C_t) = O_2(C_t)$—as required for characteristic 2 type in Definition 2.0.7.

So we define $\theta(t) := O_{2'}(C_t)$; and we set out to show that the values of the function θ must be trivial.

[3] However (as we'll see), that proof does *assume* two basic results, also fairly elementary, which were considered standard by the early 1970s when the result was formulated.

Balance and signalizer functors. Since $m_2(G) \geq 3$, G contains some elementary 2-subgroup A of rank ≥ 3. Though later we will vary over such A, for the moment we *fix* A: so that in the next few paragraphs, elements and subgroups such as t, u, B below are implicitly assumed to lie *inside* this particular A.

Using the property of "no 2-components" from our denial of component type in (2.0.10), it is a fairly elementary exercise, mainly using the properties of the generalized Fitting subgroup, to establish the symmetry property:

(2.0.13) *balance*: For commuting $t, u \in A$, we have $\theta(t) \cap C_u = \theta(u) \cap C_t$.

Gorenstein, building on balance and other ideas of Thompson going back to the Odd Order paper, developed the notions of a signalizer functor; and this was subsequently refined by Goldschmidt. We approximate the definition as:

(2.0.14) *signalizer functor*: roughly, a $2'$-valued function (on A) with balance.

The rationale for this terminology will emerge a little farther down. Since here we are working under the hypothesis that $m_2(G) \geq 3$ (so that we have such an A), and we observed in (2.0.13) that we have balance, our present choice of θ as $O_{2'}\big(C_G(-)\big)$ is then a signalizer functor in the sense of (2.0.14).

The methods now proceed by making an extension of our function θ to larger elementary abelian 2-subgroups, in a fairly obvious way: Namely for $B := \langle t, u \rangle$ with commuting t, u, set $\theta(B) := \langle \theta(v) : v \in B^{\#} \rangle$; and so on, for higher-rank groups. And now the goal is to show that this larger-θ can take only trivial values; which will imply the same desired property for our original θ defined just on t.

In particular, this construction shows why we can refer to the extended function θ as a *functor*: For we can consider the poset (i.e. partially ordered set) of nontrivial elementary 2-subgroups as the objects of a category—with morphisms given by the inclusions $\leq$, and the G-conjugations. Then the above definition via the group-span show that $B \leq C$ leads to $\theta(B) \leq \theta(C)$; while we have $\theta(B^g) = \theta(B)^g$ for our particular choice of $\theta(-)$ as $O_{2'}\big(C_G(-)\big)$. (The literature considers many other choices for $\theta(-)$.)

Signalizer-values and completeness. Now we invoke the standard, and comparatively elementary, Signalizer Functor Theorem—for a full statement, see e.g. 0.3.14 in [**ALSS11**]. The proof requires $m_2(G) \geq 3$.

One of the conclusions of that result now helps explains the terminology of "signalizer": Namely it shows that for any elementary 2-subgroup C, the $C_G(C)$-invariant value $\theta(C)$ is still a $2'$-group—this is not automatic, just from the definition of the extended-θ via the group-span. Thus the functor θ has values which are in fact "2-signalizers" in the language of Thompson.

A further important consequence of the Signalizer Functor Theorem is that θ is *complete*; roughly, it is "globally determined" by our fixed A of rank ≥ 3:

$$\text{For } t \in A^{\#},\ \theta(t) = \theta(A) \cap C_t.$$

Notice that this in particular implies the balance condition (2.0.13). Thus we could regard the various detailed conclusions of the Signalizer Functor Theorem, including the signalizer-valued and functorial properties above, as more or less logically equivalent to the definition of a signalizer functor just via balance in (2.0.14).

In practice we typically focus on the following corollary of completeness:

(2.0.15) *completeness* gives: $\theta(A) = \theta(B)$, for all B of rank ≥ 2 in A.

The rank-3 graph, and the cases for connectivity. At this point, we allow our previously-fixed A to vary—now over all elementary 2-subgroups of G of rank 3.

Notice that if A, A' are distinct such groups intersecting in B of rank 2, then completeness (2.0.15) shows that $\theta(A) = \theta(B) = \theta(A')$. Thus the separate definitions of extended-θ for subgroups of A, and of A', are compatible on their intersection—so that extended-θ is naturally defined on the union of those subgroup-posets. And so on.

Indeed this further suggests that we should consider a graph Γ: where vertices are the elementary 2-subgroups A of rank 3; and A, A' define an edge whenever we have $A \cap A' = B$ of rank 2. Then the observation in the previous paragraph extends to:

The value of $\theta(-)$ is constant on any connected component of Γ.

We then have two obvious cases: Γ is either disconnected or connected.

Suppose first that Γ is disconnected:

Here it follows, with some further work, that G has a subgroup M which is:

(2.0.16) *strongly embedded*: $|M|$ is even, but $|M \cap M^g|$ is odd $\forall g \notin M$.

Here we may quote the very-standard Strongly Embedded Theorem (for example [**ALSS11**, 0.2.3]) of Bender and Suzuki; which states that:

THEOREM 2.0.17 ((Bender-Suzuki) Strongly Embedded Theorem). *A simple group G with a strongly embedded subgroup is $L_2(q)$, $U_3(q)$, or $Sz(q)$, with $q = 2^a$.*

These conclusion-groups are often called the *Bender groups*. Note that they are the rank-1 groups of Lie type—in characteristic 2. In particular, they satisfy characteristic 2 type (including $\theta \equiv 1$); and so our proof is done in this case.

So we turn to the remaining case, where Γ is connected:

Here in particular the G-conjugacy class of A is connected; so by our earlier remark, the value of $\theta(A)$ is constant on this class. However, G permutes that conjugacy class, and hence permutes the corresponding values of θ over that class; so since these values are constant, G must normalize $\theta(A)$. Now by simplicity of G, we conclude that $\theta(A) = 1$. Finally, every involution t must lie in some A of rank 3; see e.g. the proof of 0.3.22 in [**ALSS11**]. It follows that $\theta(t) = 1$—as we wanted in (2.0.12), to in fact establish characteristic 2 type.

This completes our sketch of the proof of the Dichotomy Theorem. □

We comment that the proof actually has a *tri*chotomy structure; namely:

{ component type, disconnected, connected }.

And it happened—for our situation here with $p = 2$—that the latter two branches led to the same conclusion, of characteristic 2 type. But when we later adapt these arguments, for odd p when we are in the Even Case, the disconnected branch will remain distinct—giving our trichotomy.

Indeed the above notions related to balance and signalizer functors—including completeness and connectedness, leading to a trichotomy—were adjusted and re-used in many different situations, throughout the CFSG.

The basic Grid of subcases for the CFSG proof. The Dichotomy Theorem has given us a basic Odd/Even case division for the CFSG. But to close the Section, we further subdivide cases—by giving some more detail on the small/generic distinction that we had made, prior to stating that Theorem.

Historically, the treatment of $m_2(G) \leq 2$ was regarded as a part of the Odd Case—namely as the Small Odd Subcase. Correspondingly, the union of $m_2(G) \leq 2$ and component type is called "Gorenstein-Walter type" (GW) in [**ALSS11**].

In order to re-use, for odd primes p in the Even Case, the signalizer-functor arguments underlying the Dichotomy Theorem proof, a different small/generic case subdivision was used. For recall that characteristic 2 type focuses on 2-local subgroups; so we consider *odd* p-ranks—inside 2-locals:

Define $m_{2,p}(G) := \max\{m_p(H) : H \text{ is a 2-local subgroup of } G\}$.

The following parameter had already been of importance in Thompson's early work on N-groups [**Tho68**]:

$$e(G) := \max_p m_{2,p}(G). \tag{2.0.18}$$

And the Small Even Subcase is correspondingly given by $e(G) \leq 2$ (the *quasithin* condition, which we abbreviate by QT).

Thus the original CFSG treats a "grid" of four subcases:

REMARK 2.0.19 (The CFSG Grid). The basic subcases for the CFSG are:

	(Odd Case) GW type	(Even Case) characteristic 2 type
small	$m_2(G) \leq 2$	$e(G) \leq 2$ (quasithin)
generic	component type	$e(G) \geq 3$

◇

So we now turn to summarizing how these various subcases were handled.

2.1. Treating the Odd Case: Via standard form

Just as in the discussion before the Dichotomy Theorem 2.0.9, we continue to omit the treatment of the Small Odd Subcase $m_2(G) \leq 2$; instead referring the reader to [**ALSS11**, 1.4.6] for details.

Thus we assume that $m_2(G) \geq 3$; and that we are in the component-type branch of the Dichotomy Theorem—that is, the Generic Odd Subcase of the Grid 2.0.19.

So for some $t \in G$, the centralizer $C_t := C_G(t)$ has a 2-component L; that is, a component in $C_t/O_{2'}(C_t)$.

Fundamental contributions of Gorenstein-Walter and Aschbacher led to the notion of standard form, and the following idealized strategy—which is a further-simplified form of [**ALSS11**, 1.1.1]:

REMARK 2.1.1 (Standard-form strategy). We try ideally to proceed as follows:

(1) Show that some such L is in fact quasisimple. (That is, we want to move from a 2-component to an actual component. This is roughly the content of Thompson's fundamental B-*conjecture*—indicated below.)

(2) Show that some suitably-maximal L has certain very strong "standard form" properties. (These properties will also be described below.)

(3) Show that any simple G is (usually) a larger group of the same general type as L. (E.g. recall in the Odd Case Example 2.0.3 that linear components SL_k and SL_{n-k} of C_t arise in the linear group $G = SL_n$.) ♢

The actual path followed was more complicated; but the over-simplification above should suffice for our present expository purposes. So we'll now summarize, with only a few details, how that idealized strategy was implemented:

Obtaining components in standard form. For (1) in the strategy 2.1.1: The general form of Thompson's B-Conjecture [**ALSS11**, 1.6.1] states roughly that any 2-components of a local subgroup of X lie inside those of X. It has the corollary, in our situation where in particular $O_{2'}(G) = 1$, that the 2-components of C_t must in fact be components.

No direct proof of the B-Conjecture has been found. Instead, it was established indirectly, as a consequence of the Unbalanced Group Theorem—using a version of standard form below, adapted for 2-components. The interested reader is directed to [**ALSS11**, Sec 1.6–1.8] for details of this nontrivial process.

Thus we now assume the resulting B-Theorem, which was established by that process: so that our 2-component L is now in fact a component—i.e. quasisimple. And we turn to step (2) in the strategy 2.1.1.

Here is Aschbacher's characterization [**ALSS11**, 1.6.1] of the fundamental properties that a component L should have—on being taken maximal in a suitable sense:

DEFINITION 2.1.2 (standard form). Set $H := C_G(L)$. The component L is *standard* in G, or in *standard form*, if:

- L commutes with none of its distinct G-conjugates; and
- $H \cap H^g$ has odd order for all $g \in G \setminus N_G(L)$. (So $N_G(L) = N_G(H)$.)

The latter condition says that H is *tightly* embedded in G. This definition due to Aschbacher uses a condition resembling that for strongly-embedded in (2.0.16); but it definitely differs, since here $H < N_G(H)$. ♢

EXAMPLE 2.1.3. In the Odd Case Example 2.0.3, work in $SL_n(q)$ with even n. So if we take $k = 2$, then in effect we maximize the component $L \cong SL_{n-2}(q)$. To observe standard form: As a typical example of L^g, take the diagonal entries of -1 in the bottom two positions. Then as long as $n \geq 4$: $H \cap H^g = 1$; and the nontrivial subgroup $L \cap L^g$ is not centralized by L or L^g, so that $[L, L^g] \neq 1$. (With some work, this example could be made into an actual proof.) ♢

EXERCISE 2.1.4. Revisit some other component-type groups, for example from Exercise 2.0.5; again these should not be too small. Choose a component which seems likely to be maximal; and verify the standard-form property—at least informally, as in the Example above. ◇

The definitive verification that standard form should usually hold was provided by Aschbacher's Standard Component Theorem [**ALSS11**, 1.8.12]. That result was proved before the B-Conjecture, and so assumed that conjecture as part of its hypothesis; this restriction was of course released, after the proof of the B-Theorem. The conclusion was that: either G is one of a few explicit small classical groups in odd characteristic—and hence among the conclusions of the Odd Case; or G has an involution-centralizer component in standard form.

So we may assume that our L is standard in G; and we turn to step (3) in the strategy 2.1.1:

Treating the standard-form problems. At this point, we see that the completion of our Odd Case has been reduced to:

- considering each quasisimple L, namely a covering of some simple S; and
- determining which simple G can have that L in standard form.

The former is in fact covered by what we had earlier called the quasisimple-form of the CFSG-list 1.4.9. So we are reduced for each L to the latter, which is called the *standard-form problem* for L.

We had observed in 2.1.1(3) that the answer G in the standard-form problem for L should usually be a larger group of the same general type as L. But there are enough exceptions to complicate the problem. For example, $L_2(5)$ is standard in almost-simple $PGL_3(5)$; but we also recall[4] that $L_2(5) \cong A_5$—and we had observed in 2.0.5 that A_5 is standard in A_9. For a more nontrivial example: A_{11} is similarly standard in A_{15}; but its double cover $2A_{11}$ turns out to be standard in the Lyons group Ly.

By around 1979, the various standard-form problems had been handled—in papers by more than 20 different authors; see [**ALSS11**, Secs 1.9–1.10] for the list of authors, with a fuller description of the results.

Those results completed the treatment of the Odd Case of the CFSG. □

So we turn to the Even Case: where, roughly speaking, we will adapt essentially the same ideas as for involution centralizers in the Odd Case—but now instead for application to centralizers of elements of *odd* order, in the Even Case.

2.2. Treating the Even Case: Via trichotomy and standard type

Now we assume that we are in the Even Case, so that our simple G has characteristic 2 type. And of course we can also assume that $m_2(G) \geq 3$—since $m_2(G) \leq 2$ was covered as the Small Odd Subcase (in the language of the Grid 2.0.19).

Thus from Definition 2.0.7, we have $F^*(N) = O_2(N)$ for each 2-local subgroup N. So this holds for N given by the centralizer $C_t := C_G(t)$ of each involution t of G.

[4]E.g. from 1.5.5(1).

The groups expected in the conclusion of the Even Case should primarily be groups of Lie type in characteristic 2 (as well as a few sporadic groups).

The Small Even Subcase. We will begin with just a brief indication of the Small Even Subcase, given by $e(G) \leq 2$—the quasithin groups.

We observe first that Thompson's parameter $e(G)$ in (2.0.18), when applied to G of Lie type in characteristic 2, provides a good approximation of the Lie rank:

EXAMPLE 2.2.1 ($e(G)$ and characteristic-2 Lie rank). In $GL_n(q)$ in our Even Case Example 2.0.6, we recall by the Borel-Tits Theorem 1.3.20(5) that 2-locals are contained in parabolic subgroups. A typical parabolic is in fact provided by the centralizer C_t in that Example. Its widest odd-order subgroup is the full diagonal subgroup H; this is a Cartan subgroup, which has rank n—and that is also the *Lie* rank of $GL_n(q)$. For the determinant-1 subgroup $SL_n(q)$, these ranks are both given by $(n-1)$. ♢

EXERCISE 2.2.2. In some other even Lie-type cases, say those in Exercises 1.5.5 and 2.0.8, check the agreement of these ranks. ♢

Thus the expected conclusions in the Small Even Subcase are the given by the characteristic-2 Lie-type groups, of Lie rank ≤ 2.

Unfortunately, there are enough further examples of quasithin groups, to greatly complicate the classification problem. Even more difficult to deal with, than these extra conclusions, are the very many borderline cases: "shadows" which are not quasithin groups, but have e.g. local subgroups whose commutator subgroups are quasithin—these can be very hard to eliminate by purely local methods.

Around 1981, a substantial work on quasithin groups was left unfinished by Mason. Eventually a classification of quasithin groups was published, in a lengthy work of Aschbacher-Smith [**AS04b**]; a statement of the result appears as 3.0.1 in [**ALSS11**]. We mention that this quasithin classification was proved under a more general hypothesis of:

DEFINITION 2.2.3 (even characteristic). The condition of *even characteristic* weakens characteristic 2 type, by requiring $F^*(N) = O_2(N)$ only for 2-locals N containing a Sylow 2-subgroup of G. ♢

One useful feature of this extra generality is that the quasithin result can be used also in the "revisionism" program of Gorenstein-Lyons-Solomon—which we will indicate briefly as the new-approach (2) to the CFSG, in our afterword-Section 2.3. More precisely: using instead the revisionism-hypothesis of "even type" indicated there, a corollary of the Aschbacher-Smith result shows that the only additional quasithin group that arises under even type, beyond those of even characteristic, is the Janko group J_1. We'll say a little more about this result, in the quasithin-application subsection of later Section 8.3.

So for the rest of the section, we assume the Generic Even Subcase: $e(G) \geq 3$.

Odd signalizer functors and trichotomies. The material in this subsection is a much-oversimplified summary of the approach taken in Sections 2.2 and 2.3 of [**ALSS11**]. For full definitions and details, the reader should consult that source, or its underlying references.

From $e(G) \geq 3$ we get an odd prime p, with $m_p(N) \geq 3$ for a 2-local subgroup N.

In particular, we have $m_p(G) \geq 3$. And recalling that $m_2(G) \geq 3$ was required for the use of signalizer functors in the proof of the Dichotomy Theorem 2.0.9, we might well ask here if we can employ a suitable odd-p version of those fairly elementary arguments in signalizer functor theory.

We observe first that in the Odd Case, we had an involution t—of order 2, coprime to the odd characteristic of the Lie-type groups expected in the Odd Case conclusions. This suggests that in the Even Case, where we are expecting conclusion groups in characteristic 2, we should examine an element u of odd prime order p, and its centralizer $C_u := C_G(u)$. Let's see how this works in an example:

EXAMPLE 2.2.4 (Centralizers of odd-order elements). We mimic the development in the Odd Case Example 2.0.3: But this time our linear group $G := GL_n(2^a)$ is now of characteristic 2; and we replace the earlier diagonal involution t, which had k entries of -1, by a p-element u—which instead uses k copies of a primitive p-th root ω of unity. Then the same linear-algebra argument as before leads to block-matrix form for C_u; which again has components SL_k and SL_{n-k}.

Similarly we can mimic the development in the Even Case Example 2.0.6: We work inside the linear group $G := GL_n(p^a)$, which now is of odd characteristic; for u of order p, we can use a Jordan-matrix analogous to the one in that Example. And we obtain centralizer matrix forms analogous to the ones there, leading now to $F^*(C_u) = O_p(C_u)$. ◇

Thus we make the exact p-analogues of the earlier definitions:

DEFINITION 2.2.5 (p-component type and characteristic p type). We say that G has *p-component type*, if there is some element u of order p, with $C_u := C_G(u)$, such that $F^*\big(C_u/O_{p'}(C_u)\big)$ has a component.

And G has *characteristic p type* if $F^*(N) = O_p(N)$ for all p-locals N. (Again it suffices to have this condition just for all the centralizers C_u.) ◇

EXERCISE 2.2.6 (more odd components). Find p-components in C_u similar to the above, in other even Lie-type cases; say those in Exercises 1.5.5 and 2.0.8. ◇

With the above notions in hand, it is now straightforward to reproduce the methods of signalizer functors: including balance, completeness, and the graph Γ on rank-3 elementary p-groups. In particular, the same elementary arguments used in proving the Dichotomy Theorem 2.0.9 (plus now a much less elementary *Non-solvable* Signalizer Functor Theorem) now lead instead to a *tri*chotomy—since this time, the case of Γ-disconnected in the proof remains distinct for odd p. These re-used arguments (cf. [**ALSS11**, 2.2.1]) result in:

THEOREM 2.2.7 (Weak Trichotomy). *Assume that G is simple and of characteristic 2 type; and that $e(G) \geq 3$. Then there is an odd p with $m_{2,p}(G) \geq 3$, such that one of the following holds:*

(1) G is of p-component type;

(2) Γ is disconnected;

(3) Γ is connected—and then G has characteristic p type (as well as characteristic 2 type by hypothesis).

As the "Weak" in the name suggests, for odd p these cases are not necessarily an ideal starting point for the final classification in the Even Case. Indeed Gorenstein and Lyons, and Aschbacher, and probably others, saw that it would be advantageous to refine this trichotomy—by suitably strengthening these three cases, in order to provide substantially more information.

This resulted in the stronger Trichotomy Theorem 2.2.8 below. For our present expository purposes, we will not provide here the full technical definitions of its three cases; for those details, see [**ALSS11**, Sec 2.2]. Instead, we will try to give some background on each adjusted-case, including just a brief approximation to its definition.

A main goal was to strengthen case (1), so as to *already* contain various properties like those coming from Aschbacher's notion of standard form. The result was called *standard type*; very roughly, it includes properties of components, and their "neighbors"—defined when they intersect in a suitably-large subcomponent. And then this adjusted case (1) should in fact lead to most of the Even-Case conclusion-groups—that is, the groups of Lie type in characteristic 2.

Case (2), with Γ disconnected, is not really adjusted at all: it just provides the information that the stabilizer of a connected component of the graph Γ on p-groups is contained in some 2-local subgroup N. Now this should turn out to be impossible, since N "should" instead be a p-local. But leading this case to a final contradiction is tricky; we'll just sketch what should be a desirable path: One immediate difficulty is that for odd p, in contrast to the situation for $p = 2$ in the proof of the Dichotomy Theorem 2.0.9, this does not force N to be a *strongly p-embedded subgroup*—in the straighforward p-analogue 8.6.1 of strongly embedded in (2.0.16). So an eventual goal is to establish that N *should* be strongly p-embedded. In that situation, "uniqueness methods" show that many p-locals lie in this *unique* maximal subgroup N; for example, this property holds for rank-1 Lie-type groups in characteristic p, using 1.3.20(6). So that is basically what we will later call the *Uniqueness Case*. Correspondingly, the earlier starting-point in case (2) was called the p-pre*uniqueness type* case.

Case (3), in the original un-adjusted form stated above, is already eliminated: For Klinger and Mason [**ALSS11**, B.9.1] showed that characteristic $\{2, p\}$ type forces $m_{2,p}(G) \leq 2$, contrary to $m_{2,p}(G) \geq 3$ as assumed in the statement of the Weak Trichotomy Theorem 2.2.7. However, their methods suggested a way of expanding case (3) to have a very natural content: Notably their early result [**ALSS11**, B.9.2] showed that characteristic $\{2, p\}$ type with $m_{2,p}(G) \geq 2$ leads to the already well-known condition of $GF(2)$ *type*. We'll examine that definition and its consequences later at 8.1.2; for the moment, we'll just say that it abstracts some structure which is visible in many Lie-type groups defined over the smallest field $\mathbb{F}_2$—and also in a fair number of sporadic groups. Since these Even Case conclusion-groups could be complicated to treat from the standard-type viewpoint of (1), it was convenient to expect them instead to arise under adjusted-case (3)—now re-defined as $GF(2)$ type.

This stronger-trichotomy program, with the three cases adjusted as in the discussion just above, was implemented by Gorenstein and Lyons, in a lengthy work [**GL83**]—in fact assuming $e(G) \geq 4$. The groups with $e(G) = 3$ were handled by Aschbacher [**Asc81a, Asc83a**]—indeed all the way to the final classification of such groups. (Aschbacher did not state an explicit trichotomy; but it can be gleaned from the logic sequence of his proof.) We state the result [**ALSS11**, 2.3.9] in this simplified-for-exposition form:

THEOREM 2.2.8 ((Strong) Trichotomy Theorem). *Assume G is simple of characteristic 2 type; and $e(G) \geq 3$. Then one of the following holds:*
(1) G is of standard type for some p;
(2) G is of p-preuniqueness type, for all relevant p with $m_{2,p}(G) \geq 3$;
(3) G is of $GF(2)$ type.

We mention that for $e(G) = 3$, Aschbacher in fact allowed in (3) the more general *$GF(2^n)$ type*. We'll give that definition later at 8.1.5; it is roughly an analogue of $GF(2)$ type, made for any $\mathbb{F}_{2^n}$.

And the remainder of the CFSG then proceeded via these three subcases.

The treatment of the three subcases of the Even Case. Since our primary purpose was to introduce some of the ideas above, here we'll give only a brief indication of how these subcases were later handled.

The treatment of standard type. We recall from our discussion before the Trichotomy Theorem 2.2.8 that this case (1) was designed to contain most of the conclusion groups in the Even Case, namely G of Lie type in characteristic 2.

In striking contrast to the Odd Case—where the standard form problems were handled by many different authors—in the Even Case, all the standard-type problems were handled in a single paper: namely Gilman-Griess [**GG83**].

This was made possible essentially by the strong properties included in the definition of standard type for case 2.2.8(1). See e.g. [**ALSS11**, Ch 6] for details; here is a very rough summary: For our standard-type component L, Gorenstein and Lyons provide a theory of *neighbors*, defined by such components sharing a "standard subcomponent"—with strong information about that intersection.

Next recall as in Example 2.2.4 that in the desired conclusion-groups of Lie type in characteristic 2, the component L should roughly be a Levi complement of a maximal parabolic; hence standard subcomponents should be similar Levi subgroups inside them.

In rough outline, Gilman and Griess proceed by first determining the full Weyl group W, as in a conclusion-group G; and then identifying G as $\langle L, W \rangle$—more or less as follows: They can obtain neighbors from suitable conjugates of L under W; and then the sub-components from the intersections do have the structure of Levi subgroups, as expected from the Dynkin diagram of the desired G. In particular, since they assume $e(G) \geq 4$,[5] so that typically L has Lie rank ≥ 3 as in Example 2.2.1, Levi subgroups corresponding to rank-2 subdiagrams should be contained in components such as L (or others constructed via subcomponents), and so have the desired structure. Finally this information on rank-2 Levi subgroups, along with the desired Dynkin diagram, is the input to the standard *Curtis-Tits relations* (or more simply the Steinberg relations)—appearing in their results 2.27 and 2.30, and which we indicate at later Theorem 4.2.1. The presentation corresponding to these relations allows them to identify their group $\langle L, W \rangle$ as the desired Lie-type group G in characteristic 2.

The treatment of $GF(2)$ type. We recall next that the case 2.2.8(3) was designed to cover the sporadic groups in the conclusion of the Even Case, and also many of the Lie-type groups defined over the small field $\mathbb{F}_2$.

One advantage of this case is that $GF(2)$ type had in fact already been handled by 1978—well before the Trichotomy Theorem 2.2.8 was published. A fuller description of that work is provided in [**ALSS11**, Ch 7]; with the $GF(2)$ type classification stated at [**ALSS11**, 7.0.1].

We will just summarize a few points:

The work took place during the middle 1970s, and was not necessarily regarded (at least originally) as part of the overall plan for the CFSG: it does not use an overall induction, and in particular, no $\mathcal{K}$-group hypotheses are needed.

Various authors participated: notably Aschbacher (e.g. [**Asc76**]) early on; and Timmesfeld, who reduced the problem to a finite number of local configurations; and Smith and others, who finished off those remaining configurations.

The more general $GF(2^n)$ type classification proceeded along similar lines, with contributions also by Stroth; see [**ALSS11**, 7.5.2] for a statement.

Some aspects of this treatment are discussed in an application-subsection of later Section 8.1.

The treatment of the pre-uniqueness case. Finally we recall that case (2) of 2.2.8 was intended to be eliminated—for the final contradiction of the CFSG.

See [**ALSS11**, Ch 8] for a fuller discussion of this work. As usual we provide just a summary:

Recall that this is the pre-uniqueness case: where the stabilizer of a connected component of Γ is contained in some 2-local subgroup. The first goal was to then proceed to the "full" Uniqueness Case: meaning roughly that some 2 local is "almost" strongly p-embedded—a condition which allows for a few further possibilities, beyond the strongly p-embedded condition 8.6.1 mentioned in our earlier preliminary discussion. This step was implemented by Aschbacher-Gorenstein-Lyons in [**AGL81**]. Indeed this advance was assumed in the final statement of Trichotomy by Gorenstein-Lyons in [**GL83**]: that is, replace the pre-uniqueness case (2) of 2.2.8 above with the full Uniqueness Case.

Then, the Uniqueness Case itself was eliminated by Aschbacher; this deep and difficult work appears in [**Asc83b, Asc83c**]. We mention that proceeding from almost strongly p-embedded to strongly p-embedded does not complete the analysis: for there is no independent odd-p analogue known of the Strongly Embedded

[5]As mentioned above before 2.2.8, Aschbacher handled the cases with $e(G) = 3$; in his standard-type situation, he used various recognition theorems, including the Curtis-Tits approach as in Gilman-Griess.

Theorem 2.0.17. (Indeed the classification of the analogous strongly p-embedded subgroups was only completed using the full power of the finished CFSG; we will say more about that treatment in later Section 8.6.)

This contradiction completed the Even Case; and hence the CFSG. □

2.3. Afterword: Comparison with later CFSG approaches

The CFSG outline in the sections of this chapter so far have been concerned with:

(1) The "original" CFSG.

In the remainder of this section, we'll comment briefly on some later, alternative routes to classifying the finite simple groups:

(2) The "revisionism" program of Gorenstein-Lyons-Solomon (GLS)—see the volume [**GLS94**] and its successors:

This approach is intended to organize and improve the original CFSG proof; with the particular goal of being self-contained: namely collecting together essentially all the material, in one series of monographs.

One new feature is the replacement of characteristic 2 type with *even type*: which now allows 2-components of characteristic 2 in involution centralizers. (So in effect, the vertical partition in the Grid 2.0.19 is shifted somewhat to the left.)

So far, 6 of the planned 10 volumes have appeared; with the seventh submitted, and detailed drafts prepared for most of the rest. As we had mentioned earlier, the quasithin treatment for the GLS project will quote Aschbacher-Smith [**AS04b**]. It is also likely that the Uniqueness Case for the GLS project will appear in an anticipated volume by Stroth (a 2009 preprint can be found at his website).

(3) Meierfrankenfeld-Stellmacher-Stroth—characteristic-p methods:

An overview of this project appears in [**MSS03**]; for the current status, refer to Meierfrankenfeld's website.

The approach is intended to exploit the features of characteristic p type, for each p: That is, conclusion groups, notably those of Lie type in characteristic p, should be identified by the structure of p-local subgroups—such as parabolics. This differs from approaches (1) and (2), where for a group in characteristic p, components are produced in the centralizer of a p'-element.

The approach would classify groups of characteristic p-type, for any p; and so in particular would replace the Even Case (characteristic 2 type) of the original CFSG. But it would not replace the entire CFSG, which also covers the case of groups which are not of characteristic p type for any p (such as most alternating groups).

However, with that restriction: for the groups it treats, the approach would remove any vertical partition from the Grid 2.0.19. But there is still a horizontal partition—that is, small vs generic cases.

(4) Aschbacher's approach via *fusion systems*:

For an overview, see for example [**Asc15**]. This approach works in the context of fusion systems—the modern topological framework axiomatizing the p-local properties of finite groups: Namely given a prime p, along with a p-group P, the fusion system is roughly the category whose objects are the subgroups of P, and whose morphisms are suitable natural mappings among those subgroups. The motivating case comes from P taken as a Sylow p-subgroup of G; with the mappings given by the G-conjugations inside P—the p-fusion. (We return to the topic of fusion at later Definition (3.5.1); and to fusion systems thereafter.)

Thus the plan would be to first classify the abstract simple fusion systems; and then for each such simple system, identify the simple groups G having that system.

The approach seems most promising for groups of p-component type: since the fusion system is not affected by normal p'-subgroups, it is not necessary to avert $O_{p'}$ problems via signalizer-functor/connectivity arguments. Furthermore the approach may only be really feasible for $p = 2$; since for odd p, there are many exotic fusion systems not corresponding to finite groups G.

Applying the CFSG toward Quillen's Conjecture on $\mathcal{S}_p(G)$

In the remainder of the chapter, we turn again to applications—this time with a more topological flavor: namely "subgroup complexes", in particular the geometries afforded by the p-subgroup structure of a finite group.

As a reference for this area, I'll primarily use my book [**Smi11**] on subgroup complexes; the reader can also find there references to many further sources.

The treatment below will assume some basics of algebraic topology, notably: simplicial complexes, including the join construction; their homological algebra, including e.g. cycles, boundaries, and homology groups; and homotopy equivalences and contractibility. The reader needing a review of such material can consult [**Smi11**] and the sources it references; we mention that a particularly good topological resource for the subgroup-complexes viewpoint is Munkres [**Mun84**].

2.4. Introduction: The poset $\mathcal{S}_p(G)$ and the contractibility conjecture

Starting in the mid-1970s, topological work of Brown, Quillen, and Webb focused attention on the partially ordered set (poset) consisting of all nontrivial p-subgroups of a finite group G.

The original interest for topologists was in the group cohomology $H^*(X)$ for a finite group X; and indeed more specifically its p-part $H^*(X)_p$, when p divides the order of X. But soon, a wide range of other applications led to research of common interest across (at least):

- algebraic topology,
- finite group theory, and
- combinatorics.

For the general setup: Let $\mathcal{C}$ be some set of subgroups of a finite group G; and let $|\mathcal{C}|$ denote the set of inclusion-*chains* from $\mathcal{C}$. Since a subset of a chain is still a chain, we see that $|\mathcal{C}|$ is in fact a simplicial complex; namely:

(2.4.1) The *order complex* $|\mathcal{C}|$ of a poset $\mathcal{C}$ is its set of inclusion-chains.

This setup may seem very abstract—but you can draw pictures, at least in smallish posets. For brevity, usually we will now usually write just $\mathcal{C}$—both for the poset, and for the resulting complex $|\mathcal{C}|$; only making the notational distinction when really needed.

Normally we take $\mathcal{C}$ to be closed under conjugacy, so that the poset (and complex) admit a G-action. We also usually consider a poset which does not include 1 or G; this avoids having the poset be contractible for an essentially trivial reason—namely a unique minimal or maximal element.

The p-subgroups poset $\mathcal{S}_p(G)$. For the study of the p-structure of G, Brown around 1975 was led to consider the poset:

$$\mathcal{S}_p(G) := \{ \text{ all non-trivial } p\text{-subgroups of } G\}. \tag{2.4.2}$$

In particular, he established his "homological Sylow theorem" [**Smi11**, 0.0.1]; with the number of Sylows replaced by the Euler characteristic[6] χ of the complex:

$$\chi\big(\mathcal{S}_p(G)\big) \equiv 1 \pmod{|G|_p} \tag{2.4.3}$$

Subsequently Quillen's influential paper [**Qui78**] made an extensive further study of the topological properties of the complex; and his dramatic results stimulated considerable research activity.

To take one example: Note that subtracting 1 from both sides of (2.4.3) shows that $|G|_p$ divides the *reduced* Euler characteristic $\tilde{\chi}\big(\mathcal{S}_p(G)\big)$. And this is a property (see later 5.0.1) of the dimension of a projective G-module. Furthermore an important module related to the complex is its *Lefschetz module* $L\big(\mathcal{S}_p(G)\big)$: given by the alternating sum of the homology groups $H_i(-)$ of the complex, with coefficient $(-1)^i$ in dimension i. Because of the alternating signs $\pm$, it is actually *virtual* module, well-defined in the appropriate Grothendieck group; and its dimension is just the Euler characteristic $\chi(-)$. Then subtracting a trivial module in formal dimension -1 gives the *reduced* Lefschetz module $\tilde{L}(-)$, corresponding to *reduced* homology groups $\tilde{H}_i(-)$; and this has dimension given by the *reduced* Euler characteristic $\tilde{\chi}(-)$. So this is the dimension divisible by $|G|_p$ resulting from (2.4.3); and correspondingly Quillen, extending earlier arguments of Brown, showed (see for example 6.2.1 in [**Smi11**]) that:

(2.4.4) The *generalized Steinberg module* $\tilde{L}\big(\mathcal{S}_p(G)\big)$ is projective.

This terminology for an arbitrary finite group G indicates an analogy with the projective *Steinberg module* for a Lie-type group—we will discuss that actual Steinberg module at later Definition 5.2.8.

Webb and others pursued the area further; his survey [**Web87**] indicates the status of developments through about 1985.

Quillen's conjecture and the Aschbacher-Smith result. Quillen also showed, via a fairly easy argument, that (cf. [**Smi11**, 3.3.5]):

(2.4.5) If $O_p(G) > 1$, then $\mathcal{S}_p(G)$ is contractible.

And then he conjectured (very boldly, in my view) the far more difficult converse (cf. [**Smi11**, 3.3.8]):

CONJECTURE 2.4.6 (Quillen Conjecture). *If $O_p(G) = 1$, then $\mathcal{S}_p(G)$ is not contractible.*

[6]Recall this is the alternating sum of the dimensions of homology groups $H_i(-)$.

In fact Quillen himself established the conjecture for G solvable; we will discuss this in a moment as Theorem 2.5.7. Indeed he obtained a number of other important special cases, notably G of Lie type in the same characteristic p; some further cases are described in [**Smi11**, 8.4.2].

Later authors established various other special cases; notably Aschbacher and Kleidman in [**AK90**] explored the groups in the CFSG-list to obtain (cf. 8.4.3 in [**Smi11**]):

(2.4.7) The Conjecture holds for simple G in the CFSG-list 1.0.2.

The most general result so far is that of Aschbacher-Smith [**AS93**], which we state here in a form simplified from [**Smi11**, 8.4.4]:

THEOREM 2.4.8. *For $p > 5$, the Conjecture holds—unless $E(G)$ has certain unitary components.*

A main theme in the Aschbacher-Smith work is to exploit the central-product structure of $E(G)$: this gives the topological *join* of the complexes for the factors of $E(G)$—that is, for the components of G. The goal is to exhibit nonzero reduced homology (more specifically than in Aschbacher-Kleidman (2.4.7) above) for the possible components in the quasisimple form 1.4.9 of the CFSG-list—and try to show that these must lead to nonzero reduced homology for G.

The implementation of this approach is discussed in [**Smi11**, Ch 8]; and in the remainder of this chapter, we will sketch some of the exposition there which involves using the quasisimple CFSG-list.

2.5. Quillen-dimension and the solvable case

But before we consider material involving applications of the CFSG, it will be useful to review some background from Quillen's solvable case. (For fuller details, see [**Smi11**, Sec 8.2].)

As suggested just above, to establish non-contractibility of $\mathcal{S}_p(G)$, it will suffice to exhibit nonzero reduced homology $\tilde{H}_*$.

But where should we find that homology? More precisely, in what dimension should we find it? An answer will be suggested, in the process of examining Quillen's proof for the solvable case:

The subposet $\mathcal{A}_p(G)$ and Quillen-dimension. We note first that Quillen in fact works with a subposet of $\mathcal{S}_p(G)$:

(2.5.1) $\mathcal{A}_p(G) := \{$ all *elementary abelian* p-subgroups (> 1) of G $\}$;

This suffices, since one of his fundamental results was (cf. [**Smi11**, 4.2.4]):

(2.5.2) $\mathcal{A}_p(G)$ is homotopy-equivalent to $\mathcal{S}_p(G)$.

Furthermore, this implies that homology of $\mathcal{S}_p(G)$ vanishes in all dimensions greater than that of the subcomplex $\mathcal{A}_p(G)$. And for $d := m_p(G)$, namely the full p-rank of G, we see that any inclusion-chain of elementary p-subgroups has length $\leq d-1$—which gives the dimension of $\mathcal{A}_p(G)$.

Now often we can find nonzero homology in *exactly* that dimension; motivating:

DEFINITION 2.5.3 (Quillen-dimension). We say G has *Quillen-dimension* (at p), if we have nonzero homology $\tilde{H}_{m_p(G)-1}(\mathcal{A}_p(G)) \neq 0$; that is, in the top dimension of the complex $\mathcal{A}_p(G)$. ♢

Using Quillen-dimension is crucial for the Aschbacher-Smith approach.

Cycles from spheres. Next: How might we find—indeed construct—nonzero homology? More precisely, what group-structures can we use to exhibit $\tilde{H}_{d-1} \neq 0$?

We emphasize some crucial features of Quillen-dimension: First, $(d-1)$ is the top dimension of the complex $\mathcal{A}_p(G)$, so that the only $(d-1)$-boundary is 0—hence the $(d-1)$-cycles in fact give the homology classes $\tilde{H}_{d-1}(\mathcal{A}_p(G))$. Second, also by top dimension, we can even reduce to subgroups: because such a cycle for a subgroup H is also a cycle for G.

Now the simplest kind of $d-1$-cycle would be a sphere S^{d-1}. And recalling that the 0-sphere S^0 is just two disconnected vertices $(\{\bullet\ \bullet\})$, the simplest way to construct the sphere S^{d-1} is as the *join* of d copies of S^0 (roughly, a join of two terms makes all possible connections between them).

An easy example of a group-structure involving such spheres is given by taking $p = 2$, and considering the direct product of d dihedral groups D_{2q_i} for various odd orders q_i. For note that each dihedral group has its $\mathcal{A}_2$ given by $q_i \geq 3$ disconnected vertices—so there are multiple copies of S^0. And then it is standard (for example [**Smi11**, 8.1.2]) that $\mathcal{A}_p$ of a direct product is the join of $\mathcal{A}_p$ of the factors; so via this join-factorization we get (many) copies of S^{d-1}.

EXERCISE 2.5.4. For general p, observe similar spheres in the direct product of nonabelian groups of order pq, where p divides $q-1$. ♢

Indeed such products arise naturally in a well-known minimal-faithful situation of Thompson given in [**ALSS11**, B.1.7]:

THEOREM 2.5.5 (Thompson Dihedral Theorem). *Assume that A is an elementary 2-group of rank d, acting faithfully on a (solvable) $2'$-group L. Then LA contains a direct product of d dihedral groups $A_iL_i \cong D_{2q_i}$, where A is the product of the A_i of order 2, and $L_i \leq L$.*

Of course the solvability of L is automatic, from the Odd Order Theorem (2.0.1); but the elementary proof uses only the assumption of solvability (rather than quoting the CFSG).

The theme of such a product LA—with faithful action of A (of maximal rank d) on some L—will be prominent in the discussion that follows.

Quillen's minimal solvable case—and the solvable extension. To treat solvable groups, Quillen reduces to such an LA-situation; namely he first establishes nonzero homology [**Smi11**, 8.2.9] for:

THEOREM 2.5.6. *Assume A is an elementary abelian p-group of rank s, acting faithfully on a solvable p'-group L. Then $\tilde{H}_{s-1}(\mathcal{A}_p(LA)) \neq 0$.*

Thus in the post-Quillen terminology we have introduced, LA has Quillen-dimension. We mention that Quillen's proof is topological in flavor; however Alperin's later method, given at [**Smi11**, 8.2.9], essentially does proceed by means of the elementary join-spheres above.

Quillen then deduces the solvable case [**Smi11**, 8.2.5]) of his Conjecture:

THEOREM 2.5.7 (Solvable Quillen Conjecture). *Assume that G is solvable, with $O_p(G) = 1$, and $d := m_p(G)$. Then $\tilde{H}_{d-1}(\mathcal{A}_p(G)) \neq 0$.*

PROOF. Here is the actual process of reduction to a suitable LA: We may choose A elementary of rank d. Since G is solvable, the generalized Fitting subgroup is just the Fitting group $F(G) =: L$. By hypothesis we have $O_p(G) = 1$, so that L is a p'-group—and solvable, by that hypothesis on G. By Fitting's Theorem 1.4.2, we have $C_G(L) \leq L$; so as L is a p'-group, $C_A(L) = 1$—that is, A is faithful on L. So by Quillen's minimal-case Theorem 2.5.6, we get $\tilde{H}_{d-1}(\mathcal{A}_p(LA)) \neq 0$. Since $(d-1)$ is the top dimension of $\mathcal{A}_p(G)$, as we noted earlier, these cycles for LA also give cycles for G—as required. □

And again we see that, in our later language, G has Quillen-dimension.

It turns out that a similar reduction, though requiring a rather mild application of the CFSG, works for the p-solvable[7] case.

2.6. The reduction of the p-solvable case to the solvable case

Again see [**Smi11**, Sec 8.2] for a fuller exposition than the summary below.

Quillen mentioned that the proof for the solvable case in Theorem 2.5.7 could be immediately extended to the p-solvable case—if the "solvable" restriction could be removed from L in Theorem 2.5.6. Indeed, now assume we have p-solvable instead of solvable in that proof: so using the generalized Fitting subgroup $L := F^*(G)$, the hypothesis $O_p(G) = 1$ again leads to a p'-group—though now L is only p-solvable. And use of the self-centralizing property (1.4.6) similarly leads to A faithful on L.

So in effect, what is needed to generalize 2.5.6 is a reduction from L given by any p'-group, to a suitable A-invariant solvable subgroup $L_0 \leq L$ (with A still faithful on L_0).

And here for p odd, L can have p'-group components from $E(G)$ which are definitely not solvable:

EXAMPLE 2.6.1 (simple $3'$-groups). The Suzuki twisted groups $Sz(2^a)$ are simple $3'$-groups—indeed the *only* simple $3'$-groups, using a result of Glauberman (see [**Gor82**, 4.174]).

By the early 1990s, experts were aware that such a reduction could be obtained fairly easily—using the CFSG-list. But seemingly no one had explicitly claimed the result for the p-solvable case. It is quoted in [**AS93**, 0.5] as well-known, with a proof indicated via 1.6 (using 0.10) there. But the reduction to a solvable L_0 was, inadvertently, omitted in 0.10; later a full argument was supplied in 8.2.12 of [**Smi11**].

Since it provides an easy example of applying the CFSG inside the Aschbacher-Smith work [**AS93**] on the Quillen Conjecture, in this section we give a sketch of that reduction.

[7] Recall this means the terms of a composition series are either p-groups or p'-groups.

Notice first that the above reduction shows that we may as well assume that we have $L = F^*(L) = F(L)E(L)$: where $F(L)$ is a p'-group—which is nilpotent, and so certainly solvable; while $E(L)$ is the central product of quasisimple components, which are also p'-groups.

Coprime outer automorphisms. We now separate off the part of the argument that invokes the CFSG-list; this feature of the simple groups seems to be of independent interest, so we reproduce it from [**Smi11**, p 270]:

COROLLARY 2.6.2 (Coprime outer automorphisms). *Assume S is a simple p'-group, with P a nontrivial p-subgroup of $Out(S)$. Then p is odd, S is of Lie type, and P consists of field automorphisms; so in particular, P is cyclic.*

PROOF. Note first that p is odd: for simple S has even order by the Odd Order Theorem (2.0.1). We will now need only very basic facts from our discussion of $\mathrm{Out}(S)$ in Section 1.5:

Recall we saw there that for S alternating or sporadic, $\mathrm{Out}(S)$ is either trivial or a 2-group. But we saw in the previous paragraph that $p \neq 2$; hence S must be of Lie type.

So we recall the diagonal-field-graph Theorem 1.5.4. Now diagonal automorphisms are eliminated: since their orders are among those of the elements of the diagonal subgroup H. Similarly graph automorphisms are eliminated: those of order 2 by $p \neq 2$ above; while graph automorphisms of order 3 could occur only in types D_4 or ${}^3D_4$—whereas we saw in Example 2.6.1 that the only simple $3'$-groups are the Suzuki groups $Sz(2^a)$, which instead have type ${}^2C_2$.

Thus P consists of field automorphisms—and so is cyclic. □

The remaining argument mainly uses standard elementary facts about the "coprime action" of the p-group A on the p'-group L; see e.g. [**ALSS11**, Sec B.1] for further details.

The reduction to solvable L_0. The main point of the exposition in this area is to demonstrate how we work into a position to apply the CFSG via 2.6.2 above.

Reduction to quasisimple groups L_i. Recall we have $L = F(L)E(L)$. Let L_i (for $i \in I$) denote the components—that is, the quasisimple factors in the central product $E(L)$. Also recall from Remark 1.4.8 that the elements of A must permute the L_i, as they are the components of LA.

We will wish to define our subgroup L_0 as follows; set:

$$L_0 := F(L)\ \big(\ \Pi_{i \in I}\ \big(\ \Pi_{b \in A/N_A(L_i)}\ S_i^b\ \big)\ \big),$$

where for each i, S_i will be an $N_A(L_i)$-invariant Sylow p_i-subgroup of L_i (for some prime p_i), with:

$$C_A(S_i) \leq C_A(L_i). \tag{2.6.3}$$

We quickly check that this construction will satisfy what we need: Notice that L_0 is nilpotent, hence in particular solvable. Furthermore A normalizes the product of the $|A : N_A(L_i)|$ conjugates of S_i under A; so that A normalizes L_0. Finally to check faithful action: assume that $a \in C_A(L_0)$. Such an a centralizes $F(L)$ by definition of L_0; and also for each i, a centralizes S_i, and hence centralizes L_i

by (2.6.3). Since $L = F(L)E(L)$, and $E(L)$ is the product of the L_i, we see that $a \in C_A(L)$—but $C_A(L) = 1$ as A is faithful on L. Hence A is also faithful on L_0.

Thus it remains, for each i, to exhibit an $N_A(L_i)$-invariant S_i with the above condition (2.6.3).

Choice of suitable S_i. To move toward such a choice, we first get a general overview of all possible Sylow groups—using standard coprime-action methods:

Consider *any* prime q dividing the order of L_i; and then consider some Sylow q-subgroup denoted by Q. Now $N_A(L_i)$ must permute the set of Sylow p_i-subgroups of L_i; and the number of these is $|L_i : N_{L_i}(S_i)|$. Since L is a p'-group, this index is coprime to p; so there must be some choice of the Sylow q-group Q which is normalized by $N_A(L_i)$—and we now make that choice, for each q. Furthermore, note that L_i is the set-product of these Sylow groups Q, as we vary q.

Furthermore since elements of A must permute the L_i, $N_A(Q) \leq N_A(L_i)$; so as Q is $N_A(L_i)$-invariant, we conclude that $N_A(L_i) = N_A(Q)$.

Now we can finally begin to further refine the possible Sylow choices:

Assume first that $N_A(L_i) = C_A(L_i)$. We may take any q, Q above in the roles of "p_i, S_i": Then we have $C_A(S_i) \leq N_A(S_i) = N_A(Q) = N_A(L_i)$ using the previous paragraph; and here $N_A(L_i)$ is just $C_A(L_i)$ by our present assumption. That is, this choice of p_i and $N_A(L_i)$-invariant Sylow S_i satisfies (2.6.3).

We turn to the remaining, more interesting case, with $N_A(L_i) > C_A(L_i)$. We adopt the notation $N_A(L_i)^* := N_A(L_i)/C_A(L_i)$; this gives a nontrivial subgroup of $\mathrm{Out}(L_i)$. It is a standard feature (e.g. [**Gor80**, 5.3.5]) of coprime action that $N_A(L_i)^* \leq \mathrm{Out}(L)$ must in fact be determined in $\mathrm{Out}(S)$, where S is the simple quotient $L/Z(L)$. And this is where we can apply the CFSG, via Corollary 2.6.2: Namely S is of Lie type, with $N_A(L_i)^*$ cyclic; indeed since A is elementary abelian, we conclude that the nontrivial group $N_A(L_i)^*$ must be of order exactly p.

Now we saw earlier that L_i is the product of the $N_A(L_i)$-invariant Q over all q; so $N_A(L_i)^*$ cannot centralize all these Q. Thus we can now restrict our choice of prime p_i to one of those q for which $N_A(L_i)^*$ is nontrivial on Q; and correspondingly take S_i to be that Q. So we have $N_A(L_i)^* > C_{N_A(L_i)^*}(S_i)$; and since $N_A(L_i)^*$ is of order exactly p by the previous paragraph, we must have $C_{N_A(L_i)^*}(S_i) = 1$. This means that $C_A(S_i) = C_{N_A(L_i)}(S_i) \leq C_A(L_i)$. Thus again (2.6.3) holds—completing the proof. □

REMARK 2.6.4. We could actually be more specific about the final choice of p_i, S_i above: For once the simple quotient S is of Lie type, and $N_A(L_i)^*$ is given by a field automorphism, we can take p_i to be the characteristic prime of S—and then the full unipotent group U, over the larger field (and covered by the Sylow S_i of L_i), is not centralized by the field automorphism. ◊

REMARK 2.6.5 (The upshot). We pause to recall what has been accomplished by the argument in this section: Namely Theorem 2.5.6 holds with the "solvable" restriction on the p'-group L removed. And then it follows, by straightforward adapation of its proof, that the Solvable Quillen Conjecture Theorem 2.5.7 can be extended to p-solvable G with $O_p(G) = 1$. ◊

2.7. Other uses of the CFSG in the Aschbacher-Smith proof

In this final section of the chapter, we give a briefer indication of how the CFSG is used elsewhere in the argument, in terms of certain main features in [**AS93**]. A fuller exposition is given in [**Smi11**, Sec 8.2–8.4].

So consider the Quillen Conjecture for general G; we start with $O_p(G) = 1$.

A certain amount of argument, using induction and the p-solvable case of the previous section, leads to $O_{p'}(G) = 1$. It follows that $F(G) = 1$, so $Z\big(E(G)\big) = 1$. Thus $F^*(G) = E(G)$—which is then the direct product of *simple* components.

This suggests the main overall approach; roughly:

- In $E(G)$, find nonzero product homology—using the CFSG for the factors.
- Then show that this "propagates" to nonzero homology for G.

It turns out that Quillen-dimension is the key to establishing this propagation. And we mention that in order to obtain $O_p = 1$ in subgroups for induction, it is necessary to analyze not just simple groups, but almost-simple groups—via complements in products LA with outer automorphisms A, in Section 2 of [**AS93**].

Establishing Quillen-dimension. Consequently Section 3 of [**AS93**] is devoted to the fairly painstaking verification of Theorem 3.1 there; which roughly states (cf. [**Smi11**, 8.2.15]) that:

"Most" almost-simple groups have Quillen-dimension.[8]

The major exceptions (that is, not covered by "Most" above) are:

- Lie type groups in the same characteristic p;
- certain unitary groups $U_n(q)$ with $q \equiv -1 \pmod{p}$.

Of course the CFSG, including knowledge of outer automorphsisms, is used for the list of groups to analyze for this result.

And the main method used for establishing Quillen-dimension is to examine the detailed subgroup structure, to locate p,q-spheres S^d of the type indicated in Exercise 2.5.4. The cases are too varied to discuss here; but in general the procedure is roughly recursive—repeatedly breaking down products into sub-products, until the final factors are nonabelian with order pr for p dividing $(r-1)$.

And just a word about the exceptions: For a group G of a fixed Lie type, over fields $\mathbb{F}_{p^a}$ for various a, the root subgroups have rank a, which can be arbitrarily large—hence the value of the Quillen-dimension also becomes arbitrary large; whereas unfortunately, the known methods for sphere-construction are limited by the fixed Lie rank of G. The problem is roughly similar for the unitary groups $U_n(q)$: Quillen-dimension requires nonzero homology in dimension roughly $(n-1)$; whereas the known methods are limited by the Lie rank, which is about half that—recall the discussion of Lie rank for the twisted groups in Definition 1.3.17.

Using Quillen-dimension. We now sketch the main method in the overall proof; this is given by [**AS93**, 1.7], which completes the case where G has a component L, such that LA has Quillen-dimension—in particular, A is faithful on L, and exhibits the maximal p-rank $m_p(LA)$.

The first step for homology propagation is to adjoin $C_G(LA)$: Here the CFSG-based analysis of outer automorphisms that we mentioned above is used, to arrange that $O_p\big(C_G(LA)\big) = 1$. Then we may assume by induction (in the overall result) that $C_G(LA)$ has nonzero homology—say $\alpha \neq 0$. And now we form the usual

[8] Recall that specifying this dimension is stronger than Aschbacher-Kleidman (2.4.7).

homological product of α, with the nonzero Quillen-dimension homology (say $\beta \neq 0$) for LA which we have by hypothesis. This gives nonzero homology $\alpha\beta$ for the product group $C_G(LA)LA$.

The remaining step in the homology propagation is to proceed to G: If $\alpha\beta$ is in the image of the boundary map from G, the preimage would have to involve some elementary p-group B of G, which in particular properly contains A. Here the CFSG-based outer-automorphism analysis mentioned above can be used to force B to even fall into our product $C_G(LA)LA$. And now we use Quillen-dimension; specifically that A has the maximal rank $m_p(LA)$, so that there are no elementary $C > A$ available in the right-hand factor LA. Indeed we are able to choose a complement in B to A from the left-hand factor: namely $B = C_B(LA) \times A$. In this situation, we can apply the main technical lemma [**AS93**, 0.27], which uses only elementary product homology (the Künneth formula etc), to show that this would force α to be a boundary—contradicting our choice above of $\alpha \neq 0$.

Finishing via Robinson subgroups. So from now on, we may assume that each component L in $E(G)$ is a "non-QD" group; namely one of the exceptions in 8.2.15 of [**Smi11**]. For these cases, we use a method due to Robinson [**Rob88**] (essentially that used for Aschbacher-Kleidman (2.4.7)):

For a q-hyperelementary[9] group X, the fixed points under X in fact satisfy:

$$\tilde{\chi}\big(\mathcal{S}_p(G)\big) \equiv \tilde{\chi}\big(\mathcal{S}_p(G)^X\big) \pmod{q}. \tag{2.7.1}$$

Now if we had contractibility of $\mathcal{S}_p(G)$, the left side would be zero; so that the fixed-point count on the right side would be divisible by q.

Thus Section 5 of [**AS93**] studies the simple groups in CFSG-list, to establish Theorem 5.3 there—which roughly states that non-QD groups should have a relevant subgroup Y:

REMARK 2.7.2 (Robinson subgroups). For L a non-QD simple group—*except* certain unitary groups—we can find a 2-hyperelementary subgroup $Y \leq L$, such that $\tilde{\chi}\big(\mathcal{S}_p(L)^Y\big) = \pm 1$. We call such a subgroup Y a *Robinson subgroup*.

In fact, usually $\mathcal{S}_p(L)^Y$ is empty, so that $\tilde{\chi} = -1$; but occasionally the fixed-point set is the 0-sphere S^0, so that $\tilde{\chi} = 1$. ♢

And now we can finish the main argument:

Recall we are now assuming that $E(G)$ has all non-QD components. And at this point, we must also make use of our original hypothesis that there is no unitary component. (In fact the main result of [**AS93**] actually allows certain unitary components, which can be shown to have Quillen-dimension.)

Now for each component L of $E(G)$, using Remark 2.7.2 we take a Robinson subgroup Y. Using the product of the 2-groups from the various Y, and the cyclic group generated by the product of the generators from the cyclic subgroups in each Y, we can build a 2-hyperelementary $X \leq E(G)$. Now the fixed points for G can be shown to in fact be those for $E(G)$:

$$\mathcal{S}_p(G)^X = \mathcal{S}_p\big(E(G)\big)^X .$$

[9]This means cyclic extended by a q-group.

And for the simple factors of the central product $E(G)$, it is standard that the reduced Euler characteristic $\tilde{\chi}$ is multiplicative on the corresponding join of posets; so from the values of $\tilde{\chi}$ for the Y in Remark 2.7.1, we obtain that $\tilde{\chi}\big(\mathcal{S}_p(G)^X\big) = \pm 1$.

This is odd; so by the congruence mod 2 in (2.7.1), we conclude that $\mathcal{S}_p(G)$ has nonzero reduced homology—and in particular, is not contractible.

This completes the argument of [**AS93**] toward Quillen's Conjecture. □

CHAPTER 3

Thompson Factorization—and its failure: FF-methods

This chapter reviews some methods related to factorization of a group as a product of subgroups. The ideas were initially introduced by Thompson, with further developments by Aschbacher and others. The techniques were applied frequently throughout the CFSG; and in the latter part of the chapter, we also indicate some modern applications of the ideas in a topological context.

The exposition in the first four sections roughly summarizes some background material on the CFSG in [**ALSS11**], especially Sections B.6–B.8 there. It also draws from the quasithin work [**AS04a, AS04b**], especially background material from Chapters B, C, and E in [**AS04a**].

Introduction: Some forms of the Frattini factorization

A simple group G is not necessarily likely to admit factorizations via subgroups. Thus in the next few sections, we will consider factorizations for a more general group H (typically, a local subgroup).

A factorization setup for module-action. We first indicate a very elementary factorization, which goes back at least to Frattini around 1885; we had in fact mentioned it in the discussion after Theorem 1.6.4:

LEMMA 3.0.1. *Assume H and a subgroup X both act transitively on a set Ω. Then X is supplemented by a point stabilizer H_α ($\alpha \in \Omega$). That is:*

$$H = X \cdot H_\alpha.$$

PROOF. Let h be any element of H, and suppose $(\alpha)h = \beta$. By the transitivity hypothesis, there is also some $x \in X$ satisfying $(\alpha)x = \beta$. Then hx^{-1} is an element of H_α, say s—so that $h = sx$. This shows that $H = H_\alpha X$; and equivalently, by taking inverses, we have $H = XH_\alpha$. □

EXAMPLE 3.0.2. For a prime p, the symmetric group S_p is the product of transitive $\mathbb{Z}_p$ with a point stabilizer S_{p-1}. And S_4 is the product of transitive A_4 (or even $O_2(A_4)$) with S_3. ♢

In fact Frattini is mainly associated with the special case [**ALSS11**, A.1.5(2)]:

COROLLARY 3.0.3 (Frattini Argument). *Let $N \trianglelefteq H$, and $P \in Syl_p(N)$. Then:*

$$H = N \cdot N_H(P).$$

PROOF. By Sylow's theorem, N is transitive on $\Omega := \mathrm{Syl}_p(N)$; but H acts on N by the hypothesis of normality, and hence also on Ω—and is transitive since N is. Now 3.0.1 gives the indicated factorization. $\square$

EXAMPLE 3.0.4. Mimic Example 3.0.2: Use $p = 3$, and note that S_4 is the product of normal A_4 with the normalizer S_3 of a 3-Sylow of order 3 in A_4. $\diamondsuit$

We next further specialize to a "module" subcase; namely we assume from now on:

HYPOTHESIS 3.0.5. *H acts (not necessarily faithfully) on some elementary abelian p-group V.*

Typically H might be a p-local of a simple G, and then a very commonly made choice is $V := \Omega_1(\ Z\big(O_p(H)\big)\ )$.[1] Since we have $V \leq H$ in that case, we can then regard V as an "internal" module for H. (We also comment that much of the analysis in the early sections of this chapter was originally done just for $p = 2$, in the analysis of 2-local subgroups in the CFSG; but most goes through for any p.) It is standard in the module situation 3.0.5 that $C_H(V) \trianglelefteq H$; so:

COROLLARY 3.0.6. *Assume $P \in Syl_p\big(C_H(V)\big)$ in Hypothesis 3.0.5. Then:*

$$H = C_H(V) \cdot N_H(P).$$

EXAMPLE 3.0.7. Take $p = 2$, with $H = S_4 \times S_3$, and $V = O_2(S_4)$. Here we see $C_H(V) = O_2(S_4) \times S_3$ has 2-Sylow $P \cong E_8$;[2] and H is the product of $C_H(V) = O_2(S_4) \times S_3$ with $N_H(P) = S_4 \times \mathbb{Z}_2$. $\diamondsuit$

Some refinements, based on subgroups $Z \leq V$ and $W \leq P$. In the internal-module subcase of 3.0.6 above, the factors $C_H(V)$ and $N_H(P)$ are p-local subgroups of H. Sometimes it can be useful to work instead with suitable p-locals *containing* those original factors.

Indeed below, we introduce a version of Corollary 3.0.6, in which the new factors are the analogous centralizer and normalizer—but of suitable subgroups Z, W of V, P. After that further corollary, we will discuss more explicitly some ways of usefully choosing such Z, W.

HYPOTHESIS 3.0.8. *Assume 3.0.5, with $Z \leq V$, and W weakly closed*[3] *in P with respect to H.*

From $Z \leq V$ we get $C_H(V) \leq C_H(Z)$. Further if $h \in N_H(P)$, then $W^h \leq P$, so that $W^h = W$; that is, $N_H(P) \leq N_H(W)$. Thus from 3.0.6, we get a refined module-consequence of the Frattini Argument:

COROLLARY 3.0.9 (A module Frattini factorization (FA)). *Assume we have Hypothesis 3.0.8. Then:*

$$H = C_H(Z) \cdot N_H(W).$$

This form (FA) of the Frattini Argument arises frequently—especially in the internal-module subcase—in the analysis of p-local subgroups H of G. So we turn to exploring how the subgroups Z, W might be productively chosen.

---

[1] Recall that $\Omega_1(-)$ means the subgroup generated by elements of order p.

[2] We write E_{p^r} (or sometimes just p^r) for an elementary group of order p^r.

[3] Recall this means: whenever $W^g \leq P$, we get $W^g = W$. In particular, $W \trianglelefteq P$.

A typical choice for Z (and indeed V). *We now choose some fixed some particular Sylow p-subgroup T of H.* Take the above $P \leq T$, so that $P = C_T(V)$.

For reasons which will be emerging as we proceed through this chapter, it can be useful to use subgroups Z, W which are "independently" determined in terms of T—rather than in terms of the particular original choice of V, P. Indeed for Z, W, we will typically wish to use not just normal but *characteristic*[4] subgroups of T.

In fact, there is a standard situation within the internal-module setup, for which there is a natural choice of the module V, determined by a fairly canonical characteristic subgroup that we can use for Z:

REMARK 3.0.10 (The p-reduced context). Assume further that G has characteristic p type; recall from the Definition 2.2.5 that this means that each p-local subgroup H of G satisfies:

$$F^*(H) = O_p(H) \text{ ; and hence } C_H\big(O_p(H)\big) = Z\big(O_p(H)\big),$$

using Remark 1.4.3.

Now assume also that H contains a Sylow p-subgroup of G. Thus our Sylow T of H is in fact Sylow in G. Here it turns out to be natural to use the following[5] characteristic subgroup of T:

$$Z := \Omega_1\big(Z(T)\big) .$$

For note that by normality we have $O_p(H) \leq T$; and so:

$$Z \leq C_H(T) \leq C_H\big(O_p(H)\big) = Z\big(O_p(H)\big)$$

using the characteristic p type property indicated in the previous paragraph.

Then as our H-module, we take:

$$V := \langle Z^H \rangle.$$

For by construction, V is normal in H; and as Z is elementary abelian, and we just saw that it lies in $Z\big(O_p(H)\big)$, we get $V \leq \Omega_1 Z\big(O_p(H)\big)$.[6] In particular, V is elementary abelian. Thus we have the module-requirement of 3.0.5; and furthermore by construction $Z \leq V$, as required for 3.0.8.

Indeed, we even get that V is *p-reduced*: that is, $O_p\big(H/C_H(V)\big) = 1$; see for example [**AS04a**, B.2.13] for this standard fact. ◇

But we have not yet discussed what might be a natural choice for the weakly-closed subgroup W of the Sylow P of $C_H(V)$. Here Thompson's deeper insights will come into play—leading to Thompson Factorization and many other developments:

3.1. Thompson Factorization: Using $J(T)$ as weakly-closed "W"

We will sketch some aspects of Thompson's extremely influential analysis.

The Thompson subgroup. Thompson's study of factorizations—and their failure—led him to define the following subgroup [**ALSS11**, B.6.4]:

DEFINITION 3.1.1 (The Thompson subgroup $J(-)$). For a p-group S, we set:[7]

$$J(S) := \langle \text{ all elementary abelian subgroups } A \text{ of maximal rank in } S \rangle.$$

---

[4]Recall characteristic means invariant under all automorphisms (not just inner) of T.
[5]Though sometimes we might wish to replace this Z by some further subgroup.
[6]For brevity we often abbreviate $\Omega_1\big(Z(-)\big)$ by just $\Omega_1 Z(-)$.
[7]Originally Thompson used A abelian of maximal order, not necessarily elementary.

Observe that $J(S)$ is normal, indeed characteristic, in S. Further whenever $S \leq X$, we see $J(S)$ is weakly closed in S with respect to X. ♢

EXAMPLE 3.1.2. Take $p = 2$. For dihedral $S := D_8$, note that $m_2(D_8) = 2$, and that D_8 has two elementary subgroups E_4 of maximal rank 2; so $J(D_8) = D_8$.

On the other hand, for quaternion $S := Q_8$, $m_2(Q_8) = 1$, and there is a *unique* elementary subgroup $Z(Q_8)$ of rank 1; thus $J(Q_8) = Z(Q_8)$. ♢

EXERCISE 3.1.3 (Practice with the Thompson subgroup $J(-)$). (1) What is the generalization of Example 3.1.2 above for dihedral D_{2^n}? For the generalized quaternion group Q_{2^n}?

(2) Compute $J(S)$ for some other small (but nonabelian) p-groups S. For example, consider S extraspecial (Definition 8.1.3) of order p^3; and larger extraspecial groups; and your favorite p-groups.

(3) Show that $\Omega_1 Z(S) \leq \Omega_1 Z\big(J(S)\big)$. Hint: $\Omega_1 Z(S)$ must be *in* each A by maximal rank. ♢

We will be interested in $J(T)$, for T our Sylow p-subgroup of H; indeed we would like to use it as the weakly-closed "W" in P, in hypothesis 3.0.8. However, in general it is not even clear that $J(T)$ should be *contained* in P.

Thompson Factorization. Nonetheless, this situation $J(T) \leq P$ is our "best-case scenario"; and we get Thompson's form (cf. [**ALSS11**, B.6.5]) below, of the refined Frattini factorization (FA) in 3.0.9:

THEOREM 3.1.4 (Thompson Factorization (TF)). *Assume* 3.0.5 *holds, and also that* $J(T) \leq C_H(V)$, *for* $T \in Syl_p(H)$. *Then:*

(1) $$H = C_H(V) \cdot N_H\big(J(T)\big) .$$

Assume further that we are in the internal-module subcase $V \leq H$, *and we also have* $\Omega_1 Z(T) \leq V$. *(In particular this holds in the p-reduced setup* 3.0.10, *where* V *is constructed as* $\langle \Omega_1 Z(T)^H \rangle$.*) Then:*

(2) $$H = C_H\big(\Omega_1 Z(T)\big) \cdot N_H\big(J(T)\big) .$$

PROOF. Set $P := C_T(V)$, so that $J(T) \leq P$ by our hypothesis. Then from Definition 3.1.1, we see that $J(T) = J(P)$; and in particular, $J(T)$ is weakly closed in P with respect to H. Thus $V, J(T)$ can play the roles of "Z, W" in Hypothesis 3.0.8; and then from (FA) in 3.0.9 we get the fundamental factorization (1). Now suppose further that $\Omega_1 Z(T) \leq V \leq H$; then the more specialized factorization (2) follows from (1). Alternatively, we can take $\Omega_1 Z(T)$ for "Z" in 3.0.8, and get (2) directly from (FA). □

EXAMPLE 3.1.5. We actually begin with some non-examples—that is, situations where the above conditions fail; we will follow this direction in the subsequent Section 3.2. First consider $H := S_4$ and $V := O_2(S_4)$. Here we have that $T \cong D_8$, and we saw in Example 3.1.2 that $J(D_8) = D_8$. Now $C_H(V) = V$, so that $J(T) = D_8 \not\leq C_H(V)$—that is, the hypothesis for (TF) in 3.1.4 fails. And also the factorization-conclusion 3.1.4(1) fails: for we have $N_H(D_8) = D_8$, so the product $C_H(V) N_H\big(J(T)\big) = O_2(S_4) D_8 = D_8$ is proper in S_4. A similar argument applies to $S_4 \times S_4$; and indeed to any larger product $(S_4)^n$—this situation will re-appear in later Theorem 3.2.6.

But the hypotheses for 3.1.4 do hold, in a certain *sub*group of $S_4 \times S_4$: For let $V := E_4 \times E_4$, and $H := V \cdot S_3$, where this "diagonal" S_3 acts on each factor as in S_4 above—namely as $L_2(2)$ on its 2-dimensional natural module. Here we see that $m_2(H) = 4$, and V is the unique elementary 2-subgroup of that maximal rank in T; hence we get that $J(T) = V$, so that the hypothesis $J(T) \leq C_H(V)$ for (TF) holds. And the product $C_H(V)N_H\big(J(T)\big)$ is just $VH = H$, so that the Thompson Factorization (TF) holds—if only in a rather trivial way, by normality in H. ◇

We can ask if there are natural classes of groups H for which, under the assumption that $F^*(H) = O_p(H)$ as in the p-reduced setup 3.0.10, we always get the conditions for the form 3.1.4(2) of (TF) to hold. In fact Thompson showed in [**Tho66**, Thm 1] that this is usually the case for *p-solvable* H—aside from certain exceptions like those with $(S_4)^n$ which we saw in Example 3.1.5:

THEOREM 3.1.6. *Assume that H is p-solvable, and that $F^*(H) = O_p(H)$. Take some $T \in Syl_p(H)$. Then either the form* 3.1.4*(2) of (TF) holds:*

$$H = C_H\big(\Omega_1 Z(T)\big) \cdot N_H\big(J(T)\big);$$

or $p \in \{2, 3\}$, and $SL_2(p)$ is involved as a section of H.

But Thompson also realized that failure of factorization (typically abbreviated by FF) will often occur. Roughly:

REMARK 3.1.7 (Expecting FF in p-locals). Recall our discussion in Chapter 2 of small versus generic cases, e.g. for the Grid 2.0.19. Generically we expect, for example in the standard p-reduced situation of 3.0.10, that our simple G will have a number of different p-locals H containing T, and satisfying $F^*(H) = O_p(H)$—as in higher-rank Lie-type groups G in 1.3.20(6).

However, methods of Thompson indicate roughly (cf. [**GLS96**, 26.12.ii]) that if *all* the p-locals H over T satisfy the factorization (TF) as in 3.1.4, then we should usually expect G to contain a strongly p-embedded subgroup. Of course this last condition is mainly associated with "narrow" groups, where a Sylow is contained in a *unique* maximal p-local subgroup: For example with $p = 2$, recall the groups of Lie rank 1 appearing in the Strongly Embedded Theorem 2.0.17; for these groups, by 1.3.20(6), a Sylow group is in a unique maximal 2-local subgroup. And cf. the analogous rank-1 groups for p odd in the parallel discussion of strongly p-embedded in Section 8.6. Of course more usually, we can expect wider situations.

Thus, generically we can expect FF in at least *some* of the p-locals. ◇

An important part of Thompson's insight was that the FF-situation is tractable: indeed, his definition of $J(T)$ itself provides a tool for FF-analyis:

3.2. Failure of Thompson Factorization: FF-methods

We continue with the H-module setup 3.0.5, but assume that (TF) in 3.1.4(1) fails; that is, we have FF. Then the hypothesis there must fail, namely we get:

$$J(T) \not\leq C_H(V).$$

So from the Definition 3.1.1 of $J(T)$ via maximal-rank elementary subgroups:

Some elementary A of maximal rank in T satisfies $A \nleq C_H(V)$.

In particular, for the faithful-action quotient, we have $\overline{A} := A/C_A(V) > 1$.

FF-offenders and FF-modules. We now restrict attention to the internal-module subcase, namely $V \leq H$. Since A is elementary, we see that $V \cdot C_A(V)$ is also elementary; and now Thompson applies the hypothesis of maximal rank for A, built in to Definition 3.1.1—to get:

$$|A| \geq |V \cdot C_A(V)| = \frac{|V||C_A(V)|}{|C_{V \cap A}(V)|}.$$

However $C_{V\cap A}(V) \leq V \cap A \leq C_V(A)$, as A is abelian; so we obtain:

DEFINITION 3.2.1 (FF-offender and FF-module).

$$\frac{|A|}{|C_A(V)|} \geq \frac{|V|}{|C_V(A)|}.$$

This condition defines $\overline{A}$ as an *FF-offender* (or just offender) on V. And V is called an *FF-module* for H; or, for the faithful quotient $\overline{H} := H/C_H(V)$. ♢

Roughly, the condition says that the codimension of $C_V(A)$ in V is smallish—with respect to $|\overline{A}|$; that is, $\overline{A}$ must centralize "much" of V.

EXAMPLE 3.2.2. There are various familiar cases of such $(V,\overline{A})$; for example:

(1) Transvections: Recall (Definition B.1.2) that these are p-elements x centralizing a hyperplane of V. Thus for $\overline{A} := \langle x \rangle$ of rank 1, we have $\dim\big(V/C_V(A)\big) = 1$, giving the FF-offender condition 3.2.1.

(2) Maximal unipotent radicals of $\overline{H} := GL_n(V)$: Recall by 1.3.20(6) that the maximal parabolics are the $P_{\hat{k}}$ of Example 1.3.4. So we take the unipotent radical:

$$\overline{A} := U_{\hat{k}} = \left(\begin{array}{c|c} I_k & 0 \\ \hline * & I_{n-k} \end{array}\right)$$

This has rank $k(n-k)$, which is $\geq$ the codimension $(n-k)$ of $C_V(\overline{A})$ in V, giving the FF-offender condition 3.2.1. Also the product $H := V \cdot \overline{H}$ is a local subgroup in a larger linear group $GL_{n+1}(q)$; so we have a p-local H exhibiting FF—as we noted we should expect, in Remark 3.1.7. ♢

EXERCISE 3.2.3. Test for FF, in some small cases for $\overline{H}, V$. Hint: In Example 3.1.5, we already saw FF for $L_2(2)$ on its natural module V; note that an offender $\overline{A}$ of rank 1 comes from a transvection, as in 3.2.2(1) just above. And in the internal-module version $H = S_4$, a subgroup E_4 not contained in $V = O_2(S_4)$ gives a maximal-rank A satisfying the FF-offender condition 3.2.1.

You can see similar behavior, using a full Sylow group $\overline{A}$ of $\overline{H} = L_2(4)$, in the action on its natural module V. On the other hand: We can take W to be the orthogonal module for H, since we have $\Omega_4^-(2) \cong L_2(4)$ via a standard isomorphism (cf. 1.5.5); and now there is *no* 2-subgroup A which is an offender. (You can realize W as the 4-subspace given by even-size subsets in the full 5-dimensional permutation module for $A_5 \cong \Omega_4^-(2)$.) ♢

We mention that the offenders in Example 3.2.2 in fact exhibit *quadratic* action: that is, $[V, \overline{A}, \overline{A}] = 0$. (So the minimal polynomial for $a \in \overline{A}^{\#}$ is quadratic.) An important result of Thompson [**Tho69**] shows that an offender without quadratic action can always be replaced by another that does have the property:

THEOREM 3.2.4 (Thompson Replacement). *Every offender* contains *a quadratic offender.*

Determining the FF-offenders and FF-modules. Notice that for the group $\overline{H} = GL(V)$ in Example 3.2.2(2), the $\overline{H}$-conjugates of the offender $\overline{A}$ given there in fact generate $SL(V)$—which is most of $\overline{H}$. Indeed in studying FF-modules V for groups $\overline{H}$, we typically restrict attention to the subgroup of $\overline{H}$ generated by offenders. And since the FF-offender restriction 3.2.1 is so strong, we consider:

PROBLEM 3.2.5. *Determine* $(\overline{H}, V)$, *with* $\overline{H}$ *generated by FF-offenders on* V.

For example, Glauberman [**AS04a**, B.2.16] refined Thompson's result 3.1.6 on the p-solvable case of factorization, by explicitly specifying the $(\overline{H}, V)$ in Thompson's FF-exceptions—again they are of the type in Example 3.1.5:

THEOREM 3.2.6 (p-solvable FF). *Assume that H is p-solvable, and that we have $F^*(H) = O_p(H)$. If the factorization* 3.1.4*(2) in (TF) fails, then $p \in \{2, 3\}$, and H is essentially the commuting product of terms of form $V_i \cdot L_i$—with V_i the natural module for $L_i \cong SL_2(p)$.*

In order to treat FF for more general $\overline{H}$, it was natural to reduce to consideration of the *components* $\overline{L}$ in $E(\overline{H})$; so we now assume $\overline{H} = \overline{L}$ is quasisimple.

For $p = 2$, Cooperstein and Mason (see [**Coo78**]) listed the relevant pairs $(V, \overline{L})$—but alas, without proofs. A full proof, for all p, was later given by Guralnick-Malle [**GM02**]; roughly:

THEOREM 3.2.7 (Quasisimple FF-list). *Assume V is an FF-module for quasisimple $\overline{L}$. Then the simple quotient $\overline{L}/Z(\overline{L})$ is either of Lie type in characteristic p, or alternating (here $p = 2, 3$); with V one of certain suitably "small" modules.*

The FF-list in 3.2.7, mainly for $p = 2$, was used to pin down non-factorization situations, throughout the CFSG. We'll indicate some of the applications of FF in our later material: for example, the $C(G, T)$-Theorem 3.3.8 in the next section; and the structure of "abstract minimal parabolics" in later 4.4.1.

So we now begin in the $C(G, T)$-direction, which along the way involves FF-theory:

3.3. Pushing-up: FF-modules in Aschbacher blocks

We continue with at least one theme from the previous section: namely a focus on subgroups which normalize (or even centralize) various characteristic subgroups of the Sylow T of H. However, we "digress", in the sense that we turn from the chapter-topic of factorizations via such subgroups, to considering instead *generation* via such subgroups. Nonetheless, one further theme remains similar to the previous situation of (TF)-versus-FF: namely if the desirable result of generation fails, we will still be able to describe the possibilities in that failure-situation.

The characteristic core $C(G,T)$ and the condition (CPU). In this section, we will often restrict attention to the case of $p = 2$ used in the CFSG; though we do indicate some of the development which still goes through for general p.

For overall context, we note that much analysis in the CFSG employs:

REMARK 3.3.1 (The Thompson strategy). Given $T \in \mathrm{Syl}_2(G)$, we make an initial choice of a maximal 2-local M containing T. Next, if possible, we find an "independent" 2-local H over T—namely with the property that $H \not\leq M$. We can then exploit the intersection $M \cap H \geq T$; and hope to describe larger $\langle M, H\rangle$—which, in view of the maximal 2-local choice of M, is likely to be all of G. ◇

The phrase "if possible" above brings us back to an issue that has cropped up several times in the material up to now—namely the number of maximal 2-local subgroups above a Sylow T. If this number is ≥ 2, then we are in a position to employ the Thompson strategy. But if not, we need to:

PROBLEM 3.3.2. *Determine the possible G for which T is contained in a* unique *maximal 2-local M.*

We had mentioned that this property arises in particular for the rank-1 Lie type groups in characteristic 2, in view of 1.3.20(6). These are of course the Bender groups, namely the conclusion-groups in the Strongly Embedded Theorem 2.0.17. Indeed the property arises under the hypothesis (2.0.16) of strongly embedded, using the later variant-formulation (8.3.2). But Problem 3.3.2 is in fact more general than the strongly-embedded case.

A related problem is suggested by our continuing focus on characteristic subgroups C of T: Observe that it follows from the definition of characteristic subgroup that $N_G(T) \leq N_G(C)$. Furthermore the behavior of parabolics in Lie-type groups 1.3.20(6) suggests that for $C < T$ we should usually get $N_G(T) < N_G(C)$; that is, ideally we should be able to then "push up" to a larger 2-local $N_G(C)$. But what if not? That is, what if $N_G(C) = N_G(T)$ for *all* C? We approach this question by defining:

DEFINITION 3.3.3 (The characteristic core $C(G,T)$). For a Sylow T of G:
$C(G,T) := \langle N_G(C) : \ C \text{ a nontrivial characteristic subgroup of } T\rangle$. ◇

Notice that $\Omega_1 Z(C)$ is characteristic in C, and hence also in T; and that we also have $N_G(C) \leq N_G\big(\Omega_1 Z(C)\big)$; so in fact it suffices to consider characteristic C which are elementary abelian.

And to then describe the corresponding problem-situation, we follow the terminology of (CPU) in [**AS04a**, C.1.6], for the characteristic-core version of obstruction to pushing-up:

REMARK 3.3.4 (Obstruction (CPU) to pushing-up). We write (CPU) for the version of *obstruction* to pushing-up given by $C(G,T) < G$; it's traditional to write this in the form:

$$C(G,T) \leq M < G\ ,$$

to allow for a possibly-independent subgroup M in applications. ◇

EXAMPLE 3.3.5. An easy example for $p = 2$ is given by taking $G := S_4$. Here $T \cong D_8$ has only $Z(T)$ as proper characteristic elementary abelian subgroup. And $N_G(T) = T = N_G(Z(T))$; so $T = C(G,T)$ plays the role of M in (CPU). $\diamondsuit$

We observe that the problem of determining obstructions (CPU) in 3.3.4 in fact contains Problem 3.3.2 of T in a unique maximal 2-local M: For in the latter, we see $N_G(T)$ lies in that unique maximal M; and we saw above that $N_G(T) \leq N_G(C)$ for all C—so that $N_G(C)$ must also lie in M, giving (CPU).

In Example 3.3.5 with S_4 above, we see that T is in fact contained in unique maximal 2-local *proper* subgroup $M = T$. However, note that this M is not strongly embedded as in (2.0.16): for here, elements outside M normalize $O_2(S_4) \leq M$. And in fact we are interested in (CPU) for non-simple situations such as 2-local subgroups, not just the context of simple G.

The $C(G,T)$-Theorem(s). Suppose first we have (CPU) in a 2-*local* subgroup H over the Sylow T of G; along with the assumption $F^*(H) = O_2(H)$ of the standard p-reduced situation 3.0.10. Our assumption of (CPU) in H says that we have a proper subgroup M_H of H such that:

$$C(H,T) \leq M_H < H.$$

This is "local failure" to push up in H; and we observe that we also get failure of factorization FF for H—namely failure of the factorization 3.1.4(2) in (TF): For a fuller derivation of this, see e.g. [**AS04a**, C.1.26]; but just intuitively: We see the two factors in that form of (TF), namely $C_H(\Omega_1 Z(T))$ and $N_H(J(T))$, lie in $C(H,T)$—which by (CPU)-in-H above lies in M_H. So if H were the product of those factors, we would get $H = C(H,T) = M_H$; contrary to the assumption that $M_H < H$.

Thus we can apply the FF-list 3.2.7 in describing obstructions in (CPU). In fact, only a small subset of the cases in that list arise. For full details, see B.7.3 in [**ALSS11**]; here we just describe the crucial building-blocks via:

DEFINITION 3.3.6 (Aschbacher χ-blocks). A χ-block is (roughly) one of several very special cases $(V, \overline{L})$ of the FF-list 3.2.7: where $\overline{L} \cong L_2(2^m)$, or A_m (m odd); and where V has a *unique* nontrivial irreducible section—given respectively by the natural module, or the irreducible permutation module. $\diamondsuit$

Notice that an example of a solvable χ-block is given by S_4 in earlier Example 3.3.5: for $V = O_2(S_4)$ is the irreducible permutation module for $\overline{L} = S_3$ (it can also be regarded as the natural module for $L_2(2) \cong S_3$).

Using this language, Aschbacher's fundamental [**ALSS11**, B.7.3] is:

THEOREM 3.3.7 (Local $C(G,T)$-Theorem). *Assume that $F^*(H) = O_2(H)$, and that H satisfies local-(CPU)—that is, that $C(H,T) \leq M_H < H$ as in* 3.3.4. *Then $H = C(H,T)L_1 \cdots L_t$, where the L_i are χ-blocks.*

Further work of Aschbacher and others then led to [**ALSS11**, B.7.8] for general G; which we state very roughly in the form:

THEOREM 3.3.8 (Global $C(G,T)$-Theorem). *Assume G is simple and of characteristic 2 type, and satisfies "global" (CPU)—i.e., $C(G,T) \leq M < G$ as in 3.3.4. Then G is in a certain short list of cases: beyond the rank-1 groups in the Strongly Embedded Theorem 2.0.17, it contains notably the rank-2 Lie-type groups which admit a graph automorphism.*

EXERCISE 3.3.9. Verify (CPU), and exhibit χ-blocks in 2-locals, in extensions of $L_3(2^m)$ and of $Sp_4(2^m)$ by a graph automorphsim g.

Hint: The graph automorphism g shows that certain 2-subgroups are *not* characteristic in T: Consider $G = L_3(2)\langle g\rangle$, with $T \cong D_8\langle g\rangle$. Here the two subgroups E_4 in D_8 are not characteristic—indeed they are interchanged by the graph automorphism $\langle g\rangle$. Furthermore we had noted in Example 3.3.5 that the only other characteristic subgroup of D_8 is $Z(D_8)$, which here also gives $Z(T)$. And much as there, we have $N_G(T) = T = N_G(Z(T))$; and since we have $C(G,T) = T$, here T plays the role of M in (CPU), for G in the Global $C(G,T)$-Theorem 3.3.8. Finally, letting V denote either choice of those E_4-subgroups of D_8, we see that the 2-local subgroup $H := N_G(V) \cong S_4$ exhibits (CPU)-in-H via $M_H := M \cap H = T$; and H itself is a χ-block, as needed in the Local $C(G,T)$-Theorem 3.3.7. ♢

With the Global $C(G,T)$-Theorem in hand, as noted earlier we in particular have a solution of Problem 3.3.2: determing cases when T is in a unique maximal 2-local subgroup. So:

(in the CFSG) after 3.3.8, *we may employ the Thompson strategy* 3.3.1;

that is, we may assume T lies in at least two maximal 2-locals, say M and $H \not\leq M$.

It follows from the Global $C(G,T)$-Theorem 3.3.8 that the rank-1 Lie-type groups $L_2(2^m)$ appear as $\overline{L}$, in χ-blocks which are 2-locals in some rank-2 groups with (CPU). Correspondingly, the (CPU)-situation in 3.3.4—defined using $C(G,T)$ for the full Sylow group T—is called *rank*-1 pushing-up. And as this terminology suggests, we can in fact generalize pushing-up to situations involving higher-rank $\overline{L}$; where we use subgroups smaller than T to define the relevant (CPU)-variant:

Pushing-up using $C(G,R)$ with $R < T$. Here we give a briefer sketch: We want to generalize from our original focus on the Sylow group T, by using instead a suitable subgroup $R \leq T$. Thus (CPU), via the obvious analogous definition, will take the form:

$$R\text{-(CPU)}:\quad C(G,R) \leq M < G. \tag{3.3.10}$$

And for R to be "suitable", it turns out to be natural to assume certain restrictions—which are automatic for T, but must be verified when $R < T$:

- First, we assume that R is Sylow in $\langle R^M\rangle$. This holds e.g. if $M = N_G(R)$.
- Second, we assume that R satisfies $R = O_2(N_G(R))$.

Indeed we recall the standard definition:

DEFINITION 3.3.11 (The p-radical poset $\mathcal{B}_p(G)$). We say a p-subgroup X of a group G is *p-radical* if:

$$X = O_p(N_G(X)).$$

We write $\mathcal{B}_p(G)$ for the poset of all nontrivial p-radical subgroups of G. ♢

The terminology of "radical" is motivated by the analogous property of unipotent radicals for Lie type in 1.3.20(4); indeed see later Theorem 5.4.2 for the characterization in that case. The p-radical subgroups, and the poset $\mathcal{B}_p(G)$, are important for the topological viewpoint on group theory, especially in representation theory—see e.g. the later Alperin Weight Conjecture 5.4.3, and other uses in geometric representation theory in Section 7.5.

Various choices of $R < T$, lead to analogues of the $C(G, T)$-Theorems—which determine obstructions in the R-(CPU) situation of (3.3.10). In analogues of the Local $C(G, T)$-Theorem, again the local subgroups H are described by suitable blocks; which are more general than the χ-blocks in Definition 3.3.6, involving:

- larger groups $\overline{L}$; acting on:
- "larger" modules V—that is, with conditions weaker than FF in 3.2.1.

We mention that the work of Guralnick-Malle [**GM02**] was in fact designed for this greater generality: indeed it provides a version of the FF-list 3.2.7, which covers many of these weaker-than-FF situations.

And of course such analogues of the Local $C(G, T)$-Theorem provide the local foundation for analogues of the Global $C(G, T)$-Theorem. We will close the section by mentioning several:

Rank-2 pushing-up. Meierfrankenfeld and Stellmacher develop results (such as C.1.32 in [**AS04a**]) for the following choice of R: Roughly, R is the unipotent radical of a rank-2 parabolic, in a Sylow T from a rank-3 Lie-type group G. The more general local-blocks arising from the R-(CPU) condition in (3.3.10), and the resulting global configurations, have been used applied in various ways in the CFSG.

Pushing-up for the non-quasithin "shadow" Fi_{23}. Now we present a somewhat more detailed sketch—of a similar local-global situation, which arose in the quasithin work; namely we summarize below the initial treatment of one of the cases of [**AS04b**, 8.1.1]:

The Fischer group $F := Fi_{23}$ is *not* quasithin; so it can't be a conclusion-group G under the QT-hypothesis. However, F does contain a 2-local subgroup of the form $H \cong 2^{11} : M_{23}$, which is quasithin; and ostensibly this subgroup *could* be involved in some simple quasithin group G. Thus during the analylsis, we view Fi_{23} in Bender's language as a "shadow", to eventually be ruled out. But such shadows can be very difficult to eliminate, using purely local methods.

In the present case, we can in fact proceed by turning to other 2-local subgroups—namely involution centralizers:

We write V for $O_2(H) \cong 2^{11}$. Here V is in fact the 11-dimensional *cocode* module for M_{23}, which we had mentioned in our initial introduction of the Fischer groups in Section 1.2. This module is a section of the full cocode module for M_{24}—which is the quotient of the 24-dimensional permutation $\mathbb{F}_2$-module, modulo the 12-dimensional Golay-code submodule, generated by the code words; for fuller details, see for example the Atlas [**CCN**$^+$**85**, p 24]. In particular there is $x \in V$ fixed by the point stabilizer M_{22} in M_{23}; that is, $C_{\overline{H}}(x) \cong M_{22}$. Furthermore $C_F(x) \not\leq H$: indeed we get $C_F(x)/\langle x \rangle \cong Fi_{22}$, which is not quasithin; so that $C_F(x)$ is a non-quasithin *2-local* subgroup.

And indeed now pushing-up methods will allow us to exploit the relation of these local subgroups from the non-quasithin group F—but instead within the abstract context of a quasithin group G with a 2-local subgroup L "similar" to H:

Our abstract local L might be larger than H; but it is suitably determined by $\overline{L} \cong M_{23}$, acting on the cocode module $V \cong 2^{11}$. And an important feature of "quasithin local theory", which we will discuss in a bit more detail in Section 4.4, shows at [**AS04b**, 3.3.2(1)] that:

(3.3.12) $M := N_G(L)$ is the *unique* maximal 2-local containing LT.

We now set $R := O_2(LT)$. Then for any characteristic subgroup C of R, we get that $LT \leq N_G(C)$. But now by the uniqueness feature in (3.3.12) above, we have $N_G(C) \leq M$. So we get:

(3.3.13) $$C(G, R) \leq M < G\ ,$$

which of course is the condition we have been calling R-(CPU) as in (3.3.10).

The goal is then to use (3.3.13), to force $C_G(x) \leq M$; this differs from the situation $C_F(x) \not\leq H$ in the shadow $F = Fi_{23}$, and so will provide a *start* towards eliminating that shadow. We summarize the result of this analysis:

EXAMPLE 3.3.14 (Pushing-up applied to the shadow of Fi_{23}). Assume that the quasithin group G has a 2-local L, determined by $\overline{L} \cong M_{23}$ acting on the cocode module $V \cong 2^{11}$. Then for $x \in V$ stabilized by M_{22}, we must have $C_G(x) \leq M$. ◇

We give a quick sketch of the logic in the proof:

Assume that $C_G(x) \not\leq M$. Because $C_{\overline{L}}(x) \cong M_{22}$ is already "most" of $\overline{L} \cong M_{23}$, we can show that $C_G(x)$ also inherits the R-(CPU) condition of (3.3.13):
$$C\big(C_G(x), R\big) \leq C_M(x) < C_G(x).$$
Thus any $C_G(x)$ not contained in M should appear in the conclusion-list for a suitable analogue of the Global $C(G,T)$-Theorem, corresponding to a local-block determined by the pair (2^{11},M_{23}). But the proof of that analogue [**AS04a**, C.2.8] under the quasithin hypothesis shows that no such larger quasithin groups arise. (That is, in general we would expect $C_G(x)$ to be something like $C_{Fi_{23}}(x) \cong 2Fi_{22}$—but that is not quasithin, and so is unavailable under QT.)

This contradiction shows we must have $C_G(x) \leq M$. □

We mention that Example 3.3.14 above will be continued in later Example 3.4.7—to the *final* elimination of the shadow Fi_{23} in the QT analysis.

3.4. Weak-closure factorizations: Using other weakly-closed "W"

We now return to our chapter-theme of factorizations. The emphasis will now revert to sufficient conditions for *success* of factorizations; though in applications, treating their failure will remain significant—at least in the background.

Thompson's early analysis involving $J(T)$ also included some consideration of elementary subgroups A of one *less* than maximal rank. This approach was extended by Aschbacher; and the resulting weak-closure factorizations were used in a number of crucial places in the original CFSG.

In this section, we give an overview of some of that development; as the material is somewhat technical, our exposition will be just an over-simplified approximation.

Weak closure and candidates for Z, W. Here is a viewpoint providing some initial motivation: Continue the internal-module setup $V \leq H$ as above. We might expect (though this is not always automatic) that V lies in some A of maximal rank in T—and hence $V \leq J(T)$. And then if some conjugate V^h should fall into T, we might also expect that $V^h \leq J(T)$.

The set of all such conjugates V^h then generates the *weak closure* $W(T, V)$ of V in T with respect to H: as the name implies, it is by definition weakly closed in T with respect to H. Thus the situation of the previous paragraph would give $W(T, V) \leq J(T)$; so that $W(T, V)$ might be an alternative candidate for the weakly-closed subgroup "W" in our general factorization setup of (FA) in 3.0.9. Furthermore if $W(T, V)$ is strictly smaller that $J(T)$, then it might be easier to obtain $W(T, V) \leq C_H(V)$—in analogy with the condition $J(T) \leq C_H(V)$ for success of Thompson Factorization (TF) in 3.1.4(1).

It was in this context that Aschbacher gave further axiomatic development to Thompson's consideration of elementary subgroups of less-than-maximal rank:

DEFINITION 3.4.1 (The subgroups W_i and C_i). Assume the internal-module subcase $V \leq H$ of our module-setup 3.0.5. For $0 \leq i \leq m(V)$ set:

$$W_i(T, V) := \langle A : A \leq T \cap V^h, h \in H, \text{ with } m(V^h/A) = i \rangle \, .$$

Notice that $W_i(T, V)$ is weakly closed in T with respect to H. Indeed when $i = 0$, we see $W_0(T, V)$ is just the usual weak closure $W(T, V)$ of V in T. Also set:

$$C_i(T, V) := C_T\big(W_i(T, V)\big) \, .$$

We typically use the abbreviations W_i and C_i for these subgroups. ◊

The goal is then to find suitable values $j \geq i$, so that we may use C_j and W_i in the roles of "Z, W", in factorizations of H in the form (FA) in 3.0.9. In the discussion below of hypotheses to guarantee such factorizations, for expository purposes I will have to blur many of the details; for a fuller treatment, see e.g. B.8.6 in [**ALSS11**].

Parameter values to guarantee a weak-closure factorization. For the remainder of the section, we'll work in the context of the Thompson strategy 3.3.1: namely a maximal 2-local M over T in some simple G, and another 2-local H over T which is not contained in M. In order to follow the conventions in [**ALSS11**, B.8.6], we deviate slightly from our notation in the chapter so far: V will now denote an internal module for M, rather than H; and we will want to develop weak-closure factorizations for H—in terms of an H-module which we will denote by U. And:

We temporarily let W_i,C_i be defined from H on U (rather than M on V).

In particular, we will want a value i such that $1 < W_i \leq C_H(U)$. And this will require estimating lower bounds on certain parameters related to the groups M, H and their modules V, U. These are described only roughly in:

REMARK 3.4.2 (Weak closure parameters). The definitions are of course abstract; but they are motivated by the usual situation in the Even Case of the CFSG discussed in earlier Section 2.2, namely of a simple G having characteristic 2 type: We might expect $\overline{M} := M/C_M(V)$ and $\hat{H} := H/C_H(U)$ to be of Lie-type in characteristic 2; in particular, they should generated by root groups in the Chevalley construction, as discussed earlier in 1.3.20(1). Hence their modules V, U should be described via the 2-modular representation theory of the Lie-type groups, discussed in later Section 5.2. Notably 5.2.3(1) gives the module action in terms of weight spaces, which we had mentioned for Lie-algebra representations in 1.3.7. Of course the ranks of the root groups and weight spaces are related to the size of the fields of definition for $\overline{M} := M/C_M(V)$ and $\hat{H} := H/C_H(U)$. And we now summarize roughly: saying that many weak-closure parameters mentioned in this section are abstract approximations to the ranks of these root groups and weight spaces.

For example, $n(H)$ approximates the rank of a root group of $\hat{H}$. And the related parameter $a(H, U)$ approximates the rank of a weight space on U.

Similarly for $\overline{M}$: the ranks of a root group, and a weight space on V, are involved in the "local" parameter $m := m(\overline{M}, V)$—namely, a lower bound on the corank in V of $C_V(\bar{t})$, as $\bar{t}$ varies over the involutions of $\overline{M}$. But the parameter $r := r(G, V)$ is "global"—a lower bound on the corank in V of a subgroup A, with $C_G(A) \not\leq M$. Roughly, if r becomes larger, then centralizers of smaller and smaller subgroups of V are forced into M; and of course these centralizers should correspondingly be larger. Finally we set $s := \min(r, m)$; and we would like to maximize s—to force as many centralizers as we can into M.

And now continuing to leave aside various technical details: An important point is that these parameters can often be estimated, during case analysis in a proof. And a strong-enough set of values may be sufficient to eliminate some less-likely configurations—just numerically, without a more detailed argument. ◊

In the above language, we can roughly state Aschbacher's [**Asc81b**, 6.11.2] for large-enough s; a weaker version, with generation rather than factorization, is given at [**ALSS11**, B.8.6]. We abbreviate the result by (WC):

THEOREM 3.4.3 (Weak-closure factorization (WC)). *Assume $W_i > 1$, for some i satisfying $0 \leq i < s - n(H)$. Then $W_i \leq C_G(U)$, and:*

$$H = C_H(C_{i+n(H)})N_H(W_i).$$

We close the section by indicating two CFSG areas where (WC)-methods were used:

Weak closure in the Uniqueness Case. We recall from the discussion at the end of Section 2.2, on the Even Case, that the final contradiction of the CFSG was provided by Aschbacher's elimination in [**Asc83b**] of the Uniqueness Case—where the 2-local $M \geq T$ as above satisfies the condition "almost strongly p-embedded". How did that work proceed?

Very many subcases were in fact handled via the following overall approach: Recall that in the Thompson strategy 3.3.1, we are supplied with a second 2-local H, with $T \leq H \not\leq M$. Ideally we should produce a factorization $H = H_1H_2$, with factors H_i given by the local subgroups for (WC) in 3.4.3.

On the other other hand, we can hope to use "uniqueness methods" (cf. our discussion of the preuniqueness case (2) of the Weak Trichotomy Theorem 2.2.7), and also methods of determing obstructions to pushing-up such as (CPU) in 3.3.4—to show that many 2-locals must in fact fall into M. If this suffices to get the H_i into M, then $H = H_1H_2 \leq M$—contradicting $H \not\leq M$ in the Thompson strategy.

This argument is sometimes mostly implicit in [**Asc83b**]: for since this factorization easily finishes off a case, attention is often focused instead, in the spirit of FF in earlier Section 3.2, on controlling the relatively restricted situation where the hypotheses for the factorization in fact fail.

Weak closure in the quasithin classification. We continue in the Thompson strategy 3.3.1, with M on V, and $H \not\leq M$. The preparatory material in Section E.3 of [**AS04a**] in effect produces some further automation of the approach of [**Asc83b**] described just above. Our focus reverts to action of M on V; and:

We return to considering W_i defined by M on V (rather than H on U).

For success of factorizations, we would ordinarily expect $W_i \leq C_T(V)$; so we now examine "where" this latter condition might begin to fail, focusing on:

(3.4.4) $$w := \text{smallest } i \text{ with } W_i \not\leq C_T(V) \ .$$

Now there are natural *upper* bounds on w. For example, using methods such as (CPU)-obstruction to pushing-up in 3.3.4, we get [**AS04a**, E.3.39]; roughly:

(3.4.5) $$\text{(Certain technical hypotheses)} \implies (\ w \leq n'(\overline{M})\ )\ ,$$

where this new parameter $n'(\overline{M})$ is defined by the condition that $\mathbb{F}_{2^{n'}}$ is the field generated by an element whose order is that of the largest odd-order subgroup of $\overline{M}$ permuting with $\overline{T}$. As in 3.4.2, this is a variation on the theme of the rank of a root group in $\overline{M}$.

Thus in order to obtain a contradiction, we also want to maximize *lower* bounds on w—until they exceed the upper bound in (3.4.5) above. A useful tool in this direction is [**AS04a**, E.3.29]:

LEMMA 3.4.6 (Fundamental Weak Closure Inequality (FWCI)). *If V is not an FF module (i.e. V is not in the FF-list 3.2.7), or $w > 0$, then:*
$$w \geq r - m_2(\overline{M}),$$
where r is the parameter discussed in 3.4.2.

We will demonstrate this approach by returning to an earlier topic, namely the non-quasithin shadow Fi_{23}:

EXAMPLE 3.4.7 (Eliminating the shadow Fi_{23}). We continue the setup of earlier Example 3.3.14: Recall $V = 2^{11}$ is the cocode module for $\overline{L} \cong M_{23}$.

We find that a 3-element permutes with a 2-Sylow of M_{23}; so from (3.4.5):
$$w \leq n'(\overline{M}) = 2.$$
Now V contains a subgroup A of corank 6, containing an element x described in 3.3.14; and in $F := Fi_{23}$, we have $C_F(A) \not\leq M$—exhibiting the value of $r = 6$ in that situation. However for our quasithin G, we showed in 3.3.14 that $C_G(x) \leq M$, and hence $C_G(A) \leq M$; and this forces the "unlikely" value of $r \geq 7$ under QT. So as the module V for sporadic M_{23} does not appear in the FF-list 3.2.7, using the (FWCI) in 3.4.6 we obtain:
$$w \geq 7 - m_2(M_{23}) = 7 - 4 = 3.$$
And of course this contradicts $w \leq 2$ above.

This finishes the elimination of L with $2^{11} : M_{23}$, as in the shadow Fi_{23}. ◊

Applications related to the Martino-Priddy Conjecture

The concepts demonstrated so far in this chapter have continued to be productive in the years after the CFSG—in possibly-unpexpected directions.

In the remainder of the chapter, we indicate some fairly recent applications of the ideas: to the modern topological approach to finite group theory, via fusion systems. We will focus on Oliver's work establishing the Martino-Priddy Conjecture. I thank Bob Oliver for a number of suggestions in this area.

3.5. The conjecture on classifying spaces and fusion systems

The story starts with a basic concept within local group theory itself:

Background: fusion in group theory and cohomology. For a group G with Sylow p-subgroup T, the *p-fusion* means the data:

(3.5.1) $\{$ p-fusion: all G-conjugacies among subgroups of T $\}$.

We emphasize this is not limited to conjugacies induced just by $N_G(T)$.

EXERCISE 3.5.2. Determine the fusion in $T \cong D_8$ in $G = A_6$; A_7; etc. Hint: The two 4-groups $A \cong 2^2$ in T are not conjugate in G. But each is normalized by an S_3, not lying in $N_G(T) = T$, which fuses the three involutions of A. These 4-groups intersect in $Z(D_8)$; so we conclude that all 5 involutions in T are conjugate in G. The two elements of order 4 in T are already conjugate in T itself.

The only possible further fusion would be of the two four-groups; and indeed this does happen in $\mathrm{Aut}(A_6)$, though not in $G = A_6$. It also does not happen in A_7, which has the same Sylow T as A_6; so A_7 has the same fusion pattern as A_6. $\diamondsuit$

The fusion information has long been important in group theory; some examples:

In the CFSG itself, determining the 2-fusion was typically an initial goal; but it was also usually crucial in the final stages, namely identifying a group—the *recognition* problem, which we will be discussing in the subsequent Chapter 4.

An important tool in fusion anlysis is provided by the Alperin Fusion Theorem [**ALSS11**, B.2.6]—which shows that although p-fusion need not take place in $N_G(T)$, it does occur via a sequence of "local" conjugations. Roughly stated:

THEOREM 3.5.3 (Alperin Fusion Theorem). *A Sylow p-group T of a finite group G admits a* conjugation family*: a set of nontrivial subgroups S_i of T, with the property that: whenever $A, B \subseteq T$ with $A^g = B$ for $g \in G$, there is a sequence of elements n_i ($i = 1, \ldots, k$ for some k), with $n_i \in N_G(S_i)$, such that $A^{n_1 \cdots n_k} = B$ (and $A^{n_1 \cdots n_r} \subseteq T$ for all $r \leq k$).*

More immediately relevant to the topological focus of these sections is: The information on p-fusion provides a standard route for computing the p-part $H^*(G)_p$ of group cohomology: namely as the "stable" elements in the restriction of H^* from G to P; see e.g. [**AM04**, II.6.6].

In fact different groups can have the same cohomology: this happens at p for example if the groups have the same p-fusion pattern—we saw an example of the latter in Exercise 3.5.2 above.

Indeed, in the background here is the idea that p-cohomology should be equivalent to p-fusion. And topologists in particular have sought natural ways of making this equivalence precise—as we will see in the Martino-Priddy Conjecture below.

The category viewpoint: fusion systems. The classical topological approach to group cohomology is via a topological space, namely the *classifying space* BG. Indeed for the p-part $H^*(G)_p$, they use the *p-completed* classifying space $BG_p^\wedge$.

Also since about the 1950s, algebraic topologists have largely replaced the classical viewpoint of topological spaces (based on cell complexes) with the viewpoint of *simplicial spaces*—this is based on the language of category theory. The equivalence of the two viewpoints was established by Quillen; For more on this background material, see e.g. [**BS08a**, Sec 3.6].

Thus it is natural to ask whether we can also approach the group-theoretic fusion information by means of a suitable category. The idea and basic definitions are due to Puig—starting with his thesis published in the 1970s, and developed over the subsequent years, culminating in [**Pui06**]. For a modern survey of the area now called *fusion systems*, see e.g. the overview in Aschbacher-Oliver [**AO16**]. Earlier literature often references the survey in [**BLO03**]; there are also newer books [**AKO11**] and [**Cra11**].

We had in fact mentioned Aschbacher's project on classifying simple fusion systems, as new-approach (4) to the CFSG, in our afterword-Section 2.3. Below we will give a little more detail on general fusion systems, than we did at that earlier point; while still referring the reader to sources such as [**AO16**] for fuller details.

A *fusion system* $\mathcal{F}$ on a p-group T is roughly a category based on:

- objects: the subgroups (P, Q, ...) of T,
- morphisms: suitable injections $P \to Q$ (e.g. inner automorphisms of T);

with several further natural axioms, which roughly correspond to abstracting certain consequences of the Sylow theorems.

Of course, the standard example comes from a finite group G with Sylow subgroup T: this category, called $\mathcal{F}_{T,G}$, uses the G-conjugacies—that is, the fusion—as the morphisms.

But there are also "exotic" fusion systems $\mathcal{F}$, which do not arise from a finite group. The most famous example is the *Benson-Solomon* system. This arose originally from work of Solomon, within the standard-form branch of the Odd Case of the CFSG (recall Section 2.1): In characterizing the Conway group Co_3, with involution centralizer $2\Omega_7(2)$, he eliminated a related odd-case—showing that no finite group G could have an involution centralizer $C_G(t)$ given by the double cover of the orthogonal group $\Omega_7(q)$ for odd q. In the process, Solomon computed the 2-fusion pattern of such a hypothetical G; and this information is basically the definition of the Benson-Solomon system $\mathcal{F}$. This system is of independent interest in algebraic topology: Benson (cf. [**BS08a**, p 202]) observed that its cohomology corresponds to that of the previously-known "exotic space" $BDI(4)$ of Dwyer-Wilkerson, whose cohomology gives the rank-4 Dickson invariants $DI(4)$. For odd primes p, there are many examples of exotic fusion systems; see e.g. [**RV04**].

Fusion systems also provided the natural context for:

The Martino-Priddy Conjecture. One natural way to express the equivalence of cohomology with fusion was conjectured by Martino and Priddy, and later proved by Oliver [**Oli04**, Thm B]; it is stated in terms of the relevant equivalences for topological spaces and for fusion systems—namely homotopy equivalence, and a strong form of category equivalence:

CONJECTURE 3.5.4 (Martino-Priddy Conjecture). *Assume finite groups G, G^* have Sylow p-subgroups denoted by T, T^*. Then:*

$$BG_p^{\wedge} \simeq_{homot.eq.} (BG^*)_p^{\wedge} \qquad \Leftrightarrow \qquad \mathcal{F}_{T,G} \equiv_{fus.sys.} \mathcal{F}_{T^*,G^*}.$$

Here the equivalence "fus.sys." of fusion systems in effect requires a group isomorphism $T \cong T^*$ which further induces a category isomorphism.

In fact Martino and Priddy were able to prove the $\Rightarrow$ direction. But their hope of proving the remaining $\Leftarrow$ direction fell afoul of certain obstructions, expressed in the topological language of the "higher limit" $\lim^2$. These obstructions were later shown to vanish by Oliver:

3.6. Oliver's proof of Martino-Priddy using the CFSG

Oliver proved the $\Leftarrow$ direction in [**Oli04, Oli06**], separating p odd and $p = 2$.

His proof uses the CFSG—and we will sample some of those arguments. However, we do mention that Chermak's later proof [**Che13**] makes only a "milder" use of the CFSG; and recently, Glauberman-Lynd [**GL16**] gave a CFSG-free proof.

Oliver's general setup. We give a quick overview of the p-odd paper [**Oli04**]; again for expository purposes, definitions will be given only approximately.

The goal is to construct a supplementary *linking system*, for the fusion system $\mathcal{F}_{T,G}$ on the right-hand side of the Conjecture 3.5.4. Roughly: the morphisms in the fusion system do not really record the kernels of mappings; and the linking system in effect restores some of that lost information about centralizers in those kernels. This information in turn enables the construction of a suitable classifying space—which can then be compared to the classifying spaces on the left-hand side of the Conjecture, to complete the proof.

The linking system turns out to require consideration of a certain center-functor $\mathcal{Z}$, roughly containing centralizer-information for centers $Z(P)$ of the p-groups P; this functor is defined on the G-orbit category—essentially corresponding to the permutation representations G/P on the cosets of the p-groups P.

The obstruction to the construction of a linking system arises via the higher limits $\lim^{i>0}$ of the functor $\mathcal{Z}$—these are derived functors of the usual limit $\lim = \lim^0$. Oliver establishes suitable vanishing results on these higher limits; this removes the obstruction to the construction of linking systems, allowing completion of the proof. Here is some of the underlying group theory:

Oliver's application of the CFSG. Oliver reduces at 4.1 of [**Oli04**] to the case where G is a nonabelian simple group.

The subgroup $X(T)$, and conditions for a suitable subgroup Q. For this situation, Oliver shows it suffices to produce a subgroup $Q \leq T$ with the properties:

- Q is *p-centric*—this means that $Z(Q)$ is Sylow in $C_G(Q)$;[8]
- Q is weakly closed in T with respect to $\mathrm{Aut}(G)$; and
- $Q \leq X(T)$, where $X(T)$ is a certain characteristic subgroup of T (below).

At several later points, we will want to use the elementary observation that T itself automatically satisfies the first two conditions; so that:

LEMMA 3.6.1 (The case $X(T) = T$). *If it happens that $X(T)$ is all of T, then we can choose T as "Q"—to complete the proof of the Conjecture* 3.5.4 *for G.*

We turn to some of the discussion of $X(T)$ from [**Oli04**, Sec 3]:

DEFINITION 3.6.2 (The subgroup $X(T)$). We consider chains of normal subgroups R_i of T:

$$1 = R_0 \leq R_1 \leq \cdots \leq R_n \leq T,$$

which satisfy $[\,\Omega_1\big(C_T(R_{i-1})\big)\ ,\ R_i\ ;\ p-1] = 1$. It turns out there is a unique maximal choice for the final member R_n of such a chain; define this choice to be $X(T)$. $\diamondsuit$

We mention that for $p = 2$, we always get $X(T) = C_T\big(\Omega_1(T)\big)$; and so $X(T)$ is not really useful when Oliver treats the Conjecture for $p = 2$ in [**Oli06**].

We note also that the $(p-1)$-times-repeated commutator with $C_T(R_{i-1})$ in the definition is reminiscent e.g. of the quadratic condition in 3.2.4; and may help explain why Oliver encountered phenomena related to $J(T)$ and failure of factorization. This connection arises more clearly in the subsequent Section 3.7.

EXERCISE 3.6.3. Find $X(T)$ for some small nonabelian p-groups T, with p odd. Hint: You may wish to use some of the properties listed in Lemma 3.6.4 below. For example, an extraspecial p-group G of order p^3 has distinct abelian normal subgroups of order p^2; so $X(T) = T$ by 3.6.4(1). $\diamondsuit$

We indicate some useful properties of $X(T)$ from [**Oli04**, 3.2,3.10]:

LEMMA 3.6.4. *$X(T)$ is p-centric. Furthermore:*
(1) If A is abelian, and $A \trianglelefteq T$, then $A \leq X(T)$.
(2) If $\Omega_1(\ Z\big(X(T)\big)\ )$ has rank $< p$, then $X(T) = T$. (So we are done by 3.6.1.*)*

Also from [**Oli04**, 3.7] we give some sufficient conditions for the choice of Q:

LEMMA 3.6.5. *The following situations givea suitable Q to complete the proof:*
(a) $J(T) \leq X(T)$;
(b) T has a unique elementary abelian subgroup A of maximal rank;
(c) $T/X(T)$ is abelian.

We mention that in (a), we can take $J(T)C_T\big(J(T)\big)$ as Q. In (b), we get $A = J(T)$; so $J(T) \leq X(T)$ by 3.6.4(1), and then we are done by (a).

In fact in [**Oli04**, Conj 3.9], Oliver asks if the inclusion $J(T) \leq X(T)$ in (a) is always true for odd p, independently of the CFSG. We will return to this conjecture, in the subsequent Section 3.7.

---

[8] This centric condition had already been prominent in the topological approach; for example, to the decompositions of cohomology that we discuss in later Section 7.6.

Oliver's choice of Q for the various simple G. We next give a quick sketch of [**Oli04**, 4.2–4.4], where Oliver verifies the existence of a suitable Q, for the types of simple groups G in the CFSG.

Many cases finish via 3.6.5(b), namely T with a unique maximal elementary:

This works for G an alternating group: Recall that we saw at 1.1.1(2) that the p-rank is $\lfloor \frac{n}{p} \rfloor$. Furthermore the subgroup A, generated by the order-p elements from each part of the partition determining that rank, is unique in T—for T is determined by further grouping-together in that partition, p parts at a time.

It also works for G of Lie type, in characteristic $r \neq p$. For we commented at Example 2.2.1 that the widest elementary r'-subgroup A lies on the diagonal subgroup H of Remark 1.3.20(2). (Here "diagonal" should be interpreted so as to allow for *non*-split tori as in Example 5.2.2.) And much as in the previous paragraph, A is unique in T.

EXERCISE 3.6.6. Describe non-split tori in $L_4(2)$ for $p = 3, 5, 7$. What is the widest for $p = 3$? ♢

(The method via 3.6.5(b) would also work in some of the remaining cases below.)

Now consider cases for G of Lie type in characteristic p: Most have $X(T) = T$, and so are finished by 3.6.1. (This remark after [**Oli04**, 4.3] can be used to replace some later arguments that use the statement there that "$U_J \cap U_{J'} = U_{J \cup J'}$", which unfortunately is incorrect.) In other cases, a suitable unipotent radical, inside $X(T)$, can be chosen as "Q".

Finally for sporadic G: For $p > 3$, $m_p(G) < p$—so 3.6.4(2) completes the proof. And for $p = 3$, various methods can be used: At least 11 cases similarly have $m_3(G) < 3$, as in 3.6.4(2). Then at least 8 more cases have $T/X(T)$ abelian—and so are completed by 3.6.5(c). The few remaining cases can be finished, with some further work, via 3.6.5(b) or 3.6.1.

3.7. Oliver's conjecture on $J(T)$ for p odd

We saw above that Oliver conjectured in [**Oli04**, Conj 3.9] that the earlier sufficient condition 3.6.5(a) should in fact always hold:

CONJECTURE 3.7.1 (Oliver's Conjecture on $J(T)$). *If T is a p-group, p odd, then $J(T) \leq X(T)$.*

One motivation for Oliver was that if this Conjecture could be proved, he would be able to avoid the use of the CFSG, and obtain a much simpler proof of the Martino-Priddy Conjecture for p odd: indeed substantially simpler than that of Chermak and Glauberman-Lynd indicated earlier.

So let us examine some aspects of Oliver's Conjecture 3.7.1. In a minimal counterexample to this "J-Conjecture", we may in fact assume that:

$$X(T) \text{ is elementary abelian;}$$

so we will now abbreviate it by $V := X(T)$. Furthermore $V = C_T(V)$ by the centric property 3.6.4, so our counterexample must satisfy $J(T) \not\leq C_T(V)$—so that as in Definition 3.2.1 we have:

$$V \text{ is an FF-module under } \overline{T} := T/C_T(V).$$

Set $n := \dim(V)$. Some results in the literature show $\overline{T}$ is "not too small"; e.g.:

- Oliver [**Oli04**, 3.10] showed: $n \geq p$.
- Green-Héthelyi-Mazza [**GHM10**] showed: nilpotence class $\mathrm{cl}(\overline{T}) \geq 5$;
- Indeed Lynd showed: $\mathrm{cl}(\overline{T}) \geq \log_2(p-2)+2$.

We now proceed to a variant of the original Oliver Conjecture, discussed for example in [**GHM11**, 1.4]; cf. also [**GL13**]. Recall that the Thompson Replacement Theorem 3.2.4 shows that FF-offenders must in fact contain offenders which exhibit quadratic action. Oliver's viewpoint in fact suggests:

CONJECTURE 3.7.2 (Quadratic Conjecture). *If V is an FF-module for $\overline{T}$ an odd p-group, then $\Omega_1\big(Z(\overline{T})\big)$ contains a quadratic element.*

Notice then that for a counterexample $\overline{T}$ to this "Q-conjecture":

$\overline{T}$ has no quadratic *normal* subgroup.

Further, it seems implicit in the literature that $\overline{T}$ contains no transvections on V.

These remarks suggest some connections with material earlier in this book—which tend to show that $\overline{T}$ is "not too large". We begin by embedding our counterexample $\overline{T}$ in a full unipotent group $\overline{U}$ of $GL(V)$.

Consider the unipotent radical $\overline{U}_{\hat{k}}$ of the maximal parabolic $P_{\hat{k}}$ in earlier Example 1.3.4: observe that $\overline{U}_{\hat{k}}$ is normal in $\overline{U}$, and acts quadratically on V. So by the displayed remark above, we have:

$$\overline{T} \cap \overline{U}_{\hat{k}} = 1. \tag{3.7.3}$$

Note furthermore that a transvection $\bar{t}$ lies in some $\overline{U}_{\hat{k}}$, from the requirement that 1-dimensional $[V, \bar{t}]$ must lie in some V_k. Thus from (3.7.3), we get an easy proof that:

$\overline{T}$ contains no transvections.

Also from (3.7.3), we see that the map from our counterexample $\overline{T}$ into $\overline{U}/\overline{U}_{\hat{k}}$ must be faithful. So take $k := \lfloor \frac{n}{2} \rfloor$: from the faithful map of $\overline{T}$ into the product in $\overline{U}$ of the full unipotent groups in the linear groups on V_k and V/V_k, we conclude that:

$\mathrm{cl}(\overline{T}) \leq \frac{n}{2}$; and

$\nu_p(\ |\overline{T}|\ ) \ \leq \frac{1}{2}\binom{n}{2}$.

So, one can hope that future research will enlarge the earlier lower bounds on the size of $\overline{T}$, and shrink the upper bounds just above—until they pass each other, for a contradiction that would prove the Oliver Conjecture 3.7.2.

Ideally this brief discussion will motivate some readers to work on the Conjecture!

Oliver's work on the Martino-Priddy Conjecture demonstrates that the ideas around Failure of Factorization continue to be productive in wider areas. So although we might have been tempted to think that we were finished with Thompson Factorization, it turns out that Thompson Factorization was not finished with us.

CHAPTER 4

Recognition theorems for simple groups

This chapter reviews techniques for identifying members of the CFSG-list.

Introduction: Finishing classification problems

For contrast with recognition, we first mention an opposite situation: In the final stages of proving some classification theorem, it is often necessary to *eliminate* various configurations—which correspond to groups that do not arise in the conclusion of that theorem. For example, we sketched in Example 3.4.7 the elimination of the 2-local $2^{11}M_{23}$, corresponding to the "shadow" of the group Fi_{23}—which is not quasithin, and hence does not appear into conclusion of the quasithin classification.

Recognition. On the other hand, the more common task at the end of a classification proof is to verify that the stated conclusion-groups actually do arise; and this typically involves using structural information in the proof-so-far, to identity those specific target groups.

For this, we normally are able to use suitable "recognition theorems"; these are typically fairly generally-stated results, roughly establishing that:

$$\text{(sufficient local information)} \Rightarrow \text{(characterization of "global" } G) \ .$$

This procedure of using recognitions theorems is in the broad spirit of applications of the CFSG: for whether the relevant classification problem occurs within the CFSG proof, or in some application-result which quotes the CFSG, the recognition process normally involves inspecting the list of the CFSG for the relevant groups, in order to recognize them via suitable properties.

For some deeper general background on recognition theorems, the reader may wish to consult sources such as Chapter 3 of [**Gor82**] (which we will quote often in this chapter), and Section A.5 of [**ALSS11**].

Uniqueness. The literature in fact contains a vast number of recognition theorems. Often these have titles of the general form "A characterization of (some particular group or groups)".

But we also emphasize that the recognition problem for some group G implicitly includes the underlying *uniqueness* problem: namely that G should be defined uniquely—that is, up to isomorphism—by a suitable subcollections of its properties. Hence also relevant to our topic of recognition are many papers in the literature with titles of the general form "Uniqueness of (some particular group or groups)".

Of course for many simple groups, such as the alternating groups or linear groups, the definition (in terms of permutations or matrices) is sufficiently natural to make uniqueness immediately clear. But this was not always the case for some of the simple groups discovered after 1960—where the phrase "group of type X" was often used, to indicate a group X whose "definition" was really just a description:

believed to be unique up to isomorphism, but not proved unique until later on. Perhaps the most celebrated examples of this complication were the Ree groups $Ree(q)$: where the eventual proof of uniqueness required subtle properties of finite fields, which were finally resolved by Thompson and Bombieri; see e.g. [**Gor82**, 3.38] for further details.

A propos of uniqueness, we should also mention that the usual form of the CFSG-list 1.0.2 implicitly assumes a few small cases where several groups with different names turn out to be isomorphic; for detailed reference see [**ALSS11**, p 261]. We had seen examples, such as $A_5 \cong L_2(4) \cong L_2(5)$, in earlier Exercise 1.5.5.

Some techniques for recognition problems. In this chapter, we will sample some frequently-used recognition results. But before subdividing our discussion according to the three standard classes of simple groups in the CFSG-list 1.0.2, we first distinguish several common methods and approaches for recognition problems in general.

Here the underlying question is: What kinds of information should be sufficient—that is, as hypothesis—in recognition theorems? There are in fact many different approaches. But the information in the hypotheses is typically rather technical; so our discussion here will remain at the level of overall descriptions. The reader can consult the sources mentioned above for fuller details.

Presentations. We first briefly mention the approach via a group *presentation*: that is, definition via abstract generators and relations.

We'll see one such result for A_n, in the subsequent Section 4.1. Furthermore for the Lie-type groups, we'll examine the important Curtis-Tits Presentation in Section 4.2 immediately thereafter.

Nonetheless, we emphasize that presentation-arguments are often rather touchy; and they are not as common in finite group theory, as they are e.g. in the infinite group theory associated with geometry and topology.

Action identifications. For convenience, I'll invent a term "action" identification—to refer to recognition of G, via its action on some natural structure; such properties which are typically close to the actual definition of G. Furthermore we have mentioned earlier that the natural actions of simple groups are often highly transitive.

Here the "structure" might be only a set; e.g. 6-transitivity determines the alternating group A_n among simple groups. Or the structure might instead be:

- a graph, as in rank-3 permutation groups (cf. later 4.3.1); or
- a module, such as the natural module for $GL_n(q)$ in Remark 1.3.1; or
- a geometry on a module, e.g. projective or polar space—cf. 7.0.1; or
- a lattice, e.g. the Leech lattice for the Conway groups in Section 1.2;

and so on. We'll see various such examples of this action-approach, as we proceed in this chapter.

Internal identifications. In contrast to action-identification above, I'll also invent a term "internal" identification: to refer to hypotheses typically involving instead various internal group-structure (e.g. p-local) properties—that is, structures within G itself, rather than some external object that G acts on.

And beginning with such internal hypotheses, we typically hope (at the cost of further effort) to work toward a final identification of G—often via a more fundamental action-identification, as above.

Some common internal-identification approaches. We suggest a few:

One classic internal-hypothesis is provided by the structure of an *involution centralizer* $C_G(t)$—for reasons we had suggested around Remark 2.0.2. And indeed the literature from the 1960s and 1970s has very many such characterizations via involution centralizers. In particular, we saw in Section 2.1 that standard-form problems are in this spirit; and they provided the key to the treatment of the Odd Case of the CFSG.

But we also emphasize that the centralizer alone is not always sufficient for recognition: the most notorious example is the group $2^{1+6}L_3(2)$, which occurs as $C_G(t)$ in three different groups G—namely $L_5(2)$, M_{24}, He.

A similar internal-hypothesis is the structure of *the Sylow 2-subgroup* T of G. Again many cases of T were considered in the literature of the 1960s and 1970s. A famous example was T *abelian*—treated by Walter; and see also Bender, e.g. 4.126 in [**Gor82**]. The groups G determined by an abelian Sylow of order 2^a are:

$L_2(2^a)$; $a = 2$—$L_2(q), q \equiv 3, 5(8)$; $a = 3$—J_1 and Ree groups $Ree(q)$.

Similar classifications—for T dihedral, semi-dihedral, and wreathed—were used in the Small Odd Subcase $m_2(G) < 2$ of Section 2.1; for example see [**ALSS11**, 1.4.6].

The structures $C_G(t)$ and T above involve recognition via certain 2-local subgroups; and the literature also contains recognition theorems by various other kinds of local subgroups.

A different internal hypothesis is given by the *group order* $|G|$; again the early CFSG literature contains many characterizations by group order. And we mention here an important tool for computing $|G|$, namely the Thompson Order Formula [**Gor82**, 2.43]; a standard special case is:

THEOREM 4.0.1 (Thompson Order Formula). *For a group G with exactly two conjugacy classes of involutions—say t, u—we have:*

$$|G| = a(u)|C_G(t)| + a(t)|C_G(u)|;$$

where $a(v)$ is the number of ordered pairs (x, y) from t^G, u^G with $v \in \langle xy \rangle$.

We remark that the input to this calculation is purely "internal" to G: namely the structure of the centralizers $C_G(v)$, plus the fusion-information as in Definition (3.5.1). And of course the output of the formula can be input to characterizations by order.

EXERCISE 4.0.2 (Practice with the Thompson Order Formula). Use the formula to compute the orders of the small groups S_4, and S_5. Then for a more "realistic" example, try A_8 (this can be fairly lengthy).

Hint: Some details were provided to the class online; these now appear in appendix Remark B.2.1. ◇

A related internal hypothesis is provided by the *character table* of G. Note that this information implicitly includes the group- and centralizer-orders used above.

We turn in the next three sections to specific recognition results for the three classes of simple groups in the CFSG-list 1.0.2.

4.1. Recognizing alternating groups

Action identification. The group A_n is usually regarded as fairly easy to recognize, at least in terms of its natural action as permutations: that is, in situations where we can exhibit the natural permutation representation of degree n, we just need to establish suitable multiple transitivity.

See e.g. [**Gor82**, Sec 3.2] for some analogous recognition results for various doubly-transitive groups. Indeed in application-situations where we are *assuming* the CFSG, the determination of 2-transitive groups (which we had outlined in Section 1.6) shows that 6-transitivity suffices to recognize A_n.

Internal identification. Recognizing A_n from internal group-theoretic structures can be more difficult. Gorenstein [**Gor82**, 3.42] gives one standard approach: it is actually via a presentation—but it involves internal structures such as properties of involutions, and of 3-cycles, in a subgroup A_{n-2}.

We mention also a more general recognition-result, which is based on somewhat analogous properties of individual elements: namely the celebrated theorem of Fischer—which was originally aimed at recognition of the almost-simple group S_n, based on generation by a class of 3-transpositions: recall from (1.2.2) this means that $|xy| = 1$, 2, or 3, for pairs in the class.

In fact, various other simple and almost-simple groups arise naturally under this hypothesis; see e.g. [**ALSS11**, A.6.3]: namely some classical groups over $\mathbb{F}_2$; some orthogonal groups over $\mathbb{F}_3$; and Fischer's sporadic groups Fi_{22}, Fi_{23}, and Fi_{24}—which were discovered during the course of Fischer's work.

EXERCISE 4.1.1 (Classical 3-transposition groups). Find 3-transpositions in some of the classical groups indicated above.

Hint: Consider transvections and reflections, respectively; verification is assisted by rank-3 considerations as in the discussion leading up to later Exercise 4.3.2. Some sample details appear in appendix Remark B.2.7. ◊

Generalizations from the viewpoint of 3-transpositions led to further recognition theorems for various Lie-type groups; see the discussion of root involutions etc in the following section.

4.2. Recognizing Lie-type groups

Action identification. For most Lie-type groups G, perhaps the most natural recognition is again via an action: namely on its *building*; for details on this geometry, see e.g. [**Car89**, Sec 15.5]. For the moment, in terms of structures we had introduced e.g. at 1.3.20(4) and 1.3.11, we'll just indicate that the building is a simplicial complex—determined by the collection of parabolic subgroups of G; and in particular, "axiomatized" by the Dynkin diagram of G. We'll examine buildings in a bit more detail, in later Section 7.2; indeed see Remark 7.2.5 for one version of the definition.

We mention that the uniqueness-aspect of the recognition of G and its building relies on an underlying topological result [**Ron89**, 4.3] of Tits, which we state as (7.2.10) in our later discussion: namely that (assuming rank ≥ 3), a finite building is *simply connected*. Also for more on recognition of Lie-type G via equivalence with its building, see e.g. [**Gor82**, 3.12].

Internal identification. We turn to recognition via internal-structure hypotheses. Here the is the Curtis-Tits Presentation [**ALSS11**, A.5.1] is the standard tool, based on the earlier *Steinberg relations*.

Underlying these relation-considerations are some deeper geometric results of Tits [**Tit74**, 13.11,13.29] on buildings. To give a sample of what Tits proved, adapted for our present group-recognition purposes, below we give a rough paraphrase of the Curtis-Tits Theorem [**ALSS11**, A.5.1]. Recall from Remark 1.3.20 the setup of root subgroups, and the Dynkin diagram; we have:

THEOREM 4.2.1 (Curtis-Tits Presentation). *For an untwisted universal Lie-type group G, with simple system Π of rank ≥ 3, take generators given by root subgroups $U_{\pm\alpha_i}$ for $i \in \Pi$; and relations given by the subgroups $\langle U_{\pm\alpha_j}, j \in J\rangle$, for the rank-2 subsets $J \subset \Pi$. (The structure of these subgroups can be read off from the Dynkin diagram.)*

Then these generators and relations give a presentation for G.

Note that the G above is the "universal" form of the simple Lie-type group: that is, it is typically a quasisimple extension of the simple group over elements from the Schur multiplier—for example, it uses $SL_n(q)$ rather than $PSL_n(q)$. In effect, the generators are rank-1 groups $SL_2(q)$; with relations from rank-2 subgroups for the corresponding rank-2 subdiagrams.

EXERCISE 4.2.2. For a few rank ≥ 3 diagrams, exhibit the structure of the rank-2 subgroups given by $\langle U_{\pm j}, j \in J\rangle$, for rank-2 subsets $J \subset \Pi$, which give the relations in the presentation.

Hint: For example, for the group SL_4, the subdiagrams of A_3 have their types given by A_2, $A_1 \times A_1$, and A_2; with corresponding groups SL_3, $SL_2 \times SL_2$, SL_3. For Sp_6 with diagram C_3, the first subdiagram instead has type C_2 (for Sp_4). $\diamondsuit$

The Curtis-Tits Presentation has been applied very widely. We had already mentioned one crucial use, in our discussion of the treatment of standard type, in the Even Case of the CFSG—namely branch (1) of the Trichotomy Theorem 2.2.8: Recall this was the paper of Gilman-Griess [**GG83**], which gave the final recognition of most Lie-type groups in characteristic 2.

There are versions of the presentation applying also to twisted Lie-type groups; we won't here present those more complicated statements. An early version including the unitary groups was given by Phan; nowadays most authors instead reference the more fully detailed revision given by Bennett and Shpectorov [**ALSS11**, A.5.2].

Internal identification for small ranks. We turn to some remarks about recognizing Lie-type groups of Lie ranks ≤ 2; these are not covered by the Curtis-Tits Presentation, so that some kind of further work is required.

The groups of of Lie rank 1 were in fact mostly recognized using some pre-CFSG results on 2-transitive groups; see e.g. [**Gor82**, Sec 3.2].

For groups of rank 2, two major approaches have been used:
- early on: via *split BN-pairs*[1] *of rank 2*; see Fong-Seitz [**FS73**];
- more recently: via *Moufang generalized polygons*;[2] see Tits-Weiss [**TW02**].

[1]We had briefly suggested the BN-pair approach at the end of Remark 1.3.20(3).

[2]We sketch generalized polygons at Remark 7.3.1. For Moufang conditions, cf. Section 10.4.

But we also mention a method developed in [**AS04a**, Sec F.4], for the analysis of quasithin groups—which involves the context of *amalgams*; we'll examine this in a bit more detail in later Section 4.5. For the moment, we'll just summarize as follows: Recall first that the quasithin conclusion-groups are mainly the Lie-type groups in characteristic 2, of Lie rank ≤ 2. The amalgam for such a group consists roughly of the *quotients* of the two minimal parabolics modulo their unipotent radicals. And the main result in [**AS04a**, Sec F.4] allows recognition of rank-2 groups via the amalgam—*plus* the standard internal-structure of the involution centralizer. This recognition can be reduced to that in either of the approaches mentioned above, namely split BN-pairs of rank 2, or Moufang generalized polygons.

Some other influential internal recognitions. We briefly mention some more specialized characterizations of certain Lie-type groups, also widely applied:

Generalizing beyond Fischer's condition of 3-transpositions, Timmesfeld recognized in [**Tim73**] many of the groups defined over $\mathbb{F}_2$—using generation by a class of $\{3,4\}^+$*-transpositions*: where now product-orders $|xy|$ can be $1, 2, 3$; or 4—in which case $(xy)^2$ is also in the class. Important applications of his result were made, for example, in the treatment of groups of $GF(2)$-type—namely branch (3) of the Trichotomy Theorem 2.2.8 in the Even Case of the CFSG.

We also mention that groups over larger fields $\mathbb{F}_{2^a}$ were recognized by Timmesfeld [**Tim75**] in terms of generation via a class of *root involutions*—where the "3" in $\{3, 4^+\}$-transposition above is replaced by "any odd number".

An important early result of McLaughlin [**ALSS11**, A.6.1] classified groups G generated by *transvections* on some irreducible G-module V defined over $\mathbb{F}_2$: the groups appearing in the conclusion are:

$$G \cong SL(V),\ Sp(V),\ SO^{\pm}(V),\ S_{n+1},\ \text{or } S_{n+2}.$$

EXERCISE 4.2.3. Exhibit transvections in some small cases of the above-listed groups. Hint: Various groups of transvections appear in Remark B.2.1. ◇

We also mention Aschbacher's Classical Involution Theorem, appearing originally in [**Asc77a**][**Asc77b**][**Asc80**] (or see [**ALSS11**, 1.7.5]), which used the viewpoint of "fundamental $SL_2(q)$s" to recognize Lie-type groups defined over fields of odd order: this work was important for example in the recognition stages of the treatment of the standard form 2.1.2 problems in the Odd Case of the CFSG.

4.3. Recognizing sporadic groups

Since there is no general theory of sporadic groups, there is no really uniform approach to their recognition.

Internal identification. We mention that the literature contains many characterizations of sporadic groups via their involution centralizers. For an indication of various such results. see for example [**Gor82**, 3.50].

Action identification. There are a reasonable number of approaches to sporadic recognition via actions; we will outline a few:

There are various discussions of the Mathieu groups in terms of their actions on Steiner systems and Golay codes; and the Conway groups on the Leech lattice. The books of Wilson [**Wil09**] and Griess [**Gri98**] are good sources for this material. (And also the Atlas [**CCN**$^+$**85**]; but treatments there don't usually contain proofs.)

There is also a literature on geometries for sporadic groups, which attempt to exploit some limited analogies with buildings for Lie-type groups. We examine this topic in a little more detail in later Section 7.3.

Rank-3 permutation representations. We saw in Section 1.2 that a number of sporadic groups were discovered via rank-3 permutation representations; we now give a brief overview of this more general topic:

The condition is a weakening of double transitivity—where a point stabilizer G_α is transitive on the remaining points $\neq \alpha$. Here we assume instead:

DEFINITION 4.3.1 (Rank-3 permutation representation). In a rank-3 permutation representation, the point stabilizer G_α has exactly *two* orbits (the "G-suborbits") on the points $\neq \alpha$. $\diamondsuit$

There are various classical examples of rank-3 representations. Indeed, essentially in view of Witt's Lemma, which we had indicated at Remark 1.3.5, the situation usually arises for a classical group G, on a space V with a form: the two further orbits of G_v on 1-spaces of the same "length" are those for: (all other $w \perp v$) and (all $x \not\perp v$).

EXERCISE 4.3.2. Give the suborbits for $O_4^-(2) \cong S_5$ on the 10 non-isotropic points; and for $Sp_4(2) \cong S_6$ on the 15 points; cf. the discussion in Sections 2.1 and 2.2 of [**Smi11**]. In particular, verify that these permutation representations have rank 3. The suborbit sizes are $1, 3, 6$ and $1, 6, 8$, respectively.

Hint: Compute the index in $C_G(v)$ of substabilizers $C_G(v, w)$ etc, as suggested above. And see also Remark B.2.7—where several somewhat larger rank-3 representations are presented. $\diamondsuit$

But in addition, as we had already commented in introducing the sporadic groups in Section 1.2, certain sporadics—e.g. J_2, HS, McL, Suz, Ru, and the Fischer groups—were found (and often characterized) via rank-3 representations. For more on this topic, see e.g. [**Gor82**, Sec 2.6].

We mention that the analysis of rank-3 representations in the literature is often phrased in the language of a graph—where the edge relation is defined using one of the two suborbits.

More generally: Aschbacher in Part III of [**Asc94**] develops (and applies) a fairly uniform approach to recognition for sporadic groups; which is similarly set in the context of suitable graphs, suborbits, and connectivity—but not restricted to the rank-3 case.

Applications to recognizing some quasithin groups

The final sections of this chapter demonstrate the application of certain recognition theorems—as used to determine some of the conclusion-groups G, in the analysis of quasithin simple groups in [**AS04b**].

4.4. Background: 2-local structure in the quasithin analysis

In each of the subsequent two sections, the quasithin conclusion-group G arises essentially as $\langle L, H \rangle$: where L and H are certain 2-local subgroups, produced at the end of a process of fairly-standardized development.

The purpose of the present background-section is to provide a quick overview of the foundations for that development. But understanding those foundations is not absolutely essential for the later sections; the more impatient reader, who is prepared to accept some mystery about the origins of L and H, can skip over the background material in this section.

Implementing the Thompson strategy. We had described the Thompson strategy in earlier Remark 3.3.1, with respect to a chosen Sylow 2-subgroup T of G.

In particular we recall that in order to use the strategy, we must first deal with the case where T is contained in a *unique* maximal 2-local subgroup M of G. For the general Even Case of the CFSG, namely under the hypothesis of characteristic 2 type, this is accomplished via the Global $C(G,T)$-Theorem 3.3.8. However, the quasithin analysis instead proceeds under the weaker hypothesis of even characteristic; and so Chapter 2 of [**AS04b**] in effect gives a "quasithin $C(G,T)$-Theorem"—namely an analogue of Theorem 3.3.8, proved under the hypothesis of even characteristic.

Thus *after* Chapter 2 of [**AS04b**], we can adopt the Thompson strategy 3.3.1: where in addition to the maximal 2-local M over T, we also have a 2-local H over T which is *not* contained in M. And we can hope to either identify, or eliminate, the larger group $\langle M, H \rangle$; for the very many cases for M and H that must be considered.

In the quasithin work, we normally take H *minimal* subject to not lying in M. And typically we can take M arising as $N_G(L)$, where L is roughly an extension of a 2-group by a quasisimple group. These choices are based on what amounts to a "local theory" for quasithin groups. (Indeed such a theory can usually be given within any reasonably general classification problem.) This quasithin local theory is developed particularly in Chapter 1 of [**AS04b**], based on a substantial array of preliminary results contained in [**AS04a**]. We sketch only some salient points:

The abstract minimal parabolic H. The case of H is actually fairly quick to summarize: Our choice of H minimal subject to $H \not\leq M$ gives the technical condition called "$H \in \mathcal{H}_*(T, M)$" at [**AS04b**, 3.0.1]. On using [**AS04b**, 3.3.2(4)], we obtain that H is an "abstract minimal parabolic", in the sense of McBride—see for example [**AS04a**, B.6.1]:

DEFINITION 4.4.1 (Abstract minimal parabolics). An *abstract minimal parabolic* H roughly has a Sylow T which is not normal, but is contained in a unique maximal local subgroup. ◇

Note that this condition does hold, if H actually *is* a minimal parabolic in Lie type: for then $H/O_2(H)$ has Lie rank 1—and hence the indicated uniqueness condition holds, in view of 1.3.20(6).)

In the quasithin context, such groups are described in [**AS04a**, E.2.2]:

LEMMA 4.4.2. *For an abstract minimal parabolic H under the QT hypothesis, most of the possible $H/O_2(H)$ are rank-1 Lie-type groups in characteristic 2 (the Bender groups $L_2(q)$, $U_3(q)$, $Sz(q)$ of Theorem* 2.0.17*); but a few other cases arise.*

Indeed H typically exhibits failure-of-factorization FF on a suitable internal module, in the sense of Definition 3.2.1*; and then $H/O_2(H)$ is in the FF-list of Theorem* 3.2.7*—usually $L_2(2^m)$, as indicated in* [**AS04a**, E.2.2].

The $\mathcal{C}$-component L. For our maximal 2-local M, there will be more possibilities. We summarize various results from quasithin local theory:

First, M arises as $N_G(L)$, for a suitable "$\mathcal{C}$-component" L: namely L, modulo a solvable normal subgroup, is a simple group which is described in Section 1.2 of [**AS04b**].These $\mathcal{C}$-components have some of the properties of ordinary quasisimple components—such as commuting.

Next, the maximality is expressed via the condition called"$L \in \mathcal{L}_f^*(G,T)$"; here the subscript f indicates that $L/O_2(L)$ acts faithfully on a suitable module V inside $O_2(L)$.

And usually we even have the Fundamental Setup of [**AS04b**, 3.2.1]; where subsequent results in Section 3.2 of [**AS04b**] determine the case-list for possible pairs given by $\overline{L} := L/O_2(L)$ and the internal module V. Typically $\overline{L}$ is a Lie-type group of rank 1 or 2 in characteristic 2, with V one of just a few possible "small" modules for $\overline{L}$.

Overview of the treatment of cases. The upshot of the above is basically that the main quasithin analysis must treat a fairly large case-list: indexed primarily by the possible pairs $(\overline{L}, V)$ in M—but for each of these, considering also the various possibilities for $H \not\leq M$ given above in Lemma 4.4.2.

As we had already previewed in the analysis for Example 3.3.14, that treatment is assisted by pushing-up considerations as in Section 3.3: Notably we saw in (3.3.12) that by [**AS04b**, 3.3.2(1)]:

M is the *unique* maximal 2-local subgroup over LT.

And then for $R := O_2(LT)$, we saw in (3.3.13) that we get the R-(CPU) condition in (3.3.10):

$$C(G,R) \leq M < G.$$

Hence we can exploit $C(G,R)$-theorems such as [**AS04a**, C.2.8].

The "majority" of the cases for $(\overline{L}, V)$ are treated in several chapters following [**AS04b**, Ch 4]:

First, most quasithin conclusion-groups are of Lie-type and Lie-rank 2, in characteristic 2; these are recognized in [**AS04b**, Ch 5]—as we will outline, in our subsequent Section 4.5.

Next, in most cases of $(\overline{L}, V)$ which do *not* lead to a quasithin conclusion-group, the module V does not satisfy the FF-condition; these "shadows" are eliminated in Chapters 7–9 of [**AS04b**]. In fact we had already outlined the elimination, using weak-closure and pusing-up methods, of the case $\overline{L} = M_{23}$ on $V = 2^{11}$, in Example 3.4.7.

So: Chapters 10–16 of [**AS04b**] are then devoted to treating only a comparatively few small cases of $(\overline{L}, V)$—but those small cases are disproportionately difficult.

4.5. Recognizing rank-2 Lie-type groups

We now outline Chapter 5 of [**AS04b**]—to demonstrate the use of recognition theorems, in identifying the main quasithin conclusion groups, namely the Lie-type groups of rank 2 in characteristic 2.

That chapter works under Hypothesis 5.0.1 there, which has $\overline{L}$ is $L_2(2^n)$, for some $n \geq 2$. Thus a number of different subcases for the module V will be considered. However, we note that most cases for $n = 1$, corresponding to the small field $\mathbb{F}_2$, involve extra difficulties—and so are postponed to later chapters of [**AS04b**].

Constructing parabolics. Recall from Lemma 4.4.2 that $H/O_2(H)$ is usually also a Lie-type group of rank 1 in characteristic 2. Consequently under the present hypothesis, both L and H resemble rank-1 parabolics—which might be expected to generate some rank-2 Lie-type group X. In this interpretation, $T \cap L$ should play the role of a full unipotent group of X; and so $N_G(T \cap L)$ should play the role of a Borel subgroup. Of course, substantial work is required to implement this outline:

Thus Section 5.1 of [**AS04b**] first develops various further restrictions on the possibilities for V and H. For example, the "Borel" $N_L(T \cap L)$ should be roughly an extension of the 2-group $T \cap L$ by an odd-order subgroup, and in particular solvable; so that a Hall $2'$-subgroup should play the role of a Cartan subgroup.

Then Section 5.2 of [**AS04b**] determines the corresponding "amalgams"—essentially the possibilities for $L/O_2(L)$ and $H/O_2(H)$, along with specific possibilities for the Sylow $T \cap L$. This argument proceeds by verifying conditions of a preliminary result [**AS04a**, F.1.1]: which is a version of the "weak BN-pairs of rank 2" condition, where the possibilities had been determined in celebrated work of Delgado-Goldschmidt-Stellmacher [**DGS85**]. Very roughly:

- The input gives the rank-1 parabolics and their intersection—*only* mod O_2.
- And the output gives the (previously-unknown) cases for the Sylow 2-group.

We emphasize that this determination of the amalgam does *not* yet identify the group generated by the parabolics. This is roughly because, in contrast to buildings of rank ≥ 3 as in (7.2.10), buildings of rank 2 are *not* simply connected—instead, they have an infinite universal cover, whose automorphism group would be some infinite group, with our finite G as a quotient.

The possible amalgams are of course basically those for the rank-2 Lie-type groups in characteristic 2. However among those, the $L_3(4)$-amalgam—based on two $L_2(4)$'s as quotients of parabolics—in fact admits an extension, corresponding to enlarging one $L_2(4) \cong A_5$ to A_7; and that extension also arises in the list of amalgams here. In the case of that extended-amalgam, we identify the quasithin conclusion group $G \cong M_{23}$; with recognition accomplished using the uniform method of [**Asc94**, 37.10].

Recognizing most conclusion groups. Section 5.3 of [**AS04b**] now implements the hybrid method described in Section 4.2: Namely the initial lemmas establish, for each of the remaining amalgams, the structure of $C_G(t)$ for 2-central t.[3]

[3]Recall that this means that t lies in the center of some Sylow 2-subgroup of G.

And this information, together with the amalgam, provides the hypothesis of the recognition theorem [**AS04a**, F.4.31], identifying G as the corresponding rank-2 Lie-type group in characteristic 2. We mention that in that method, the structure of the subgroup $C_G(t)$ provides sufficient further relations beyond the amalgam, to "collapse" the infinite automorphism group of the universal cover of the building down to the desired finite G.

Hence at the end of Chapter 5 in [**AS04b**], the infinite families of groups satisfying the quasithin hypothesis have now arisen.

But alas, the analysis of the remaining "small" cases for $(\overline{L}, V)$ (which may or may not lead to conclusion-groups) will require difficult and detailed work—occupying another 11 chapters there.

In fact the last case of $(\overline{L}, V)$ within the Fundamental Setup to be thus treated is $\overline{L} \cong L_3(2)$, in Chapter 14 of [**AS04b**]—with V a natural module; the reduction to this V holds after [**AS04b**, 12.4.2]. And next we will summarize the treatment of the very last subcase of the case ($L_3(2)$,natural V):

4.6. Recognizing the Rudvalis group Ru

In Section 14.7 of [**AS04b**], as noted above we have $\overline{L} \cong L_3(2)$ on natural V.

Further Sections 14.3 and 14.4 of [**AS04b**] had shown for 2-central z that the subcase in which $U := \langle V^{C_G(z)} \rangle$ is nonabelian leads to $G \cong HS$ or $G_2(3)$. In fact [**AS04a**, I.4.8, I.4.5] indicates a number of different possible recognition theorems for these identifications: including the methods of rank-3 graphs, or involution-centralizer characterizations, or weak BN-pairs of rank 2—techniques we had already mentioned in earlier sections of the present chapter.

Constructing parabolic-like subgroups. So Section 14.7 of [**AS04b**] begins with the final subcase, where U is abelian; and it remains to construct local subgroups resembling those in the Rudvalis group Ru.

Taking $C_G(z)$ for "H", subsequent lemmas reduce to $H/O_2(H) \cong S_5$; and obtain specific action of H on sections of U—of dimensions $1, 4, 6$. Further $O_2(H)$ is shown to consist just of U of order 2^{11}.

This information in turn leads to determination of $O_2(L)/V$ as the 8-dimensional adjoint-module for $L_3(2)$; so that $O_2(L)$ also has order 2^{11}.

Thus at this point, H and L now have the structures of two standard well-known 2-local subgroups of Ru.

Recognizing the Rudvalis group. These subgroups are then the hypothesis for applying the preliminary result [**AS04a**, J.1.1]—which identifies G as Ru. We mention that this preliminary recognition result proceeds via uniqueness of the rank-3 permutation graph.

CHAPTER 5

Representation theory of simple groups

Applications of simple groups often proceed via their linear *representations*. One standard reference for general representation theory is Curtis-Reiner [**CR90**]. As usual we will confine ourselves to a limited number of salient features; the experienced reader can skip over the basic review in the introductory section below.

Introduction: Some standard general facts about representations

A representation is a group homomorphism $\rho : G \to GL(V)$, where V is a vector space over a field F. And then V is the FG-*module* corresponding to that representation. Typically we can reduce to the case where V is *irreducible*: that is, where no proper nonzero subspace W of V is G-invariant.

"Ordinary" representation theory. We first consider the case $\mathrm{char}(F) = 0$. (The theory is similar when $\mathrm{char}(F)$ is a prime p that does not divide $|G|$.)

In this situation, the group algebra FG is *semisimple*; and then representations are *completely reducible*—that is, they decompose into a direct sum of irreducible representations.

Much of the information about a representation ρ in in effect encoded in its *character*: namely the values of the traces $\mathrm{Tr}\big(\rho(g)\big)$ of the representation matrices over $g \in G$. The set of characters for the irreducible representations ρ gives the *character table* of G. The character of ρ determines for example its decomposition into irreducibles. See for example Isaacs [**Isa06**] for basic character theory.

Modular representation theory. We turn to $p := \mathrm{char}(F)$ which *does* divide $|G|$. See e.g. Feit [**Fei82**] for fuller details of the modular theory. We only mention in rapid summary:

In this situation, the group algebra FG is not semisimple; and representations need not decompose into irreducibles—that is, there can be *non*-split extensions among the irreducible sections.

So some attention is also focused on representations which are *indecomposable*—that is, which cannot be further decomposed as a direct sum. These are of course less precisely described than irreducibles; typically just in terms of a composition series: The multiplicities of the various irreducible sections in such a series are determined— by the character; but if we proceed up an ascending series of submodules, the irreducibles might appear in various different orders. So now the character of the representation only describes those irreducibles and their multiplicities in a composition series.

Some restrictions on the possible non-split extensions are given by Brauer's theory of *p-blocks*: these blocks are the terms in a maximal decomposition of FG into two-sided ideals. Non-split extensions are only possible between irreducibles associated to the same block.[1] In particular, each irreducible module I is visible in a unique block, by means of its *projective cover* $P(I)$—an indecomposable summand of the block, which has I as its unique irreducible quotient; and which is also projective, in the sense of the standard definition:

DEFINITION 5.0.1 (projective module). There are many equivalent definitions of a *projective* module; perhaps the most common is, as a direct summand of a *free* module. A free module is in turn a direct sum of copies of the group algebra FG. In particular, a projective module has a basis admitting free (i.e. regular) action by a Sylow p-subgroup of G—and so has dimension divisible by $|G|_p$. ◇

Knowledge of the projective indecomposables, and not just the irreducibles, is a crucial part of the modular representation theory. A very special situation for a projective indecomposable is given by defect 0:

REMARK 5.0.2 (Defect groups and defect 0). To a block B is associated a certain p-group D, called the *defect group* of the block. An extreme case is given by a block of *defect 0*: Here $D = 1$; and the block contains a single ordinary irreducible—which when read mod p, remains irreducible, giving the single p-modular irreducible I of the block. Now I in addition to being irreducible is also in fact projective; so that I is in fact equal to its indecomposable projective cover $P(I)$. This situation holds for an ordinary irreducible iff its dimension is divisible by $|G|_p$. ◇

The p-modular theory is important for many applications—especially for modules involved in the structure of p-local subgroups of G.

Some other features. For simple groups (especially in the p-modular case), it is often important to study modules which are "small" in some suitable sense. For example, in Theorem 3.2.7 we had mentioned that the FF-condition 3.2.1 leads to one important notion of smallness.

In fact for many groups, the nontrivial irreducible module N of smallest dimension can often be considered the "natural" module or G—in analogy with the terminology of V as the natural module for $GL(V)$.

There are various constructions by which smaller modules V, W lead to larger modules: notably their tensor product $V \otimes W$. For $V = W$, this leads to the standard theory of symmetric and exterior powers $S^k(V), \Lambda^k(V)$. And often a "natural" module V leads to a suitable "adjoint" module—typically related to $\Lambda^2(V)$. This is in analogy with behavior in classical Lie-type groups, where the adjoint module (defined by the underlying Lie algebra, as in Example 1.3.12) has such a relation with the exterior square.

Given an extension of a normal subgroup A by a group B, Clifford's Theorem [**CR90**, 11.1] describes how representations of the extension AB are assembled from representations of A—using suitable representations of B (or perhaps of suitable central extensions of B).

In the next few sections, we consider more specific aspects of representation theory, for each of the three standard classes in the CFSG-list 1.0.2:

[1] Furthermore the ordinary characters are also associated with particular p-blocks.

5.1. Representations for alternating and symmetric groups

Typically it is most natural to first discuss representations of almost-simple S_n; and then pass down to representations of the normal subgroup A_n, via Clifford's Theorem indicated above. So in this section we focus on S_n.

Ordinary representations. We first make some naive remarks about the smallest irreducible: Let P denote the natural permutation module; this has basis given by the n points permuted by S_n. Then P decomposes as $T \oplus N$, where: T is a 1-dimensional trivial submodule—spanned by a vector with coefficient 1 in all n places; and N is the $(n-1)$-dimensional *natural* irreducible submodule—whose vectors have coefficient-sum 0.

The *general* theory of the irreducibles for S_n is a very classical topic—going back to the 19th-century theory of Young diagrams; and it is prominent not just in the algebra literature but also in combinatorics. For more modern treatments, see for example James-Kerber [**JK81**] and Sagan [**Sag01**].

REMARK 5.1.1 (Some features of ordinary irreducibles for S_n). We give just a rapid overview of some salient points:

• A conjugacy class in S_n is determined by a cycle-type, which in turn is determined by a partition of the n points—customarily denoted by λ.

• It is standard that the number of irreducible characters of a group is equal to the number of conjugacy classes (so that the character table is a square matrix, indeed invertible); hence for S_n, this value is the number $p(n)$ of such partitions.

• Furthermore there is a natural 1:1 correspondence, with the partition λ determining an irreducible I_λ. This I_λ has a basis indeed by *standard Young tableaux*: where, given boxes arranged in rows of lengths given by the parts of the partition λ, we insert the values of $1, \cdots, n$—*increasing* in each row and column. The dimension of I_λ is given by the celebrated "hook-length formula", which we give in the Appendix at (B.3.2). ♢

Here are some easy sample calculations, not even requiring that formula: The trivial module T above corresponds to the "trivial" partition—into just one part; there is only one way to fill the corresponding single row in increasing order; so $\dim T = 1$. The natural module N above corresponds to the partition with parts of size $(n-1), 1$: The column-increasing conditions means that we cannot put 1 into the single box in the second row; but the other $(n-1)$ choices are possible for that box—and each allows just one increasing way to fill the top row. Thus N has dimension $(n-1)$.

EXERCISE 5.1.2. Using the formula, write out the details for the above remark; and explore some other examples, say for $n = 4$.

Hint: Some sample details appear in appendix Remark B.3.1. ♢

Modular representations. Consider the permutation module P above, but defined now with coefficients over $\mathbb{F}_p$. If p does not divide n, the discussion of $T \oplus N$ is much as before. But assume instead that p does divide n: Then T is a submodule of N—and P is now indecomposable. Furthermore N is now reducible; so we refer to the irreducible $(n-2)$-dimensional quotient N/T as the "natural" irreducible.

In much the same way, the above general theory has to be further refined—since the earlier modules I_λ may be reducible, when read mod-p.

For example, there will now be fewer irreducibles: It is standard that their number is the number of conjugacy classes of p'-order; and these in turn correspond with p'-partitions—namely where the part-sizes are not divisible by p. The latter in turn correspond with p-regular partitions, in which no p successive parts have the same size; and it is these p-regular partitions λ which are customarily used to index the modular irreducibles. For such λ, the irreducible will be a suitable quotient $\overline{I}_\lambda$ of the characteristic-0 irreducible I_λ above.

This p-modular theory is still under vigorous development; again see for example James-Kerber [**JK81**]. For example, some features still not determined include: the dimensions of the irreducibles $\overline{I}_\lambda$; and the *decomposition matrix*, expressing the characteristic-0 module I_λ, mod-p, via a composition series in terms of $\overline{I}_\mu$ for p-regular μ which are suitably below λ in a natural ordering $\leq$.

REMARK 5.1.3 (Connections between S_n and $GL_n(q)$). We continue an ongoing theme, introduced after earlier Example 1.3.1: Recall that S_n is the Weyl group of $GL_n(q)$, as in 1.3.20(3). It turns out that the representation theory of S_n in fact leads naturally to some parts of the representation theory for $GL_n(q)$. Due to space limitations, such connections will remain mainly implicit, in the following section on representations. But the connections-theme will recur much more explicitly—when we discuss maximal subgroups, in the following Chapter 6; and indeed elsewhere. $\diamondsuit$

5.2. Representations for Lie-type groups

The reader may wish to review material from Section 1.3, e.g. the overall Lie-type setup in Remark 1.3.20.

Ordinary representations. A good source here is Carter [**Car93**]. I'll rapidly sketch some classical *Deligne-Lusztig theory*; note that more recent developments also use the later *Lusztig induction*.

We indicated in Remark 1.3.23 that we can obtain a finite Lie-type group G as the fixed points in an algebraic group $\overline{G}$ defined over the algebraic closure $\overline{\mathbb{F}_p}$, under a suitable "Frobenius" automorphism F; where F involves some power of the field automrophism $x \mapsto x^p$, and possibly some graph automorphism.

Now inside $\overline{G}$, fix some maximal torus $\overline{T}$ (i.e. Cartan subgroup), which is stable under the action of F. There is a relation with the Weyl group W:

$$\text{(5.2.1)} \qquad \text{classes of } F\text{-stable } \overline{T} \quad \stackrel{1:1}{\leftrightarrow} \quad \text{“}F\text{-conjugacy” classes in } W.$$

And roughly: The fixed points $\overline{T}^F$, as we vary $\overline{T}$, give rise to the various tori in finite G—not just the split Cartan subgroup, but also the non-split tori; as in:

EXAMPLE 5.2.2 (split and nonsplit tori in $GL_n(q)$). First let V denote a vector space of dimension d over the prime subfield $\mathbb{F}_p$. Then $GL_d(p)$ contains a cyclic subgroup T of order $p^d - 1$; and any subgroup of T which remains irreducible on V is an example of a *nonsplit* torus. This is in contrast to a split torus, namely a diagonal subgroup of order $(p-1)^d$. And notice that a nonsplit torus becomes split—if the field of definition for the matrix action is extended from $\mathbb{F}_p$ to $\mathbb{F}_{p^d}$.

Now consider $GL_n(q)$, for $q = p^a$: Here the various choices for full tori correspond to the possible partitions of the n diagonal positions: For given a diagonal block of of size $k > 1$, by the previous paragraph we can find a nonsplit size-k torus of order $q^k - 1$. A split full torus, of order $(q-1)^n$, arises from the trivial partition $1, 1, \cdots, 1$; but all other partitions give nonsplit full tori.

And we had already seen in Remark 5.1.1 that the partitions in turn correspond, via cycle-types, to conjugacy classes in the Weyl group $W = S_n$. This in particular continues the theme of connections in Remark 5.1.3 above.

We return to nonsplit tori in Section 9.3; especially Exercise 9.3.1. ♢

Now choose a prime different from the characteristic p of $\overline{G}$; this prime is typically denoted by ℓ.

Fix an irreducible (in particular, 1-dimensional) character θ of $\overline{T}^F$. Define $R_{\overline{T}}^{\overline{G}}(\theta)$ as the alternating sum of the θ-components of the ℓ-adic cohomology, with compact support, of the variety $\{g \in \overline{G} : g^{-1}F(g) \in \overline{U}\}$, where $\overline{U}$ is a full unipotent group $\overline{U}$ normalized by $\overline{T}$.

Deligne and Lusztig showed that these sums determine much of the ordinary representation theory for G, in the following sense: Namely $R_{\overline{T}}^{\overline{G}}(\theta)$ is a virtual representation of G; and further, each irreducible for G is contained in an $R_{\overline{T}}^{\overline{G}}(\theta)$, for some $\overline{T}, \theta$. Finally the sets of irreducibles in the $R_{\overline{T}}^{\overline{G}}(\theta)$ are disjoint, up to "geometric conjugacy" of θ.

Subsequent research has analyzed the $R_{\overline{T}}^{\overline{G}}(\theta)$ for many classes of Lie-type G; and of course development is continuing.

Modular representations. Here for Lie-type G in characteristic p, the modular theory is most interesting for the natural-characteristic prime p; and correspondly, most of the literature considers this p-modular case.

However we briefly mention some work on the "cross-characteristic" case; that is, q-modular representations for $q \neq p$. Suitable "small" representations are studied for example in Landazuri-Seitz [**LS74**]. But see also a number of papers of Guralnick and Tiep in this area.

So we now focus in the p-modular case, for p the characteristic prime of G.

Some general features of weight theory. Here the main idea is to mimic the "highest weight" properties of representations of the underlying Lie algebra $\mathcal{G}$ of G; cf. Humphreys [**Hum78**] for details of the latter. We had sketched elements of the weight-theory, mainly in the special case of the roots which arise on the adjoint module $\mathcal{G}$, earlier in Remark 1.3.7.

We now recall, and expand somewhat on, that earlier discussion of more general weights: The roots arise as irreducible (1-dimensional) characters of the Cartan subalgebra $\mathcal{H}$. We fix a simple system $\Pi = \{\alpha_1, \ldots, \alpha_n\}$, affording a basis for the space generated by the roots inside $\mathcal{H}$. Using the natural inner product on $\mathcal{H}$ given by the Killing form, we obtain the corresponding basis $\{\lambda_1, \ldots, \lambda_n\}$ of $\mathcal{H}^*$ which is dual to the simple co-roots—namely those roots divided by their squared-lengths. The resulting characters λ_i of $\mathcal{H}$ are called the *fundamental weights*. These are determined and listed at e.g. [**Hum78**, p 69], for the various types of simple $\mathcal{G}$.

For the corresponding Lie-type group G, the weights—the integral lattice which is spanned by the fundamental weights—play the role of characters of a Cartan subgroup H. This statement in fact requires certain technical adjustments for dealing with finite fields; for example, note that $H = 1$ when G is defined over the smallest field $\mathbb{F}_2$. We summarize some of the features emerging from this theory:

REMARK 5.2.3 (Some weight theory for natural-characteristic modules). An irreducible module V for G mimics many of the standard properties of Lie-algebra representations; for example:

(1) The action of a Cartan subalgebra H on V is completely reducible: into *weight spaces*, namely 1-dimensional subspaces on which the character of H is given by one of the weights. The action of a root group U_α on a weight space is determined via exponentiation from the action of the root subspace $\mathcal{U}_\alpha$ in (1.3.9).

(2) The fixed points V^U, for a full unipotent group U, are 1-dimensional. Furthermore $N_G(V^U)$ is a parabolic subgroup of G.

(3) The weight λ on V^U is *highest* in V, in the natural ordering: where for any other weight μ on V, $\lambda - \mu$ is a positive linear combination of simple roots.

(4) The weight λ on V^U is *dominant*: that is, it has the form $\sum_{i=1}^{n} a_i\lambda_i$—with $0 \le a_i \le q-1$, for an untwisted group G defined over $\mathbb{F}_q$. Notice that the number of these weights (called *q-restricted*) is q^n, for n the Lie rank.

(5) There is a bijection between the dominant weights λ and the irreducible modules $V(\lambda)$. ◊

EXAMPLE 5.2.4 (Weights for the natural module of $L_3(2)$). We consider for example the natural module V for $L_3(2)$; whose underlying root system we had examined at earlier Example 1.3.10. Explicit computation of suitable inner products shows that the highest weight is in fact λ_1—which is fundamental, and in particular dominant. Note as at [**Hum78**, p 69] that it can be expressed in terms of the simple roots, as $\frac{2}{3}\alpha_1 + \frac{1}{3}\alpha_2$. And the two remaining, lower weights on V are $\lambda_1 - \alpha_1$ and $\lambda_1 - (\alpha_1 + \alpha_2)$—exhibiting via (1.3.9) the action of negative-root subgroups. ◊

EXERCISE 5.2.5. Mimic this discussion of weights in some more examples of small G and V: for example, in the 4-dimensional natural module for $Sp_4(2)$; and the 8-dimensional adjoint module for $L_3(2)$.

Hint: Some details appear in appendix Remark B.3.3. ◊

One approach to this high-weight theory proceeds via restriction from the irreducible representations of the algebraic groups $\overline{G}$. This is the viewpoint used in the celebrated lecture notes of Steinberg [**Ste68**]. We next mention some further features which are usually deduced in that context:

Assume G is an untwisted group Lie-type group $G(q)$ of rank n, defined over $\mathbb{F}_q$ where $q = p^m$. It is standard that the number of p-modular irreducibles is equal to the number of conjugacy classes having p'-order. One can show group-theoretically that this number is q^n—and this agrees with the number of q-restricted dominant weights λ in 5.2.3(4). This underlies the correspondence of such λ with irreducibles $V(\lambda)$ in 5.2.3(5).

Furthermore the irreducibles can be studied more finely using the viewpoint of the groups $G(p)$ defined over the prime field $\mathbb{F}_p$, and its irreducibles—these are the "basic" irreducibles, corresponding to the p^n dominant weights which are p-restricted. Indeed given a q-restricted dominant λ, which has a decomposition via p-powers, of the form $\sum_{i=1}^{m} p^i \mu_i$—so that each μ_i is p-restricted—we have the important expression (see e.g. [**Jan03**, 3.17]):

THEOREM 5.2.6 (Steinberg Tensor Product Theorem). *We have:*

$$V(\lambda) = V(\mu_1) \otimes \sigma\big(V(\mu_2)\big) \otimes \cdots \otimes \sigma^{m-1}\big(V(\mu_m)\big),$$

where σ generates the Galois group $\mathrm{Gal}(\mathbb{F}_q/\mathbb{F}_p)$ *of order m.*

EXERCISE 5.2.7. Express as products the irreducibles of $SL_2(4)$ and $SL_3(4)$. Hint: Some further details appear in appendix Remark B.3.4. $\diamondsuit$

Some finer points of module structure. We first single out one module of particular significance:

DEFINITION 5.2.8 (Steinberg module). The largest q-restricted dominant weight, of form $\sum_{i=1}^{n} (q-1)\lambda_i$, corresponds to the irreducible called the *Steinberg module St* for G. This module has a number of very special properties:

- It has dimension $q^{|\Phi^+|}$—equal to the order $|U|$ of a full unipotent group.
- It is the unique irreducible which is also projective—of defect 0 as in 5.0.2.
- It is the product of the conjugates of the basic-Steinberg modules for $G(p)$. $\diamondsuit$

EXERCISE 5.2.9. Here are some very unusual relationships, for certain small Steinberg modules: They can be verified using the 2-modular character tables in the Modular Atlas [**JLPW81**].

(1) The Steinberg module of $L_3(2)$ is given by its 8-dimensional adjoint module. Hint: The tensor product of the natural 3-dimensional module V with its dual $\hat{V}$ affords the space $\mathrm{gl}_3(2)$ of 3×3 matrices—with a trivial submodule given by scalar matrices. So from the product character, remove a trivial character: the result is afforded by the adjoint module $\mathrm{sl}_3(2)$.

(2) For $Sp_4(2)$, the 16-dimensional Steinberg module is given by the tensor product of the natural 4-dimensional module with its conjugate under a graph automorphism. $\diamondsuit$

Next we turn to a general irreducible $V(\lambda)$ for some high weight λ:

As an initial approximation to the module $V(\lambda)$ for G, we can consider the characteristic-0 *Weyl module* $W(\lambda)$, defined for the underlying Lie algebra $\mathcal{G}$—but in fact read mod-p; the result has $V(\lambda)$ as its unique irreducible quotient. From *Weyl's formula* [**Hum78**, p 139], which we state as (B.3.6), we know the dimension of $W(\lambda)$; but determining the typically-smaller dimension of the irreducible $V(\lambda)$ has been a major open problem in the theory.

For example, for $L_4(2)$, the adjoint module $W(\lambda_1+\lambda_3)$ for the Lie algebra has dimension 15; but the irreducible module $V(\lambda_1+\lambda_3)$ is the 14-dimensional quotient modulo a trivial submodule. (We will revisit this irreducible in later Exercise 5.6.1.)

EXERCISE 5.2.10. For $G = Sp_4(2)$ of type C_2: Apply Weyl's formula to $W(\lambda_1)$, to describe the natural 4-dimensional symplectic module—which is irreducible. Then apply the formula to $W(\lambda_2)$, to obtain the 5-dimensional orthogonal module. This is the natural module for $\Omega_5(2)$ of type B_2—which is isomorphic to $Sp_4(2)$, as we had observed toward the end of Remark B.2.7. This time the Weyl module has a trivial submodule, and so is reducible. The symplectic module can be regarded as the "spin" module for the orthogonal group $\Omega_5(2)$.

Hint: Some further details appear in appendix Remark B.3.5. ♢

Lusztig in 1979 stated a conjecture for the dimensions of the $V(\lambda)$. Later Andersen-Jantzen-Soergel [**AJS94**] showed it must hold for sufficiently large p. But recent examples of Williamson[2] show that a lower bound for p must be fairly large. (E.g. larger than *Coxeter number* h that Lusztig had originally hoped for.)

A different approach to high-weight theory, using only *finite* G, was developed by Curtis and Richen; see [**Cur70**]. The basic context is that of a split BN-pair, which we had mentioned at Remark 1.3.20(3), and in the discussion of rank-2 in the later part of Section 4.2.

Here a *weight* includes not just λ (as a character of the *finite* subgroup H), but also certain scalars μ_i—roughly recording the effect of the $U_{-\alpha_i}$ for $i \in \Pi$ on a high-weight vector. This compensates for "small" H—for example, $H = 1$ for groups defined over $\mathbb{F}_2$. Of course, the final results are the same as those arising from the algebraic-groups approach; but sometimes the purely finite-group context of Curtis-Richen can be more convenient.

One result that fully generalizes from the Lie-algebra situation is given by 10.1.7 in [**Smi11**]—which extends to all parabolics the irreducibility under the Borel subgroup $B = UH$ that is a consequence of the 1-dimensionality of $V(\lambda)^U$ in Remark 5.2.3(2):

THEOREM 5.2.11. *For irreducible $V(\lambda)$ and parabolic $P_J = U_J L_J$, the fixed points $V(\lambda)^{U_J}$ are irreducible under L_J; they afford the module $V(\lambda|_{L_J})$ for L_J.*

Examples of this property have already appeared implicitly; e.g. in Remark B.1.1, for the action of the maximal parabolics $P_{\hat{k}}$ indicated there, on the natural module V: Namely we saw that the usual k-subspace V_k of V arises as $V^{U_{\hat{k}}}$—and affords the natural irreducible module for a suitable factor of $L_{\hat{k}}$.

Indeed in the Ronan-Smith presheaf-viewpoint [**Smi11**, 10.1.8], which we return to at Remark 7.5.5, the irreducibles—or equivalently, their highest weights—are in 1:1 correspondence with irreducible presheaves on the building: Here the presheaf is defined at all parabolics; but each is determined already by the values at the *minimal* parabolics. These values are irreducible by Theorem 5.2.11.

EXERCISE 5.2.12. Explore such presheaves, in some small modules for groups such as $L_3(2)$; $Sp_4(2)$; $L_4(2)$. Hint: We already mentioned Remark B.1.1, for the case of maximal parabolics on the natural module studied there. And see also [**RS89**], which methodically explores the presheaves for several examples. ♢

[2]See `http://people.mpim-bonn.mpg.de/geordie/Torsion.pdf`

Some remarks on "small" representations. For classical matrix groups, the *natural* representation is the obvious action on the vectors of the defining space. For exceptional groups, the term "natural representation" is sometimes used informally to refer to the *smallest* module.

EXAMPLE 5.2.13 (The Cayley-algebra module for $G_2(q)$). For example, the group $G_2(q)$ acts on the 7-dimensional Cayley algebra—see e.g. [**Hum78**, p 105]; this has a 6-dimensional quotient in characteristic 2. ♢

A further comparatively small representation is the adjoint module—arising from the underlying Lie algebra $\mathcal{G}$ for G, in the Chevalley construction in Section 1.3. For classical groups, the action on the n-dimensional natural module is usually described via $n \times n$ matrices; and the adjoint module corresponds roughly to the *conjugation* action of these square matrices on the underlying space of not-necessarily-invertible $n \times n$ matrices

The weights that arise on an irreducible $V(\lambda)$ fall into orbits under the Weyl group W of G. A corresponding further notion of smallness is given by: The dominant-weight λ is called *minimal* (or *minuscule*) if there is just *one* W-orbit in Weyl module $W(\lambda)$. Such weights for all Lie types are listed at [**Hum78**, p 105]. In this case, $W(\lambda)$ must remain irreducible, when read modulo any prime p; that is, $W(\lambda)$ gives the irreducible $V(\lambda)$ for the finite group. Note in type A_m that each fundamental weight λ_i is minimal; in particular this includes λ_1, which gives the natural module. In type B_m, for groups $O_{2m+1}(q)$, the weight λ_1 for the natural module is not minimal; but λ_m is minimal, and gives the irreducible spin-module of dimension 2^m. We saw this distinction for type B_2 in earlier Exercise 5.2.10.

The FF-context of Definition 3.2.1 gave still another notion of "small" module; we mention that the work of [**GM02**] establishing the FF-list 3.2.7 made heavy use of the above weight-theory.

5.3. Representations for sporadic groups

Of course there is no common *theory* of the structure of sporadic groups; much less, for their representations.

But often, it is possible to get information about some of their smaller representations: from their construction; or from their containments in other groups, whose representations are more familiar. Many such details be found using the Atlas [**CCN**$^+$**85**]; or especially the Modular Atlas [**JLPW81**] for modular representations. We quickly indicate the dimensions and fields for several such modules:

- From Golay codes: M_{12} in $6/\mathbb{F}_3$; M_{24} in $12/\mathbb{F}_2$.
- From the Leech lattice: Co_1 in $24/\mathbb{F}_2$; $3Suz$ in $12/\mathbb{F}_3$; $2J_2$ in $6/\mathbb{F}_5$.
- Since $J_1 < G_2(11)$, J_1 is represented in $7/\mathbb{F}_{11}$.
- From $J_2 < G_2(4)$, we find the group J_2 in $6/\mathbb{F}_4$.
- From $3J_3 < U_9(2)$, we get $3J_3$ in $9/\mathbb{F}_4$; etc ...

However, some groups have no really small irreducibles; For example, the Baby Monster BM has minimal dimension $4370/\mathbb{F}_2$; and the Monster M has minimal dimension $196882/\mathbb{F}_2$.

Applications to Alperin's conjecture

By way of full disclosure: I find the p-modular theory for the Lie-type groups a particularly appealing area. And this is behind my choice here of Alperin's conjecture, to demonstrate applications in this chapter: the conjecture is motivated by aspects of modular Lie theory, and allows for some specially elegant uses of that theory in verifying the Lie-type case.

5.4. Introduction: The Alperin Weight Conjecture (AWC)

In early 1985, Alperin [**Alp87**] stated a bold conjecture on the modular theory for *any* finite group. In this section, we briefly review some of the background.

Alperin isolated, from the high-weight theory for a Lie-type group G, the following subset of features—from among those which we had indicated earlier at Remark 5.2.3 and thereafter:

An irreducible module $V(\lambda)$ determines a high-weight 1-subspace X.

- Further $N_G(X)$ is a p-local: say $P_J = N_G(U_J)$—where $U_J = O_p(P_J)$.
- $K_J := \langle U_{\pm i} : i \in J \rangle$ is of Lie-type—with projective irreducible Steinberg St_J.
- Also $\lambda \cdot St_J$ is projective irreducible module for $HK_J = L_J \cong N_G(U_J)/U_J$.

And conversely: The pair $(U_J \ , \ \lambda \cdot St_J)$ in turn determines $V(\lambda)$.

EXERCISE 5.4.1. Exhibit such pairs $(U_J \ , \ \lambda \cdot St_J)$, for irreducibles of some smaller groups, such as $L_3(2)$; $Sp_4(2)$; $L_4(2)$. Hint: Some further details appear in appendix Remark B.3.7. ♢

Alperin focused on generalizing these pairs to an arbitrary finite group H. He defined an "abstract" *weight* as a pair (P, S): where P is a p-subgroup of H (so that the normalizer $N_H(P)$ is a p-local subgroup); and S is a projective irreducible module for $N_H(P)/P$.

We note that it is elementary to show that such an S can exist only in the case that $P = O_p\big(N_H(P)\big)$; so in fact we may restrict attention to such P. This is the important *p-radical* condition (recall Definition 3.3.11) on P—such subgroups have proved crucial in many other places in the literature. And when H is actually of Lie-type, as we had mentioned earlier, a standard result (e.g. [**Smi11**, 4.4.1]) on the unipotent radicals motivated the terminology of "p-radical" for general groups:

THEOREM 5.4.2. *For H of Lie-type in characteristic p, $\mathcal{B}_p(G)$ consists of the unipotent radicals of parabolics.*

Alperin then conjectured that these abstract-weights in fact determine the number of modular irreducibles:

CONJECTURE 5.4.3 (Alperin Weight Conjecture (AWC)). *For a finite group H, and any prime p dividing $|H|$:*

#(p-modular irreducibles for H) = #(Alperin-weights—up to conjugacy) .

EXAMPLE 5.4.4. We summarize a verification for $H = A_7$ at $p = 2$: There are 6 conjugacy classes of odd order—so 6 is also the number of the 2-modular irreducibles. We can check that the 2-radical subgroups P are represented up to conjugacy by: the Sylow D_8; and two subgroups E_4—one contained in a subgroup A_4, the other not. The first and third of these representatives have $N_G(P)/P$

given by the trivial group, and $L_2(2)$, respectively; each of these quotients has just one projective irreducible, namely the trivial module and the Steinberg module—so together they contribute 2 Alperin-weights. Finally the second E_4, that lying in an A_4, has $N_G(P)/P$ given by an E_9 inverted by an involution. The ordinary character table of this quotient has 2 characters of degree 1, and 4 of degree 2. Those latter 4 are of defect 0 via Definition 5.0.2—and so this P contributes 4 more Alperin-weights. So we also have $2 + 4 = 6$ Alperin-weights, as desired. $\diamondsuit$

EXERCISE 5.4.5. Verify the conjecture for some other small groups. $\diamondsuit$

The Alperin Weight Conjecture is of course fascinating in its own right. It has also various further consequences, such as conjectures of Broué; see for example Schmid [**Sch07**, p 202].

We mention that for general groups H (as opposed to Lie-type G), there is not necessarily any "natural" bijection between the irreducibles and the Alperin-weights; for example, given an irreducible V, one can't necessary hope to find the projective irreducible S visibly embedded in the restricted module $V|_{N_H(P)}$.

Various important special cases H of the Conjecture were verified early on; for example Alperin in [**Alp87**] mentions: solvable groups; S_n; $GL_n(q)$; and some others. Seemingly Alperin himself did not originally check all the possible Lie types; soon Cabanes [**Cab88**] gave a general argument—using the viewpoint of modular *Hecke algebras*. We will revisit some aspects of the motivating Lie-type case, in Section 5.6 below.

Various authors have studied the AWC from the viewpoint of general groups H; and a number of topological viewpoints are mentioned for example in Chapter 11 of [**Smi11**]. But for the remainder of this chapter, we'll focus instead on the substantial literature which approaches the AWC via reduction to simple groups—this of course in particular requires verification for each group in the CFSG-list 1.0.2.

5.5. Reductions of the AWC to simple groups

I thank P. H. Tiep for suggestions in this area; much more detail on applications in representation can be found in his survey [**Tie14**].

Some earlier history of reductions. On the more specific topic of reductions, for more details see e.g. Navarro-Tiep [**NT11**].

The reduction-approach was pioneered by Dade: who gave a number of variants of the Conjecture (e.g. the "projective" conjecture); and also announced an anticipated proof of the relevant reductions.

This motivated verifications of the conjectures for many of the simple groups, by a number of researchers, including many of Dade's students. Various papers of An include an overview of much of this verification-literature. But unfortunately, no complete proof of Dade's reduction has yet appeared.

We do however mention one feature of verification-proofs for simple G, from that period: We had already commented in the previous section that in order for $N_G(P)/P$ to have a projective irreducible S, necessarily $P = O_p(N_G(P))$—the p-radical condition of Definition 3.3.11. So in order to implement the verification of the AWC for G, it is first necessary to determine the poset $\mathcal{B}_p(G)$ of p-radical subgroups. The determination of this poset is also of interest for applications in geometry and topology; see e.g. [**Smi11**, p 121], and later Section 7.5. The literature of that early period contains such determinations for many simple G, by various authors. The reader is particularly directed to various relevant papers of Yoshiara, Sawabe, O'Brien, and An.

More recent approaches to reduction. In the last decade or so, there has been a resurgence of interest in reductions for the AWC and related conjectures.

We first summarize a reduction given by Navarro-Tiep [**NT11**]: For the AWC, it suffices to show that all simple groups G are *AWC-good*; this is defined by conditions (1.a) $\cdots$ (3.d) in Section 3 of their paper. Indeed we mention that (1.b), together with the bijection in 3.2 there, roughly requires a "partition" of the AWC, indexed via the p-radical subgroups; and in fact further requires verification for central p'-extensions of G.

The proof of their main reduction Theorem 5.1 relies repeatedly on consequences of AWC-goodness established in their Theorem 3.2. And 3.2 in turn parallels Theorem 13.1 of Isaacs-Malle-Navarro [**IMN07**], which gives an analogous goodness-condition the *McKay Conjecture*—a topic we will mention later, as remark (1) in the brief concluding section of this chapter.

Since AWC-goodness has stronger requirements than those of the AWC itself, it is not sufficient to just quote the above-mentioned literature verifying the AWC for various simple G. For example, Section 6 of [**NT11**] now checks AWC-goodness for G of Lie type in natural characteristic p—going beyond the earlier-indicated verification of AWC for such G by Cabanes.

In fact, Cabanes himself in [**Cab13**] gave an additional modern reduction-proof for the AWC. Furthermore Puig in [**Pui11**] stated a variant of the AWC, and likewise reduced it to checking central p'-extensions of simple groups.

5.6. A closer look at verification for the Lie-type case

In the section, we will focus on Alperin's motivating case of Lie-type G in characteristic p. We won't try to outline Cabanes' general verification of the AWC for this case; instead, we'll just consider a few special aspects, which can be checked using just a few features of the p-modular theory in Section 5.2.

We had seen at Remark 5.2.3(4) that for untwisted G of rank n over $\mathbb{F}_q$, the number of p-modular irreducibles is q^n; this followed e.g. by calculating the number of q-restricted dominant weights. So to verify the AWC, we want to obtain this same number q^n, by calculating the number of Alperin-weights (P, S).

We had mentioned earlier that for G of Lie type, the p-radical condition necessary for the existence of the projective irreducible S of $N_G(P)/P$ is by Theorem 5.4.2 satisfied exactly by the unipotent radicals U_J; recall such a J varies over a subset of the simple system Π. Now for U_J, we saw at Remark 1.3.20(4) that the normalizer $N_G(U_J)$ is the parabolic P_J; and we have $N_G(U_J)/U_J \cong L_J = H \cdot K_J$, where K_J is generated by the $U_{\pm\alpha}$ for $\alpha \in J$. Since K_J is also of Lie type, it has a projective irreducible St_J—which is unique (as we mentioned at Definition 5.2.8). Consequently the possibilities for S are the extensions $\lambda \cdot St_J$ from K_J to L_J, where λ denotes a 1-dimensional character of H/H_J, for $H_J := H \cap K_J$.

Thus to count the pairs (P, S), we must count the pairs $(U_J, \lambda \cdot St_J)$: varying J over subsets of Π; and for fixed J, counting extensions $\lambda \cdot St_J$ from K_J to L_J.

The case of general q. In the general case, q is some power of the prime p; and in particular, we have $|H| = (q-1)^n$.

Since $L_J = HK_J$, where $H_J = H \cap K_J$ is the product of the H_i with $i \in J$, we can obtain a semidirect product $L_J = K_J H_{\hat{J}}$: where $H_{\hat{J}}$ is the product of the H_i with $i \in \hat{J} := \Pi \setminus J$. Hence for the character λ of $H/H_J \cong H_{\hat{J}}$ indicated earlier, the number of choices is $|H_{\hat{J}}| = (q-1)^{|\hat{J}|}$; and so this gives the number of extensions $\lambda \cdot St_J$.

And now combining J-terms over fixed $|J| = i$, for the total number of Alperin-weights, using just the binomial theorem we get:

$$\sum_{i=0}^{n} \binom{n}{i} (q-1)^{n-i}\, 1^i = \big((q-1)+1\big)^n = q^n \ ;$$

as desired. □

The subcase $q = 2$. We end the section with a slightly different viewpoint on the subcase corresponding to the smallest field $\mathbb{F}_2$—where the analysis has some particularly elegant features.

In this case, the target value for our count of Alperin-weights will be 2^n; and this value is of course immediate, from setting $q = 2$ in the binomial-theorem expression above.

But let us examine a little more deeply the assumption $q = 2$: We get $H = 1$ for the Cartan subgroup. Hence $L_J = K_J$, and the only possible value for λ is the trivial character 1. Thus St_J is the only possible choice for the extension $\lambda \cdot St_J$. So the number of Alperin-weights $(U_J, \lambda \cdot St_J)$ is just the number of subsets J of Π, a set of size n; so we get 2^n for our count—just as the size of the power-set of Π.

Let's go even a little farther: As in our introduction to the AWC in Section 5.4, we recall from Remark 5.2.3(2) that each irreducible $V(\lambda)$ has a high-weight 1-subspace X which is stabilized by a parabolic P_J. From the discussion in the previous paragraph, including triviality of λ as a character of $H = 1$, we see that in the present situation with $q = 2$, we may as well replace the notation of $V(\lambda)$ with $V(J)$—where $V(J)$ now denotes the unique irreducible in which P_J is the stabilizer of a high-weight space. And now the module-weight correspondence has been simplified to the form $V(J) \leftrightarrow (U_J, St_J)$. We cannot expect such a natural bijection in more general groups.

Indeed, we can even shift attention to the "complementary" parabolic $P_{\hat{J}}$, where $\hat{J} := \Pi \setminus J$. This has the advantage, for our present purposes, that the U_{-i} for $i \in \hat{J}$ do *not* stabilize the high-weight space X. And then, using the viewpoint of presheaves in [**Smi11**, 10.1.8] which we had mentioned after Theorem 5.2.11, it follows that the fixed subspace $V(J)^{U_{\hat{J}}}$, which is in fact generated by the action of $L_{\hat{J}}$ on X, has dimension $2^{n-|J|}$—and affords the Steinberg module $St_{\hat{J}}$ of $L_{\hat{J}}$.[3] So we could even replace the correspondence of the previous paragraph with the form $V(J) \leftrightarrow (U_{\hat{J}}, St_{\hat{J}})$—where this time the restriction $V(J)|_{P_{\hat{J}}}$ in fact contains the subspace $V(J)^{U_{\hat{J}}}$ affording $St_{\hat{J}}$. And as noted above, we cannot expect such a natural "internal" correspondence with subspaces, for general groups and modules.

[3]Roughly because minimal parabolics $P_{\hat{i}}$ have Levi complements $L_{\hat{i}} = K_{\hat{i}} \cong L_2(2)$; with Steinberg module given by the 2-dimensional natural module of $L_2(2)$.

EXERCISE 5.6.1. Explore these features, for some small Lie-type groups such as $L_3(2)$; $Sp_4(2)$; $L_4(2)$.

Hint: Again the examples in [**RS89**] will be helpful. Note also that in the various modules for $L_4(2)$ and $Sp_4(2)$ considered in appendix Chapter B, the parabolic P_J stabilizing a high-weight space was always a maximal parabolic. Hence in those cases, the complementary parabolic $P_{\hat{J}}$ is a single minimal parabolic, with Levi complement $L_2(2)$; and necessarily $St_{\hat{J}}$ is then the natural module.

So in addition, consider also the 14-dimensional irreducible V for $G = L_4(2)$; this is afforded by the 15-dimensional Lie-algebra adjoint module $\mathrm{sl}_4(2)$, modulo its trivial submodule of scalars. Here the highest weight is given by the highest root $\alpha_1 + \alpha_2 + \alpha_3 = \lambda_1 + \lambda_3$; it is afforded by the root subspace which underlies the root group for 2-central z, in the notation of Remark B.2.1. This space has stabilizer P_J given by the involution centralizer $C_G(z)$ there, namely the minimal parabolic $P_2 \cong 2^{1+4}L_2(2)$; so we have $J = \{2\}$. Hence the complementary parabolic $P_{\hat{J}}$ is the maximal parabolic $P_{\hat{2}}$ as in Remark B.1.1, with unipotent radical $U_{\hat{2}} \cong 2^4$. Here $V^{U_{\hat{2}}}$ is the 4-subspace given by the root subspaces underlying the 4 root subgroups generating $U_{\hat{2}}$; and we can see that this space affords the Steinberg module $St_{\hat{2}}$ for $L_{\hat{2}} \cong L_2(2) \times L_2(2) \cong \Omega_4^+(2)$. $\diamondsuit$

A glimpse of some other applications of representations

We end the chapter with a briefer indication of some other applications of representation theory.

(1) The McKay Conjecture. This conjecture of McKay, from around 1971, concerns ordinary characters: Indeed let $\mathrm{Irr}_{p'}$ denote the set of characters which have p'-degree. The conjecture says, for a finite group H with Sylow group P, that:

$$|\mathrm{Irr}_{p'}(H)| \stackrel{?}{=} |\mathrm{Irr}_{p'}(N_H(P))|$$

EXERCISE 5.6.2. Check the conjecture at $p = 2$ for A_5; A_6; $L_3(2)$; etc. $\diamondsuit$

Isaacs-Malle-Navarro in [**IMN07**, Thm B] in effect reduce the McKay Conjecture to simple groups: namely it suffices to show that all simple groups are *good* (for p)—goodness is defined by conditions given in Section 10 there.

Various cases of simple groups G are in fact treated in that paper, and in further work in the subsequent literature. (We had mentioned in Section 5.5 that this reduction largely inspired the reduction in [**NT11**] for the AWC.)

Recently Malle-Späth announced the full verification for $p = 2$; see [**MS16**].

(2) The Brauer Height 0 Conjecture. This conjecture of Brauer from around 1955 concerns the theory of ordinary characters, via p-blocks and their defect groups in the sense of 5.0.2. It asserts that, for a p-block B of a finite group H, with defect group D, that:

$$[\text{Degrees of irreducibles in } B \text{ have } p\text{-part } \tfrac{|G|_p}{|D|}] \quad \stackrel{?}{\Leftrightarrow} \quad [D \text{ is abelian}].$$

First $\Leftarrow$ was reduced to quasisimple groups by Berger-Knörr [**BK88**]. Various quasisimple cases were then treated—see e.g. the history in Kessar-Malle [**KM13**]. And in fact 1.1 of that paper completed the treatment. We mention that for characters of Lie-type groups, they use the later method of Lusztig induction, which we only indicated without details in Section 5.2.

The $\Rightarrow$ direction would follow, if we knew a certain strong form, in Navarro-Späth [**NS14**], of the Alperin-McKay Conjecture (this last is a blockwise version of the McKay Conjecture).

The "would follow" argument in fact comes via Navarro-Tiep [**NT13**] and Giudici-Liebeck-Praeger-Saxl-Tiep [**GLP$^+$16**]; and that in turn uses Aschbacher's work on maximal subgroups of classical groups—which we indicate at Theorem 6.2.1.

(3) Dade's projective conjecture. Recently Späth announced a similar reduction to simple groups, for Dade's projective version of the AWC, which we had mentioned early in Section 5.5.

(4) The Ore Conjecture. This is a conjecture in general group theory, dating from about 1951. It asserts that:

CONJECTURE 5.6.3 (Ore Conjecture). *For a nonabelian simple group G, every element should be a commutator $a^{-1}b^{-1}ab$.*

A substantial literature on this conjecture has developed over the intervening years, and many simple groups had been covered.

Recently Liebeck-O'Brien-Shalev-Tiep in [**LOST10**] completed the analysis of all remaining cases. Their methods are character-theoretic (and use numerical computation); for example, they use the standard lemma of Frobenius that:

$$g \text{ is a commutator } \Leftrightarrow \sum_{\chi \in \mathrm{Irr}(G)} \frac{\chi(g)}{\chi(1)} \neq 0.$$

EXERCISE 5.6.4. Explore, either directly or using the lemma, some small cases of the conjecture; for example, the groups A_5, $L_3(2)$, A_6. $\diamondsuit$

For Lie-type groups, some use is also made in [**LOST10**] of the Deligne-Lusztig theory sketched in Section 5.2.

CHAPTER 6

Maximal subgroups and primitive representations

For further reference, we mention that Wilson's recent book [**Wil09**] has a good discussion of maximal subgroups for various kinds of simple groups. Indeed Wilson has been a major contributor in the determination of maximal subgroups, especially in sporadic groups. But also see the surveys of Kleidman-Liebeck[**KL88**] and Liebeck-Saxl [**LS92**].

I thank the referee for a number of valuable suggestions in this chapter.

Introduction: Maximal subgroups and primitive actions

To get started, let's first explore how we might approach the problem of finding maximal subgroups—particularly in a simple group G.

I'll begin by indicating some overall features for the almost-simple symmetric group $G = S_n$. This will first of all provide background for our more formal treatment of maximal subgroups of S_n in the subsequent Section 6.1. But in addition, these features will have suitable analogues for maximal subgroups in a number of other possible simple groups—notably classical groups in Section 6.2.

In fact, we'll start with essentially a more methodical version of the analysis which had been briefly sketched in earlier Remark 1.1.1(3):

A structures-list (S). Our eventual goal is a list of the maximal subgroups M of G. But in fact we want more—namely to understand the maximals, ideally via some role that they play in the context of G. Indeed, often a subgroup H of G will preserve some natural *structure*; in the present case of S_n, meaning some additional substructure on the n points. So here is a possible initial step, on the way to a final list—which I'll call.

(S): Obtain a list of the possible substructures (and their stabilizers)

Indeed if we can show that any proper subgroup $H < G$ stabilizes one of those structures, then the stabilizers in the structures-list (S) at least give possible *candidates* for the maximal subgroups M. So this structures-viewpoint will be a main theme, in the present chapter on maximal subgroups.

Below are several easy examples of such structures. First for $1 < k < n$:

(1) (intransitive:) If H fixes a k-subset, then $H \leq S_k \times S_{n-k}$.

Now with (1) in hand, we are reduced to considering subgroups H which are *transitive* on the n points. And we next recall the setup we indicated in Exercise 1.1.3:

DEFINITION 6.0.1 (blocks of imprimitivity). Assume for some $1 < k < n$ that k divides n. If G preserves a partition of the points into parts of size k, we say those parts form a system of *blocks of imprimitivity* for G. ♢

Hence we have:

(2) (imprimitive:) If H permutes such k-blocks, then $H \leq S_k \wr S_{\frac{n}{k}}$.
Most readers will be familiar with this concept; but in case not:

EXERCISE 6.0.2. In Exercise 1.1.3, give blocks for $n = 4$ and $n = 6$. The centralizer of a regular involution should suggest examples. ♢

And now with (2) in hand, we are reduced to considering subgroups H which act *primitively* on the n points. This in turn suggests a related kind of list, which I will call:

An actions-list (A). Permutation-group theorists also need the classification of all possible types of primitive actions. See e.g. Cameron [**Cam99**] for background on permutation-group theory.

So we now let H denote some "general" subgroup of S_n acting primitively on the n points—where H is not necessarily assumed maximal in S_n. But in fact it is standard that:

(6.0.3) A transitive group H is primitive iff H_α is maximal in H.

So in seeking general primitive actions, we have not completely escaped maximal considerations. Indeed a natural intermediate goal, on the way to our final maximals-list for S_n, might be:

(A): Obtain a list of the possible actions for a primitive group H.

Here I'll just say, rather vaguely, that this list should give a suitable "qualitative" description of H: The possibilities will typically depend on features such as the structure of the socle:

REMARK 6.0.4 (socle). The *socle* soc(H) is the product of the minimal normal subgroups of H. Each minimal normal subgroup is the product of isomorphic simple groups: these may be nonabelian simple; or abelian—these latter giving normal subgroups of suitable prime-power orders. ♢

Let's consider how we might use these two lists, toward obtaining our final maximals-list: The actions-list (A) should be essentially a refinement of the structures-list (S). Roughly, we should be able to fit each primitive group in (A) into one of the structure groups in (S). Indeed, we can consider chains of proper inclusions among members of (A): in any chain that cannot be extended, the final member should determine a structure in (S), hence a candidate for a maximal subgroup, which in particular is primitive.

I have tried here to emphasize the distinction between (S) and (A): because it seems to me (as an outsider to permutation-group theory) that this distinction is sometimes blurred in the literature.

6.1. Maximal subgroups of symmetric and alternating groups

I thank Cheryl Praeger for assistance with this section.

For further detail on this specific material, see for example Cameron [**Cam81**], or Praeger [**Pra83**], or Wilson [**Wil09**, Sec 2.6]. In addition, this area was the topic for the 2008 Venice Summer School lectures of Michael Aschbacher; these can be found online, at URL:

`http://users.dimi.uniud.it/~mario.mainardis/scuolaestiva2008/venotes.pdf`

Structures: The O'Nan-Scott Theorem. Much of a structures-list (S) for S_n was known classically: mainly "obvious" structures—starting with the intranstive and imprimitive cases (1)(2) in the introductory section above, and continuing with various fairly straightforward primitive cases.

And in addition, it was known that any such list must also be subject to at least one limitation: since a further case, namely almost-simple groups, can arise "unpredictably"—that is, outside of any methodical list of structures.

However, *completeness* of any suggested structures-list was not known until:

Some history of the structures-list (S). At the Santa Cruz AMS Summer Symposium in 1979, O'Nan and Scott announced a completeness proof for a suitable structures-list (S) for S_n. This result is usually called the *O'Nan-Scott Theorem*; and we will state it below as Theorem 6.1.1. But first, we metion a complication (which did not in fact affect its proof).

The proof was deduced, as suggested in the introductory section above, using a primitive actions-list—stated by Scott [**Sco80**, p 328]. Unfortunately, that list was found to be incomplete; a correction was given by Aschbacher [**AS85**, App]. The final result is often called the *Aschbacher-O'Nan-Scott Theorem*: we state it as Theorem 6.1.3 below; and we examine that final, corrected actions-list (A), somewhat informally, by means of the table in Remark 6.1.4.

Perhaps confusingly, some of the literature seems to use the term "O'Nan-Scott Theorem" to refer to *both* the above results. However, to try to prevent such confusion, I will follow the naming convention indicated above—namely the O'Nan-Scott Theorem for the structures-list (S), and the Aschbacher-O'Nan-Scott Theorem for the final, corrected actions-list (A).

The structures in the O'Nan-Scott Theorem. As indicated above, the original statement of the O'Nan-Scott Theorem nonetheless remained correct: This is because the action-type originally omitted in [**Sco80**, p 328] could not in fact be terminal in an inclusion-chain; and thus could not lead to a new maximal-stucture. So with that observation, Scott's analysis at [**Sco80**, p 329], fitting the various primitive types into possible structures, delivered the correct final structure-list. In particular, the resulting maximal structures which are primitive appear as (3)–(6) in Theorem 6.1.1 below. The result had been obtained independently by O'Nan.

We state it essentially in the form given by Wilson [**Wil09**, Thm 2.4]:

THEOREM 6.1.1 (O'Nan-Scott Theorem). *A proper subgroup H of S_n, other than A_n, lies in one of following subgroups (which stabilize the structures indicated in parentheses on the right):*

(1) $S_j \times S_k$, *where* $n = j + k$; *(intransitive: j-set; partition j, k)*
(2) $S_j \wr S_k$, *where* $n = jk$; *(imprimitive: blocks, giving array $j \times k$)*
(3) $S_j \wr S_k$, *where* $n = j^k$; *("product": k-hypercube of side j)*
(4) $AGL_d(r) \cong r^d : GL_d(r)$, $n = r^d$; *(affine d-dimensional space over $\mathbb{F}_r$)*
(5) $L^k(S_k \times Out(L))$, *L simple,* $n = |L|^{k-1}$; *($soc(H)_\alpha = L$ diagonal in L^k)*
(6) an almost-simple group H. *(?—no "predictable" structure)*

Cases (3)–(6) are primitive.

We note that as H is primitive in cases (3)–(6), the point stabilizer H_α is maximal in H by (6.0.3). So it may seem ironic, that in determining maximal subgroups in the Theorem for the particular almost-simple group S_n, we are in effect "reducing" in (6) to determining maximal subgroups such as H_α for *all* almost-simple groups.

EXERCISE 6.1.2. Which of (1)–(6) arise for S_4? S_5? $\cdots$ S_8? Use for example the Atlas [**CCN**$^+$**85**] to decide case (6). Some details appear in Remark B.4.1. ◇

We remark that the "candidates" in cases (1)–(6) are typically, but not always, maximal; the details of verifying which are actually maximal were handled (for A_n as well as S_n) by Liebeck-Praeger-Saxl [**LPS87**]. Their result includes partial, but still reasonably general, restrictions on the "unpredictable" almost-simple case (6). We also mention (as briefly suggested earlier) that the verification of maximality involves analyzing possible proper *inclusions* among pairs of members of the cases (1)–(6); for example, if $j = k$ in case (2), the group is properly included in case (3). The analysis of this inclusion-problem was often difficult—notably, with pairs in case (6).

Actions: The Aschbacher-O'Nan-Scott Theorem. We now resume our earlier discussion of the actions-list (A), namely the Aschbacher-O'Nan-Scott Theorem 6.1.3 below. This result also has been very heavily applied in the literature on permutation groups.

Below we give one fairly standard statement, that of Liebeck-Praeger-Saxl as in [**LPS88**, Thm, Sec 2]; in particular, we follow the convention there of first giving only the *names* of the cases, with discussion of those fairly intricate cases given elsewhere:

THEOREM 6.1.3 (Aschbacher-O'Nan-Scott Theorem). *Any finite primitive permutation group is permutation-equivalent to one of the types I, II, III(a), III(b), and III(c)—briefly described in our Remark* 6.1.4 *below.*

Some remarks about the statement and proof. Before trying to summarize that specific list of actions, we'll first comment further on several general features relevant to the result.

We add a detail to our earlier discussion of the procedure which corrected Scott's original form of the actions-list: Namely Aschbacher saw that Scott had omitted the action usually called *twisted* wreath products; it is the case III(c) in Theorem 6.1.3 above. Equivalently, it is abbreviated by TW in Remark 6.1.4 below; we will indicate there how this case was already in fact included in case (3) of the the maximal structures-list—so that its omission in the original action-context did not affect the original proof of the structures-list in the O'Nan-Scott Theorem 6.1.1.

We had remarked that the CFSG has been extensively applied to problems in permutation-group theory; see e.g. Cameron [**Cam81**]. Seemingly it was not originally clear if the CFSG was needed for the (S) and (A) lists; this point is perhaps not crystal-clear in the literature. However, the usual modern published proofs *do* seem to make a fairly mild use of the CFSG—via the Schreier Conjecture 1.5.1, which of course is common elsewhere in permutation-group theory. But see also the discussion of Wilson's approach to (S) in appendix Section A.1.

Next we mention several such proofs of the specific actions-list given in the O'Nan-Scott-Aschbacher Theorem.

We caution that the various approaches indicated below may make different subdivisions of the primitive-types, as they appear in their final statements. To avoid confusion among those statements, it may be helpful for the reader to first look at one or more sources in the literature, which make a comparison of the cases in those approaches: e.g. Table 2 in Baddeley-Praeger-Schneider [**BPS07**], which we essentially reproduce in our Remark 6.1.4 below; or Table 6.1 in the forthcoming Praeger-Schneider book [**PS**].

Now to indicate those various approaches:

Cameron's view [**Cam81**, 4.1] of the proof has been influential; e.g., it led to a later short self-contained proof, in Liebeck-Praeger-Saxl [**LPS88**] (from which we quoted Theorem 6.1.3).

A version of the result for actions of *quasiprimitive* H (meaning that all normal subgroups of H are transitive) appears in Praeger [**Pra93**]; this version does not use the CFSG, even via the Schreier Conjecture 1.5.1.

A later subdivision into 8 basic action types appears for example in Baddeley-Praeger [**BP03**]; and a more extended and elementary discussion is given in Section 6 of [**PLN97**]. We include this approach in Remark 6.1.4 below; and in later applications, often follow this case division, rather than the one in Theorem 6.1.3.

Finally the texts of Dixon-Mortimer [**DM96**], Cameron [**Cam99**], and Praeger-Schneider [**PS**] all have chapters on O'Nan-Scott theory (structures and actions).

A glimpse of the actions in the Aschbacher-O'Nan-Scott Theorem. For later reference, I'll roughly state the actions-list (A), via the table in Remark 6.1.4 below—using an 8-type viewpoint from [**PLN97**, Sec 6].

REMARK 6.1.4 (Actions in the Aschbacher-O'Nan-Scott Theorem). I won't here try to give real details on those cases. However, column 2 of the table gives the abbreviations of the names for the 8 cases—followed by a parenthetical comment intended to give a brief suggestion of the underlying structure. As in Table 2 of [**BPS07**], column 1 gives the correspondence of the column-2 cases with the case-divisions in [**LPS88**]; while column 3 gives containments of column-2 actions within the cases of the S_n-structures-list in the O'Nan-Scott Theorem 6.1.1. (Note that the term *holomorph* of X below means $X \cdot \mathrm{Inn}(X) \cdot \mathrm{Out}(X)$.)

cases in [**LPS88**]	cases in [**PLN97**, Sec 6]	$\leq$ 6.1.1-cases
I	HA (holomorph of abelian-group)	(4)
II	AS (almost-simple group)	(6)
III(a)(i)	SD (simple-diagonal group—as H_α)	(5)
III(a)(ii)	HS (holomorph of simple-group)	$<$ (3),(5)
III(b)(i)	PA (wreath-product action)	(3)
III(b)(ii)	CD (compound-diagonal—i.e. of SDs)	$<$ (3)
if 2 components	HC (holomorph of compound-group)	$<$ (3)
III(c)	TW (twisted wreath-product)	$<$ (3)

(In SD, H_α is the full diagonal in a product of isomorphic simple groups.) ♢

Notice that from the completeness of the primitive-list (A) of the Aschbacher-O'Nan-Scott Theorem 6.1.3—as given in column 1 (or 2) of Remark 6.1.4—along with the inclusions in column 3 there, we immediately deduce that the cases for the maximal-structures list (S) which are primitive are exactly the cases (3)–(6) of the O'Nan-Scott Theorem 6.1.1.

We also recall our earlier remark that deducing (S) from (A) in this way makes use of the CFSG, via the Schreier Conjecture 1.5.1; where the deduction of the weaker quasiprimitive types [**Pra93**] would not.

We won't outline the proof of the actions-list in the O'Nan-Scott-Aschbacher Theorem, as given in Remark 6.1.4. However, there are some perhaps subtle points, underlying the above argument leading from it to the O'Nan-Scott Theorem 6.1.1, which we discuss in appendix Section A.1; notably analyzing inclusions among the cases, in the spirit of the remark after Exercise 6.1.2.

6.2. Maximal subgroups of Lie-type groups

Just as with the symmetric groups, the maximal subgroups of the classical groups had been a topic of research already from the early days of group theory. Indeed the analogous problems had been studied in the context of Lie algebras, with applications to Lie groups, notably by Dynkin; applications to algebraic groups were taken up by Seitz and others, as we will be indicating later in the section. For a survey of this area, see e.g. Liebeck [**Lie95**].

The study of the maximal subgroups for the finite Lie-type groups was subdivided into two cases: namely the classical matrix groups; and the "exceptional" groups—where now this term combines our previous usage for groups corresponding to actual exceptional Lie algebras (types E_6, E_7, E_8, F_4, G_2), with the non-classical twisted groups: namely the types ${}^2C_2$(i.e. Sz), ${}^2G_2$(i.e. Ree), ${}^3D_4$, ${}^2F_4$, and ${}^2E_6$.

Maximal subgroups of classical groups. A good discussion of this case appears in Wilson [**Wil09**, Sec 3.10]; and here we will largely parallel his viewpoint.

Aschbacher [**Asc84**] continued, for the classical matrix groups, the viewpoint of structures for S_n which we saw in the O'Nan-Scott Theorem 6.1.1. A first case:

Maximals for linear groups. We begin with the standard example given by the linear subcase, namely $G := GL_n(q)$. This time, the structures will of course not be for n permuted points, but instead will be described inside the n-dimensional natural module V for G. In particular, we will continue the theme of connections between S_n and $GL_n(q)$ which we had indicated at Remark 5.1.3. Also much as for S_n with respect to A_n, we work primarily with $GL_n(q)$, which mod-center is almost-simple; since results can usually be easily translated to the simple section $L_n(q)$.

We essentially follow Wilson's "elementary" version [**Wil09**, Thm 3.5],[1] presented as an analogue of a structures-list (S) for $GL_n(q)$. Indeed we make our statement largely parallel to the that of the O'Nan-Scott Theorem 6.1.1—even largely preserving the case-numbers from that earlier result; but at the left, we also indicate (most of) Aschbacher's names C_i for those "families" of structures.

For brevity, we will usually suppress the indication of the field $\mathbb{F}_q$—writing just GL_n, instead of $GL_n(q)$ for $q = p^a$. The result can be stated roughly as:

---

[1]A technical point: The "structured" subcases in the conclusion we denote by C_+ in the Theorem below seem to have been inadvertently omitted in case (vi) of Wilson's statement; and those subcases might *not* be the preimage of an almost-simple group (some are even solvable).

THEOREM 6.2.1 (Aschbacher). *A proper subgroup H of GL_n, not containing SL_n, lies in one of the following subgroups (preserving, in the n-dimensional natural module V, the structure indicated in parentheses on the right):*

C_1: *(1)* $q^{jk}(GL_j \times GL_k)$, *where* $n = j + k$; *(reducible (parabolic!)—j-subspace: J)*
C_2: *(2)* $GL_j \wr S_k$, *where* $n = jk$; *(imprimitive—direct sum: $V = \oplus^k J$)*
C_4: *(2′)* $GL_j \cdot GL_k$ *(commuting),* $n = jk$; *(tensor product: $V = J \otimes K$)*
C_7: *(3)* $GL_j \wr S_k$, *where* $n = j^k$; *(k-hypercube tensor-product: $V = \otimes^k J$)*
C_6: *(4)* $r^{1+2d} Sp_{2d}(r)$, $n = r^d$, $r \neq p$ *prime;* *(lift of affine, to extraspecial)*
C_+: *(6)* $F^*(H)$ *quasisimple—plus certain structure-cases; see below.*
In (4), if $r = 2$, *replace* $Sp_{2d}(r)$ *by the relevant orthogonal group* $O^{\perp}_{2d}(2)$.

We'll provide further discussion of the final conclusion C_+ in a moment.

EXERCISE 6.2.2. Which of the above cases arise for $L_3(2)$? for $L_4(2)$? As in earlier Exercise 6.1.2, again use e.g. the Atlas [**CCN$^+$85**] to decide the corresponding case (6) here; and note that this case is further subdivided below—cf. later Exercise 6.2.3. Also recall $L_4(2) \cong A_8$, where A_8 appeared in 6.1.2. Some details appear in appendix Remark B.4.2. ♢

To recover the further structures in Aschbacher's actual statement, we now subdivide the conclusion we had denoted by C_+ above: That conclusion certainly contains the the unpredictable-case analogue of (6) in the O'Nan-Scott Theorem. But it also contains some predictable groups, corresponding to preserved structures; and these groups also are "usually" quasisimple. Thus Aschbacher defined three further structure-families within what we called C_+, beyond the five others we indicated in our statement of 6.2.1 above. These further families are:

C_3: $GL_m(q^r)\mathbb{Z}_r$, where $n = mr$, for r a prime; (extension field)
C_5: $GL_n(q^{\frac{1}{r}})$, for r a prime dividing $\nu_p(q)$; (subfield)
C_8: $Sp_n(q)$, $O_n(q)$, $U_n(q)$. (subgroup for a form)

It was convenient to include these at first within our notation C_+; but it was then necessary to use the word "plus" there: since as we had mentioned earlier, these classical subgroups might not be quasisimple—for certain small dimensions and fields. (Some examples arise in appendix Remark B.4.2 mentioned below.)

EXERCISE 6.2.3. Exhibit C_3 and C_8 for $L_4(2)$; and C_5 for $GL_2(4)$. Some details for $L_4(2)$ appear in Remark B.4.2; and $S_3 \cong L_2(2)$ is maximal at least in $A_5 = SL_2(4)$ in Remark B.4.1. ♢

Finally we emphasize that, beyond the classes C_3, C_5, and C_8 just indicated, conclusion C_+ still also contains quasisimple groups which arise "unpredictably"—that is, not from any obvious substructure. Many authors refer to this as a further subcase "C_9".

Maximals for the other classical groups. In fact, Aschbacher also gave a similar treatment for other classical groups, with forms: there the families C_i now also contain a few further substructures for those forms (for example, distinguishing isotropic and non-isotropic subspaces). See especially Wilson [**Wil09**, Ch 3] for statements of these results for Sp_n, U_n, O_n.

EXERCISE 6.2.4. Explore families for S_6—viewed as classical $Sp_4(2)$, or $O_4^-(3)$. Hint: The Atlas [**CCN$^+$85**] has some details, at the bottom of page 4. The array-format there makes it easy to compare with the cases for S_6 in the viewpoint of the O'Nan-Scott Theorem 6.1.1, in Remark B.4.1. ◇

Again, the "candidates" in the families C_i are usually, but not always, maximal. Many such details are handled by Kleidman-Liebeck [**KL90b**].

Maximals via the algebraic-groups viewpoint. Furthermore, maximal subgroups for classical G were *also* described via the Lie theory: this involved first studying the overlying infinite algebraic group $\overline{G}$. I thank Gary Seitz for suggestions about this aspect of the maximal-subgroups program.

One reference for families C_i in algebraic groups is Liebeck-Seitz [**LS98a**]. Here we will single out just a few aspects of that exposition:

The analogue of the inclusion-problem indicated after Exercise 6.1.2, indeed for pairs within the unpredictable almost-simple family C_9, is the crucial analysis of:

$$X < Y < \overline{G} \text{ with } X, Y \text{ irreducible on the natural module } \overline{V}. \tag{6.2.5}$$

This was treated by Seitz [**Sei87**]—building on analogous work of Dynkin for Lie algebras. (That problem for exceptional groups was treated by Testerman [**Tes88**].)

The families for algebraic groups then led naturally to analogous results for the finite Lie-type groups G. In particular, this helps describe much of the subcase for characteristic p in the almost-simple case C_9.

For further reference, see also the overview in Malle-Testerman [**MT11**]. The topic is also part of the material in the lectures planned by those authors at the Venice Summer School in September 2017; see URL:

`//users.dimi.uniud.it/~mario.mainardis/summerschool2017/programme.html`

Maximal subgroups of exceptional groups. For a survey of this area, see for example Liebeck-Seitz [**LS03**]; we largely summarize that exposition, in our fairly informal sketch below:

The analysis of maximal subgroups remains focused on structures—where now for non-classical groups, the families arise even more predominantly from the Lie theory than in the classical-via-algebraic case indicated just above. (But also compare with the approach in Wilson [**Wil09**, Ch 4] and its references; as well as Aschbacher's approach to types G_2 and E_6 via analysis of a "natural" module.)

Maximals for exceptional algebraic groups. Again just as in the classical-via-algebraic treatment above, the Lie approach first treated maximals for the corresponding overlying algebraic groups. Here we provide a brief summary, intended to be suggestive, but with only an approximation of the details.

REMARK 6.2.6 (Background from characteristic 0). For exceptional Lie groups over $\mathbb{C}$, Dynkin in [**Dyn52**] gave maximal *connected* subgroups. The main cases emerge from underlying Lie-algebra substructures:

- maximal parabolic subalgebras: roughly, from the building—cf. C_1;
- maximal-rank reductive[2] subalgebras (certain root subsystems)—cf. C_8;
- scattered cases—mostly simple, and smallish (but note e.g. $F_4 < E_6$). ♢

EXERCISE 6.2.7. Describe some maximal-parabolic structures in G_2; in E_6. Some details appear in appendix Remark B.4.3. ♢

For the exceptional algebraic groups, Seitz in [**Sei91**] gave a result analogous to Dynkin's; this was extended to the maximal positive-dimensional case by Liebeck-Seitz [**LS90**]. The work was originally done for characteristic $p > 7$; it was later extended to small p in [**LS04**].

Finite groups of exceptional Lie type. The algebraic-group results led to similar results for finite exceptional Lie-type groups: Indeed [**LS90**, Thm 2] gives a structures-list, which we informally summarize in the form:

REMARK 6.2.8 (Maximal subgroups in exceptional Lie type). First there are cases correspdonding to structures indicated earlier; that is:
- groups from cases for Lie algebras (and algebraic groups), as in Remark 6.2.6.

In addition, for a group defined over the finite field $\mathbb{F}_q$, we expect also further cases, of the "same" Lie type—compare with Aschbacher's subfield-family C_5 (though there is no non-classical analogue of the extension-field case in C_3):
- groups for a subfield (including twisted groups).

Like the root-subsystem situation from Lie algebras as above, these form part of the almost-simple subcase. In addition, we get:
- a few "exotic" local subgroups—for example, $2^3L_3(2)$ in $G_2(\text{odd})$.

Finally, any group not arising from one of the structures above comes from:
- unpredictable almost-simple groups;

where of course this case compares with earlier "C_9". ♢

In fact, in further pursuing this last unpredictable-subcase of the almost-simple case: those in characteristic p are called *generic* (cf. [**LS98b**]); and those in other characteristics are called *non-generic*. Non-generic possibilities are listed in Theorem 4 of Liebeck-Seitz [**LS03**]; and generic cases in Theorems 5 and 7 there.

Finally Theorem 8 of [**LS03**] gives an overall summary of the work indicated above. Also used in the program are further papers of Testerman [**Tes88**]; and Liebeck-Seitz [**LS99a**].

Some details of the overall program remain unfinished: for example, conjugacy of some of the almost-simple cases.

6.3. Maximal subgroups of sporadic groups

Many maximal subgroups of sporadics are described in the Atlas [**CCN$^+$85**], but without proofs; that source roughly records the state of knowledge circa 1985. A summary with fuller references appears in [**Wil86**].

[2]Reductive means that the solvable-radical is given just by the center.

Wilson's recent book [**Wil09**] contains tables of maximal subgroups for each of the 26 sporadic groups. These give the status essentially as of the present (2017). In summary: Those tables are known to be complete—except possibly for the Monster M. A still more-recent survey of maximal subgroups of the sporadic groups, including corrections to some errors in the literature, will soon be appearing in [**Wil17**].

Sometimes those maximal subgroups are visible from the viewpoint of preserved structures; for example, in Steiner systems, or Golay codes, or the Leech lattice. A standard example is Co_2—which arises in Co_1, as the stabilizer of a "length-2" vector in the Leech lattice.

In other cases, suitable local subgroups are maximal; these may or may not preserve any obvious structure external to the group. E.g., the maximal 2-local subgroup 2^4A_8 in M_{24} is also visible as the stabilizer of an "octad" in the Steiner system $\mathcal{S}(5, 8, 24)$.

But often, other methods are needed; especially, to determine maximals which are almost-simple.

Some applications of maximal subgroups

Results on maximal subgroups, some of which we've sketched in the earlier sections of this chapter, have been applied in a very wide variety of problems.

So this may be a good point at which to expand a little on our brief introduction to applications, in the introductory remarks before Section 1.4—especially since we noted there that the applications in the chapters before this one were largely focused on more specific questions related to group structure.

6.4. Background: Broader areas of applications

So we'll now briefly describe (in no particular order) a number of important more general areas of application; again with the caveat that there are many other areas that could be mentioned here.

(1) *Random generation and probabilistic group theory.* There has been considerable research activity studying groups from a probabilistic viewpoint; for example, the probability of generation by sets of randomly-chosen elements of suitable types. Frequently such questions reduce to problems about simple groups, and their permutation representations—hence maximal subgroups. The interested reader may wish to consult Liebeck's survey [**Lie13**]. Some sub-topics include: probability of generation by some special subset—e.g. by a pair of elements, as in the applications indicated in Section 6.7; random walks on generators, and the size of a minimal generating set, as in Section 6.5; probabilistic properties of representation varieties—namely mappings of suitable finitely generated groups (e.g. Fuchsian groups, surface groups, triangle groups, ...) into linear groups; diameter and growth properties of the Cayley graph determined by a generating set; and more.

(2) *Actions of finite linear groups.* For a linear group $H \leq GL_n(p)$, one can study special properties of H-orbits on nonzero vectors v in the natural module V for $GL_n(p)$; such as the number or length of those orbits, For example, the case of all orbit lengths coprime to p is the topic of Section 6.6. Furthermore the first part of Theorem 6.6.1 in that section includes one notion of a "long" orbit, namely a single

regular orbit: these were determined by Liebeck in [**Lie87a**, App 1].[3] Similarly one can study the situation of comparatively few orbits: the cases of one and two orbits appear in Hering [**Her85**] and Liebeck [**Lie87a**].

(3) *Algebraic combinatorics.* When combinatorial structures (such as graphs, designs, association schemes, etc) have a high degree of symmetry, they can be very naturally studied via the structure of the underlying symmetry groups. There are many corresponding applications of the CFSG, in particular of maximal subgroups; and a number of surveys of that area: for example, Praeger [**Pra97**] overviews work on graphs whose automorphism group is quasiprimitive in the sense of Section 6.1. Examples of such applications to distance-transitive graphs are indicated in later Section 9.1; and to the expander graphs of computer science, in Section 10.6.

(4) *Computational group theory.* This very active area is concerned with algorithms for computing with groups—and consequently, also with the theoretical efficiency of those algorithms, in the sense of computer science. These in turn require various statistical measures in simple groups; such as estimating the proportions of elements with various properties, so that they can be found efficient with essentially random methods. The interested reader might wish to consult the survey of Niemeyer-Praeger-Seress [**NPS13**]. Very frequently these algorithms involve permutation representations of the underlying groups, and consequently maximal subgroups. An example of such an application, to the proportion of p-singular elements, is discussed in later Section 9.2; while algorithms for computing structural features such composition factors are indicated in Section 9.8.

(5) *Degrees of primitive permutation groups.* For applications in the more general area of permutation groups, we had already mentioned such sources as Cameron's early survey [**Cam81**]. We mention in particular the sub-area of primitive groups—and in particular properties of the degrees of the corresponding primitive representation. There are many results in the corresponding literature, and we indicate some in Section 9.5, including prime-power index and odd index.

And now in the remainder of the chapter, we sketch a few particular applications of maximal subgroups. These were mainly suggested to me by expert colleagues.

6.5. Random walks on S_n and minimal generating sets

I thank Persi Diaconis for suggesting this topic.

We first summarize some background from Whiston [**Whi00**]:

Maximal independent generating sets. In [**NP92**], Neumann and Praeger gave an algorithm for recognizing, in a linear group G, the span of a set S of elements. The span is viewed as being built up from a *random walk* on G based on S—that is, building up words in S, via successive multiplication of previously built-up words by further random choices from S.

Holt and Rees [**HR92**] made some adjustments, to improve the convergence behavior of the algorithm to the final span of S—typically, to the full original group G.

[3]That result was in turn further applied in the Robinson-Thompson approach [**RT96**] to the $k(GV)$ problem—a conjecture of Brauer, now a theorem, asserting, for a p'-group G faithful on an $\mathbb{F}_p$-module V, that the number of conjugacy classes in the GV is $\leq |V|$.

Diaconis and Saloff-Coste [**DSC98**] then gave more specific results on the convergence of the span to G, for more general groups. They worked with bounds (such as that indicated below) on the parameter $m(G)$—namely the size of a maximal independent generating-set S for G; where "independent" has the obvious meaning, that no element of S is in the span of the remaining elements.

Thus it is of interest, in particular for efficiency of computations such as those above, to determine the value of $m(G)$—especially for almost-simple groups G.

Whiston's result for S_n. For G given by the symmetric group S_n, Whiston [**Whi00**, Thm 1] showed that $m(S_n) = (n-1)$; of course this value achieved by the standard generating set S given by adjacent transpositions.

The proof quotes the actions-list (A), namely that of the Aschbacher-O'Nan-Scott Theorem given in Remark 6.1.4. (Thus this part of the argument seems to require the CFSG, via the Schreier Conjecture 1.5.1.)

In overall summary, the flow of the argument in effect actually follows the structures-list (S), that is, the O'Nan-Scott Theorem 6.1.1. And to finish the proof, Whiston eventually quotes the full strength of the CFSG, to determine the possibilities for almost-simple groups in case (6) of 6.1.1.

Use of the maximal subgroups in the structures-list. We roughly sketch the logic flow:

To summarize first the strategy: From an independent generating set S of S_n, remove an element to obtain a subset S'. The span H of S' is by independence a proper subgroup of S_n, hence lies in some maximal subgroup M. One can avoid the trivial case of $M = A_n$, by re-defining S' via the removal of an even permutation. Hence M can be described using the structures-list in 6.1.1.

And now show: Either $m(M) \leq (n-3)$; or $m(M) = (n-2)$, and any further indepndent element added to H will generate the full S_n.

The implementation of this strategy based on a maximal M roughly follows a deduction of the structures-list in 6.1.1 from the actions-list in 6.1.4; namely:

First reduce to M transitive on the n points: otherwise we are in the intransitive structure-case given by (1) $S_j \times S_k$; and here we can apply induction to S_j and S_k, together with some further argument, to finish by establishing the desired bounds on $m(M)$ indicated above.

Next, reduce to M primitive on the n points: otherwise we are in the imprimitive case (2) $S_j \wr S_k$; where we can similarly apply induction to S_j, S_k to finish.

Now we can quote the primitive actions-list for the Aschbacher-O'Nan-Scott Theorem in Remark 6.1.4. However, since we are in the process of proving a purely group-theoretic assertion, it is really only the *group*-structures for the primitively-acting M that matter; it is for this reason that in effect we really just need the corresponding structures-list from the O'Nan-Scott Theorem 6.1.1. (Recall we indicated the relevant inclusions in column 3 of the table in 6.1.4.) So we can summarize the remaining argument for the 8 types of primitive actions:

Types PA,CD,HC,TW: These lie in structure (3) $S_j \wr S_k$; this is the same *group* as in (2) considered above, and so can be finished with that earlier argument.

Type HA: This lies in (4) $r^d GL_d(r)$. Here we can finish using arguments based on possible lengths of subgroup-chains, determined in Cameron-Solomon-Turull [**CST89**].

Types SD,HS: These lie in (5) $L^k(S_k \times \mathrm{Out}(L))$. But numerical bounds coming from induction, using the overgroup $S_{|L|} \wr S_k$ much as in (3), can be used to finish.

Type AS: This lies in (6) almost-simple. Here we use the CFSG to determine the list of possible simple groups. And now to finish, we can quote results in the literature on minimal degrees of permutation representations, along with bounds in Kleidman-Liebeck [**KL90b**].

6.6. Applications to p-exceptional linear groups

I thank P. H. Tiep for suggesting this topic.

We first sketch some material from the paper of Giudici-Liebeck-Praeger-Saxl-Tiep [**GLP$^+$16**]: (We had also briefly mentioned this paper in regard to the Brauer Height 0 Conjecture, in the final (afterword-glimpse) section of Chapter 5.)

The examples of p-exceptional groups. We say H is *p-exceptional* if all orbit lengths are coprime to p; that is, if each v is fixed by some Sylow group of H.

Of course, we may as well assume that p *does* divide $|H|$. Furthermore, we may as well assume that H is irreducible on V—avoiding the family C_1 of Aschbacher's Theorem 6.2.1. With regard to the family C_2, corresponding to an "imprimitive" direct-sum decomposition of V which might be preserved by H, there are some subtleties—which we will be mentioning below.

There are various natural situations giving examples of p-exceptionality:

We might in particular have all H-orbits of the *same* size; recall that the total number of nonzero vectors $|V^\#|$ is not divisible by p. This subcase is called a $\frac{1}{2}$-*transitive* action; these are described in Theorem 6 of [**GLP$^+$16**]. (Furthermore Corollary 7 there discusses the related notion of $\frac{3}{2}$-*transitive*.)

Of course a special case of $\frac{1}{2}$-transitive action is fully transitive action of G on $|V^\#|$. Some standard examples include: $SL_n(V)$; $Sp_n(V)$ for even $n = 2m$; and $G_2(2^a)$ when $n = 6$ for the Cayley-algebra module of earlier Example 5.2.13; as well as a full nonsplit torus T of order $p^n - 1$, as in Example 5.2.2.

The main result Theorem 1 of [**GLP$^+$16**] shows that the above cases are in fact almost all the examples of the p-exceptional condition which have a primitive action on V:

THEOREM 6.6.1. *Any subgroup $H \leq GL_n(p)$ which is irreducible, primitive, and p-exceptional, must be one of:*

(i) H is transitive on $V^\#$. *(These are known—see* [**Lie87a**, App 1]*.)*

(ii) $H \leq \Gamma L_1(p^n)$. *(Nonsplit tori—described in* [**GLP$^+$16**, 2.7]*.)*

(iii) $p = 2$; n-dimensional V is the natural irreducible for $H = S_c$ or A_c, where $c = 2^r - 1$ or $2^r - 2$.

(iv) the inclusion $H < GL_n(p)$ arises from one of the inclusions in the following list: $SL_2(5) < GL_4(3)$; $L_2(11) < M_{11} < L_5(3)$; or $M_{23} < L_{11}(2)$.

EXERCISE 6.6.2. Explore some of the orbit sizes in (iii) and (iv) above. Hint: Some details are given in appendix Remark B.4.4. ♢

The paper also contains a result [**GLP$^+$16**, Thm 3] covering much of the imprimitive case: namely if $H = O^{p'}(H)$, then one has transitivity as in (i) above on the nonzero vectors of the components of a direct-sum decomposition of V under H.

Maximal subgroups in the proof. Of course we may as well assume that we have $H < GL_n(p)$; otherwise we get transitivity as in conclusion (i).

Consequently H lies in some maximal subgroup M of $GL_n(p)$; and we can make use of the list of C_i-structures in Aschbacher's Theorem 6.2.1.

More precisely, the proof in [**GLP**$^+$**16**, Sec 12] actually parallels the logic-sequence in Section 11 of Aschbacher [**Asc84**], by which he deduces his result on the families C_i. We give a rough sketch of this deduction below.

Note first that the hypotheses of irreducibility and primitivity eliminate the cases C_1 and C_2. (Irreducibility also eliminates the subcase $Sp_n(p)$ in C_8.)

Next the consideration of field extensions in case C_3 is roughly automated: We choose a maximal such extension: namely d dividing n, so that $q := p^{\frac{n}{d}}$ is maximal subject to $H \leq \Gamma_d(q)$. We may assume that $d \geq 2$, since if $d = 1$ we get conclusion (ii). Then [**GLP**$^+$**16**, 12.1] shows, using irreducibility of H over $\mathbb{F}_p$, that the intersection $H_0 := H \cap GL_d(q)$ is *absolutely* irreducible.

Furthermore the case C_5 would require realizing H (modulo scalars) over a proper subfield of $\mathbb{F}_q$; but [**GLP**$^+$**16**, 6.1] shows that this situation does not lead to any p-exceptional examples.

Now consider the case where H preserves a tensor decomposition $J \otimes K$ of V, by means of factors of dimension ≥ 2, where the action is that arising in the C_4-case. Then [**GLP**$^+$**16**, 4.1,4.2] show this leads to a contradiction to absolute irreducibility above. Thus we may assume there is no such decomposition.

Analysis now focuses on the socle structure $\mathrm{soc}(H/Z)$, where Z is the subgroup of scalars in H_0.

The situation where the socle contains some minimal normal subgroup of H which is an elementary abelian r-group would lift in H to an extraspecial r-group, as in case C_6. But this would lead by [**GLP**$^+$**16**, 7.1] only to imprimitive examples.

The situation where the socle contains some minimal normal subgroup of H which is a product of more than one nonabelian simple group L would lead to a k-hypercube tensor decomposition as in C_7. But this by [**GLP**$^+$**16**, 5.1,5.3,2.8] would lead either to reducible or imprimitive cases;[4] or to the cases in (i) and (ii); or to a case in (iv), namely $SL_2(5) \trianglelefteq H \cap GL_2(9)$ (where the latter is embedded in the group $GL_4(3)$ on V).

This has reduced to case of a simple socle; so that H itself is almost-simple modulo center; in effect "C_9". Here we apply the full force of the CFSG, to determine the list 1.0.2 of simple groups, giving the possibilities of $F^*(H/Z)$. And below we briefly summarize the work of Sections 8–11 of [**GLP**$^+$**16**] analyzing those cases:

For $F^*(H/Z)$ of Lie type in the same characteristic p, [**GLP**$^+$**16**, 8.1] shows that only the transitive conclusion (i) can hold. (Notice this completes case C_8, by eliminating $O_n(q)$ and $U_n(q)$.)

For $F^*(H/Z)$ alternating, [**GLP**$^+$**16**, 9.1] allows either conclusion (iii), with S_c or A_c on its natural irreducible; or (iv) with $F^*(H) = SL_2(5) = 2A_5$.

For $F^*(H/Z)$ sporadic, [**GLP**$^+$**16**, 10.1] shows that we must have just cases in (iv), from either $M_{11} < GL_5(3)$, or $M_{23} < GL_{11}(2)$.

Finally for $F^*(H/Z)$ of Lie type in characteristic other than p, we obtain via [**GLP**$^+$**16**, 11.1] either the transitive conclusion (i); or cases in (iv)—arising from $F^*(H/Z) \cong A_5 \cong L_2(4) \cong L_2(5)$; or $A_6 \cong Sp_4(2)' \cong L_2(9)$; or $L_2(11)$.

[4] It seems implicit that this is the argument for eliminating the case 2.8(ii) there.

6.7. The probability of 2-generating a simple group

In this section, we provide an application of maximal subgroups, beyond just the symmetric and linear groups considered in the previous two sections.

We first summarize some exposition from Liebeck-Shalev [**LS95**]:

2-generation for simple groups. In [**AG84**, Thm B],[5] Aschbacher and Guralnick showed (using the CFSG) that any simple group G *can* in fact be 2-generated. That is, there exists some pair $x, y \in G$ with $\langle x, y\rangle = G$.

For the special case $G = S_n$, Netto had conjectured in 1882 (see [**Net64**]) that "most" pairs (x, y) work: that is, that for *randomly* chosen x, y, we find that the probability($\langle x, y\rangle \geq A_n$) $\to 1$, as $n \to \infty$. Dixon [**Dix69**] later proved this conjecture—in fact, before the CFSG and the O'Nan-Scott Theorem 6.1.1. And then he ventured the same conjecture for all other inifite families of simple G.

Since then various authors, using the CFSG, have now covered all G, establishing the general result—which we phrase as:

THEOREM 6.7.1. *prob(2-generating simple G) $\to 1$, as $|G| \to \infty$.*

Notice this statement is already about simple groups; so the CFSG is really being used just for the list 1.0.2 of the simple groups. Of course, the detailed properties of those groups are heavily used in the proof.

In fact, the proof proceeds via the complementary probability, showing:

probability(random x, y *not* generating G) $\to 0$.

Indeed if generation fails, then $\langle x, y\rangle < G$, and hence $\langle x, y\rangle$ falls into some maximal subgroup $M < G$; hence we can apply the lists of candidates for maximals, from the first three sections of the chapter.

Use of maximal subgroups, especially of exceptional groups. First notice that the sporadic groups G are automatically excluded from the statement: for $|G| \to \infty$ doesn't make sense for the 26 sporadics, which are not members of infinite families.

So we now summarize how the various infinite families of simple groups G were handled:

(alternating:) We already mentioned that the symmetric groups were handled by Dixon [**Dix69**]—work done before the availability of the maximal subgroup list in the O'Nan-Scott Theorem 6.1.1. So we won't here try to describe Dixon's proof.

(classical:) These groups were treated by Kantor-Lubotzky [**KL90a**]: The cases are subdivided via Aschbacher's families $\mathcal{C}_i$ in Theorem 6.2.1. We followed such a subdivision in the previous section; so we won't here follow this similar subdivision.

("exceptional":) The remaining infinite families of simple groups are the types called exceptional in this chapter: namely untwisted groups for exceptional Lie algebras, and non-classical twisted groups. These groups were handled by Liebeck-Shalev [**LS95**]. Their argument followed the case-subdivision that we summarized in Remark 6.2.8. So below, we will illustrate some of those subdivisions.

The main idea: Compute the probability that generation by a random pair x, y fails, as the sum—over all maximal subgroups M—of the terms:

$$\text{prob}(\text{random } \ x, y \in M) = (\tfrac{|M|}{|G|})^2 = \tfrac{1}{|G:M|^2}.$$

[5]We'll explore that paper a bit more in later Section 9.6.

Since the value for M occurs for each member of its conjugacy class of size $|G:M|$, in practice we just compute the contribution $\frac{1}{|G:M|}$ from the whole class.

A first reduction appears as [**LS95**, (c), p 110]; this can be compared with earlier [**KL90a**, (**), p 69]. Roughly: For G of Lie type over $\mathbb{F}_q$, and a maximal-type M coming from a natural "structure" as in 6.2.8, the index $|G:M|$ is a non-constant polynomial in the field-size q. So for such M, the sum of terms $\frac{1}{|G:M|}$ should indeed tend to 0 as $q \to \infty$. This now leaves only the sum over "unpredictable" almost-simple M in 6.2.8.

In fact, in [**LS95**], the first and more easily-handled sum can be made over terms in a larger class, called $\mathcal{K}$ ("known"). It includes first the structured-types of M from 6.2.8; we recall these are (in highly abbreviated and possibly cryptic wording): maximal parabolics; maximal-rank-as-reductive subgroups; the indicated scattered cases; the cases for subfields and extension fields; twisted subgroups; and the indicated exotic local subgroups. But it also includes some fairly large almost-simple M not predictable from any structure: namely those of order $\geq$ about $|G|^{\frac{5}{13}}$. (Also used in this area is Liebeck-Saxl-Testerman [**LST96**].)

Finally for the sum over the remaining M in the class $\mathcal{U}$ ("unknown"), namely small almost-simple M—of order below the bound of $|G|^{\frac{5}{13}}$)—the computations use estimates for the probability of 2-generation of M using a pair of elements specifically containing an involution—quoted from Malle-Saxl-Weigel [**MSW94**].

CHAPTER 7

Geometries for simple groups

From the earliest history of group theory, there has been an important theme of the study of groups via their actions on suitable *geometries*: where the term "geometry" can be fairly broadly interpreted.

The theme has been particularly prominent in geometric topology—where the geometries are typically those in the usual continuous context, namely configurations defined by lengths, angles, etc. This holds to some extent also in algebraic topology; although weaker notions of geometry are reasonably common there. For some background in this area, see e.g. Adem-Davis [**AD02**].

However for finite groups, instead of such continuous geometries, it is usually natural to instead study *discrete* (indeed, finite) geometries. For a permutation group like S_n, the geometric context can be very weak—e.g. just the combinatorics of a finite set. But more commonly, finite groups act naturally on geometries exhibited by various *simplicial complexes*.

Often these complexes may arise from inclusion-chains in a partially ordered set; we had introduced these briefly in Section 2.4: for example, we saw in (2.4.2) the poset $\mathcal{S}_p(G)$, defined by the nontrivial p-subgroups of any finite group G. And sometimes it is of interest to study this particular geometry, especially for the case of simple G.

In that general direction, we focus in this chapter on the viewpoint of set-theoretic *projective* geometry: here the model case is given by a linear group $GL_n(p)$, with the projective space 7.0.1 given by the poset of proper nontrivial subspaces of the natural module V. This in turn provides an initial context for:

Introduction: The influence of Tits's theory of buildings

For the more general class of Lie-type groups G, the value of this approach was cogently demonstrated by Tits's theory of buildings—which provided for any such G a suitable simplicial complex, generalizing the model case of projective space. We had only briefly suggested buildings, in discussing action-identifications early in Section 4.2; we'll soon be saying considerably more, in Section 7.2 (especially Remark 7.2.5).

But first, we'll provide some historical context on the study of buildings:

The classification of semisimple algebraic groups (~1955) led to a unified understanding of the various different groups of Lie type. But that did not yet provide a unified *geometric* analysis of those groups. This was roughly because: A classical group G with natural module V acts on a natural geometry determined by the subspaces of V—isotropic, in the case of a form: namely the underlying projective or polar space. But a group G of exceptional type may not have such a "natural" V.

Tits ($\sim$ 1965) defined a common structure, the building, providing a suitable analogue of projective space—for both classical and exceptional G. (We'll examine it more closely at later 7.2.5.)

To overview this unification, we expand a little on the background introduced above:

EXAMPLE 7.0.1 (projective space and polar space). Consider the projective space $\mathcal{P}(V)$ a vector space V; this poset consists of the proper nontrivial subspaces W of V. In particular, projective points and lines are given by the various 1- and 2-subspaces. The order complex of this poset, in the sense of (2.4.1), is the simplicial complex of inclusion-chains $W_1 < W_2 < \cdots < W_r$ among such subspaces.

Similarly for a space V with a form, the *polar space* arises from the poset of totally isotropic (or totally singular) subspaces. ♢

It is easily checked that the stabilizer of some chain c is a parabolic subgroup P_c; and that a chain-inclusion $c < d$ determines the reversed inclusion, namely $P_c > P_d$, in the poset of parabolics. The beginning of Tits's theory was the observation that this simplicial complex based on the parabolics (which is one view of the building; compare 7.2.5) is available for *any* Lie-type group G.

In particular, this parabolic-complex is common to both classical *and* exceptional (including twisted) groups. Tits then proceeded to give a uniform axiomatization of the building-geometries, with a detailed analysis of their properties.

Since then the theory of buildings has had very far-reaching applications; see for example Abramenko-Brown [**AB08**, Ch 13,14] for a survey of many of these.

In this chapter, after introducing buildings, we will restrict to one direction of further development: namely "similar" geometric approaches to *other* types of groups—notably the sporadic simple groups in Section 7.3.

7.1. The simplex for S_n; later giving an apartment for $GL_n(q)$

A geometric viewpoint on S_n will provide a useful introduction to various features of the building in the subsequent Section 7.2. Notice that once again, instead of considering simple A_n, we prefer to work first with the almost-simple group S_n.

Given the n-transitive action of S_n, the obvious simplicial complex for S_n is the full simplex Γ on the n permuted points; this has dimension $(n-1)$. Admittedly the geometric simplex-structure of Γ does not really add any new content, beyond the combinatorics of the n-set. Nonetheless, the geometric view of the simplex will provide a useful initial example—demonstrating many geometric concepts of Tits which will be prominent later in the chapter.

We will in fact focus attention on the boundary $\Sigma := \partial\Gamma$ of Γ.

A set-theoretic approach to the simplex-boundary Σ. This viewpoint is particularly valuable, when we regard S_n as the Weyl group of the Lie-type group GL_n; this continues our theme of connections between these groups (cf. 5.1.3).

For note that the boundary Σ of the simplex Γ (which we may informally think of as a "hollowed-out" simplex) is topologically a sphere; and of course it still admits the induced action by S_n. We'll soon be seeing that such a sphere gives an *apartment* for the building Δ of GL_n—and indeed that the building can be regarded as being assembled from these spheres. To develop this viewpoint, we'll want to examine this explicit sphere in a more abstract framework:

REMARK 7.1.1 (The set-theoretic hollowed-simplex Σ). We proceed essentially by analogy with projective space in Example 7.0.1: This time we take as our poset all the nonempty proper sub*sets* of the n points. Note that we take *proper* subsets—in order to obtain the boundary Σ, rather than the filled-in simplex Γ. And then the simplices of the order complex Σ of our poset are just the inclusion-chains among those subsets. This view of Σ is essentially the *barycentric subdivision* of our original view of the boundary Σ. We mention that Σ is a triangulation of a sphere of dimension $(n-2)$. $\diamondsuit$

For the above abstract setup, we examine some features in a small explicit example:

EXAMPLE 7.1.2 (The apartment-hexagon Σ defined by S_3). We consider the case $n = 3$: thus we work with the group S_3. This is also the Weyl group for the Dynkin diagram A_2, corresponding to linear Lie-type groups GL_3. Let a, b, c denote the 3 points permuted by S_3.

These three points determine a 2-dimensional simplex Γ: namely the filled-in triangle on the 3 points a, b, c. The boundary $\partial\Gamma$ is then just the triangle itself. However in the barycentric-subdivision view of Remark 7.1.1, we have vertices given not just by the three points, but also by the 3 edges such as $\{a, b\}$ among them. Thus our boundary Σ in fact becomes a hexagon.

For later reference, let's sketch this hexagon. For brevity of notation, I omit braces—writing just a and a, b for the vertices. The edges are of course the $\subset$-inclusions among them:

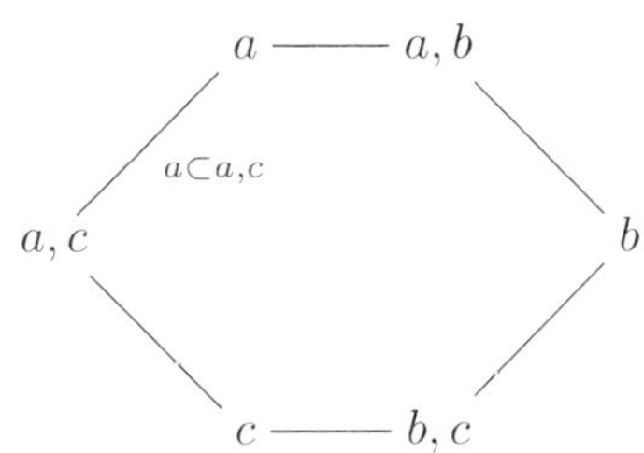

$\diamondsuit$

Notice that the next-larger case of S_4 would be much harder to draw: The subdivision of the surface of a 3-simplex, namely a tetrahedron, has 24 maximal faces—each a 2-simplex, i.e. a filled-in triangle.

We now use the explicit hexagon in Example 7.1.2 above, to demonstrate a number of geometric properties of the abstract $(n-2)$-sphere Σ in Remark 7.1.1—using the language developed by Tits in the context of buildings.

Observe first that the hexagon in 7.1.2 is a triangulation of a circle—that is, of a sphere having dimension $(n-2) = 1$.

REMARK 7.1.3 (Type for vertices and simplices). Next we note that since Remark 7.1.1 for general n works with a barycentric subdivision, each of its vertices corresponds to a k-subset of the n points—for some $k \in \{1, \cdots, n-1\} =: \Pi$. We refer to this value k as the *type* of the vertex. Furthermore each simplex σ in Σ also inherits a type—since σ is an inclusion-chain among vertices, that type is just a nonempty subset $J \subseteq \Pi$; with the dimension of σ given by $|J| - 1$. Thus in the hexagon of Example 7.1.2, the vertex types for a and a, b are 1 and 2; and edges such as $(a \subset a, c)$ have full type $\{1, 2\}$.

In the general language, the simplices of full type Π, hence having dimension given by $(n-2)$, are called *chambers*; while those of the next-lower dimension $(n-3)$ are called *panels*. ♢

A group-theoretic approach to Σ. We turn to another abstract approach to Σ—this differs from the approach in Remark 7.1.1 in being group-theoretic, rather than set-theoretic. For that reason, we will able to more easily see the generalization to Weyl groups W other than just S_n. Furthermore this alternative characterization of Σ can be useful in various theoretical arguments.

REMARK 7.1.4 (The coset-complex viewpoint on Σ). Instead of identifying the indexing set Π with the sizes $1, 2, \cdots, (n-1)$ of proper subsets of our n-set, we can instead identify Π in effect with the simple system of the underlying root system: more precisely, with the simple reflections w_i corresponding to those simple roots. Thus in fact we re-define our set, namely as $\Pi := \{w_1, \cdots, w_{n-1}\}$, via the usual generating set for S_n—consisting of the adjacent-transpositions $w_i := (i, i+1)$.

Further we want to define the *parabolic* subgroups of W: For a subset $J \subseteq \Pi$, we define the subgroup W_J as that generated by the w_j for $j \in J$. (Cf. the conventions in 1.3.20(4).)

For example, we have $W_\emptyset = 1$; and $W_\Pi = W$; and note for $\hat{1} := \Pi \setminus \{1\}$ that $W_{\hat{1}}$ is a point stabilizer, isomorphic to S_{n-1}. Indeed we can check that $W_{\hat{k}}$ is the stabilizer of a k-subset of the n points. So those k-subsets are in 1:1 correspondence with the cosets of $W_{\hat{k}}$.

By means of this correspondence, we get a simplicial complex isomorphic to Σ: where now for a set $J := \{k_1, \cdots, k_r\}$ of distinct vertex-types, a set of cosets of $W_{\hat{k}}$ for $k \in J$ will determine a simplex of type J, precisely when those cosets have nonempty intersection. Note however that a simplex of type J will have stabilizer isomorphic to $W_{\hat{J}}$ indexed by the complement $\hat{J}$.

This setup gives the *coset complex* version of Σ. ♢

We use this group-theoretic view to demonstrate some further geometric features:

Note that the panel stabilizer $W_k = \langle w_k \rangle$ of order 2 switches the chambers containing that panel—there are in fact exactly two such chambers. Tits calls this condition a "thin" geometry. By contrast, for an untwisted Lie-type group defined over some finite field $\mathbb{F}_q$, the number of chambers over a fixed panel will be $q + 1 \geq 3$—the "thick"-geometry situation for the building in 7.2.5.

We can define *paths* between chambers (and hence distance)—where adjacency of chambers is defined by sharing a panel. Such an adjacency has a type, given by the single member of Π not in the type of the panel; and consequently a path between chambers also has a type, now given by a word in the generators of W. The maximal value of this distance for S_n is given by $\binom{n}{2}$. When we view S_n as the Weyl group of the root system Φ of type A_{n-1}, this value is equal to the number of positive roots $|\Phi^+|$.

EXERCISE 7.1.5. Exhibit-path types in the hexagon for S_3 in Example 7.1.2. ♢

Some local aspects of "diagram" geometry. In particular, in 7.1.4 above we are in effect identifying the indexing set Π—viewed as the generators of S_n—with the nodes of the Dynkin diagram of Lie-type A_{n-1}:

$$\overset{1}{\circ} - \overset{2}{\circ} - \overset{3}{\circ} - \cdots - \overset{n-2}{\circ} - \overset{n-1}{\circ}$$

We turn to a "local" or "residual" feature of of this association, which is important for the viewpoint of Tits, namely:

REMARK 7.1.6 (Residual structures in Σ). Put your finger over node k; recall this node is for the stabilizer $W_{\hat{k}}$ of a k-subset X of the n points. Now notice that the "residual" diagram, that is, the part not covered, in fact has the Lie type $A_{k-1} \times A_{n-k-1}$. This diagram first of all describes the isomorphism type of the group $W_{\hat{k}}$, namely $S_k \times S_{n-k}$. But in fact, it also describes structure of the geometric *link* (or *residue*, in the language of Tits) of X—that is, the geometry of all simplices "adjacent" to our k-set X: This is the topological join, of: (the geometry Σ_k, given by the subsets of X), with (the geometry Σ_{n-k} of supersets of X). These Σ_i are spheres like Σ, but of smaller dimensions; and of course they admit the action of the factors S_k, S_{n-k} of $W_{\hat{k}}$.

The example above was for a singleton $\{k\}$ from Π; but similar remarks hold for simplices of more general types $J \subseteq \Pi$. ♢

EXERCISE 7.1.7. Describe the links of simplices of various types J, in various groups S_4; S_5; etc. ♢

This local view of the diagram is a fundamental inductive feature of Tits's approach.

Geometric properties similar to those in this section in fact hold for all Weyl groups W; for details see e.g. [**Car89**, Ch 2] and [**AB08**, Ch 1–3].

7.2. The building for a Lie-type group

Before commenting on more theoretical aspects of buildings, we will first exhibit some Lie-type buildings explicitly. (For an approach which begins directly with the diagram-geometry viewpoint of Remark 7.1.6, see Buekenhout-Cohen [**BC13**].)

In particular, we will largely parallel our discussion of the sphere Σ for S_n in the previous section: First we indicate a small explicit example for $n = 3$; and then we turn to the group-intrinsic construction of the building via parabolic subgroups—and in that context, we review various corresponding geometric properties.

The projective plane for $L_3(2)$. We recall our theme of connections between S_n and GL_n; where the natural explicit geometry for $GL_n(q)$ is the projective space 7.0.1 for its natural module. And corresponding to our earlier small example of S_3, we consider the smallest Lie-type group with this Weyl group, i.e. $G = L_3(2)$. Here the projective space is just a projective plane; and we can exhibit the geometry in some detail:

EXAMPLE 7.2.1 (The projective plane for $L_3(2)$). Choose a basis a, b, c for the 3-dimensional natural module V of $L_3(2)$.

Then there are 7 projective points (namely 1-dimensional subspaces); we can denote these in abbreviated form as: a, b, c, ab, ac, bc, abc. Furthermore there are also 7 projective lines (given by the 2-dimensional subspaces); these are $\langle a, b\rangle$ and $\langle a, c\rangle$ etc.; we typically abbreviate these as a, b; a, c; etc.

Our simplicial complex Δ is determined by taking the points and lines as vertices, with their inclusions giving edges. Notice the "local" feature that there are 3 points per line, and 3 lines per point. Consequently there are 21 edges, in our 1-dimensional complex Δ.

Often (e.g. [**Smi11**, 2.1.13]) the plane is drawn on a barycentrically-subdivided triangle with corners a, b, c: where the 7 points are the vertices; and the 7 lines are given by the 6 line-segments, together with an additional circular "line" joining the points ab, ac, bc. But for our present approach, it is more natural to draw the plane as a bipartite graph: with the two parts being given by the point-vertices and the line-vertices. Notice that when we do this, the subgraph determined by the points a, b, c alone, with their 3 lines, just gives us a (folded-over) copy of the hexagon Σ of Example 7.1.2.

Indeed just as in Σ, our vertices have types in the set $\Pi := \{1, 2\}$: with projective points corresponding to linear dimension 1, and lines to linear dimension 2. We could instead name the vertex types via projective dimension, say as $\{P, L\}$. $\diamondsuit$

EXERCISE 7.2.2. Draw the bipartite graph for the projective plane of $L_3(2)$. $\diamondsuit$

We now examine some crucial geometric features, within this example:

The subgraph Σ induced by a, b, c is an example of an *apartment* in the building Δ. We claim in fact for the chambers (i.e. edges—of the form $a \subset a, b$) that:

(7.2.3) Any two chambers of Δ lie in some apartment.

You can explore this visually, in the graph drawn in Exercise 7.2.2. Or, to proceed more abstractly: The maximal-distance case occurs if $a \notin \langle c, d\rangle$ and $c \notin \langle a, b\rangle$. Then $\langle c, d\rangle \cap \langle a, b\rangle$ is a point x, which cannot be a or c. So here we can use the apartment generated by a, c, x. The shorter-distance cases proceed by similar and indeed easier proofs.

EXERCISE 7.2.4. Similarly draw the bipartite graph Δ for the polar space of $Sp_4(2)$, in the language of 7.0.1: namely using only the 2-subspaces which are isotropic. Work in terms of a symplectic basis with $(a \not\perp d)$ orthogonal to $(b \not\perp c)$. This time there will be 15 points, and 15 isotropic lines. Here the apartment, as before in the subdivision-view with both points and lines as vertices, will be an octagon; and again check that any two chambers lie in an apartment. $\diamondsuit$

The reader can find more small-dimensional exercises in [**Smi11**, Sec 2.2].

The general building. For $GL_n(q)$ with any n, the easiest approach to the building Δ is as projective space 7.0.1; in particular, we saw that this is parallel with the set-theoretic approach to the apartment Σ in 7.1.1.

However, we can also proceed in parallel with the group-theoretic approach to Σ in 7.1.4; cf. [**Smi11**, Sec 2.2] and its references. This time, we have a correspondence: of k-spaces, with cosets of the parabolic subgroup $P_{\hat{k}}$ stabilizing a k-space in Example 1.3.4.

Indeed we may as well work in the case of general Lie-type G, over a field $\mathbb{F}_q$, where $q = p^a$. In fact, we will restrict to the subcase of G *un*twisted; similar remarks will hold for twisted groups, but at the cost of somewhat more complicated statements.

So we recall various structural features related to parabolics, from earlier Remark 1.3.20: such as the root system Φ, with simple subsystem Π, and Weyl group W; the unipotent subgroup U, Cartan subgroup H, and Borel subgroup given by $B = UH$; and parabolic subgroups $P_J = U_J L_J$ for $J \subseteq \Pi$; along with the monomial subgroup $N := N_G(H)$, where $N/H \cong W$. Then our analogue of the group-theoretic approach to the apartment in 7.1.4 is:

REMARK 7.2.5 (The building for a Lie-type group). We obtain the *building* Δ for the Lie-type group G as the complex given by the cosets of the maximal parabolic subgroups $P_{\hat{i}}$, over $i \in \Pi$; where a simplex of type J is determined by a set of cosets of the $P_{\hat{i}}$ for $i \in J$, which have nonempty intersection.

We recall from Remark 1.3.20(4) that the parabolics have the self-normalizing property $N_G(P_J) = P_J$; hence the permutation representation on their cosets is equivalent to that on their conjugates. So we can equally well view Δ as the simplicial complex defined by *conjugates* of the maximal parabolics: where a simplex is defined by a set of conjugates, of different types—which intersect at least in a Borel subgroup. This conjugate-version can be convenient in some calculations. $\diamondsuit$

EXERCISE 7.2.6. Exhibit Δ for $L_3(2)$ in the coset/conjugate view; and check the isomorphism of this complex with the set-theoretic projective plane in 7.2.1. $\diamondsuit$

Some geometric properties of the building. We'll now indicate some of the fundamental properties of Δ; a number of these had been previewed, for the apartment Σ, in Section 7.1.

The simplicial complex Δ has dimension $|\Pi| - 1$. It is in effect built up from many apartment-spheres Σ of that dimension; we'll say more about this below.

Each simplex has a type $J \subseteq \Pi$; those of maximal and just-submaximal dimensions are called chambers and panels. For G defined over $\mathbb{F}_q$, we have panel stabilizer $P_{\hat{k}} \geq L_k \cong SL_2(q)$; and this group is transitive on the $q+1$ chambers on the panel. Thus Δ is a thick geometry,[1] since $q + 1 > 2$.

Again we have paths and distances between chambers: with path-types still given by words in W; and maximal distance $|\Phi^+|$—exhibited in an apartment Σ.

To continue the theme of geometric relations between W and G: A chamber stabilizer is given by $P_{\emptyset} = B$. The N-orbit of B is a single sphere: namely the complex Σ for W as in 7.1.4, called an *apartment* of the building Δ.

The common stabilizer of the chambers in the apartment is the Cartan subgroup H. One crucial property apartments is the analogue of earlier (7.2.3):

(7.2.7) Any two chambers are contained in some apartment.

[1]Setting $q = 1$ in expressions for Δ can give analogues for the thin geometry of the apartment Σ for W.

In fact the chamber "B" (that is, the chamber stabilized by B) is in $q^{|\Phi^+|}$ apartments; and each such apartment is determined by its unique chamber at the maximal distance $|\Phi^+|$ from B, and the corresponding paths to it from B. In topological language, Δ is a "bouquet" of such apartment-spheres.

The above geometric facts in turn reflect some group-theoretic relations:

We have a double-coset decomposition, written $G = BWB$: where BWB is shorthand for BNB—since $H \leq B$, and $N/H \cong W$. This in turn leads to a canonical form for elements of G, namely the *Bruhat decomposition*. (It is roughly a generalization of the Jordan canonical form from $GL_n(q)$.)

Indeed for parabolics we similarly have a decomposition $P_J = BW_JB$, in terms of the parabolic subgroups W_J of the Weyl group in earlier Remark 7.1.4.

The diagram-inductive ("residual") property. We particularly emphasize the analogue for Δ of the earlier local property 7.1.6 for Σ:

This time in the Dynkin diagram of type A_{n-1} for $GL_n(V)$, we associate dimensions of subspaces of V to nodes of the diagram:

$$\overset{1}{\circ} - \overset{2}{\circ} - \overset{3}{\circ} - \cdots - \overset{n-2}{\circ} - \overset{n-1}{\circ}$$

REMARK 7.2.8 (Residual structures in Δ). And now on covering up the node for the stabilizer $P_{\hat{k}}$ of a k-subspace V_k of V in Example 1.3.4, the remaining subdiagram of type $A_{k-1} \times A_{n-k-1}$ is the diagram for $P_{\hat{k}}/U_{\hat{k}} \cong GL_k \times GL_{n-k}$. Indeed it also describes the geometry of the link in Δ of the vertex V_k—as a join:

$$(\Delta_k\text{=subspaces of } V_k) * (\Delta_{n-k}\text{=superspaces of } V_k = \text{subspaces of } V/V_k).$$

And the GL-factors in $P_{\hat{k}}/U_{\hat{k}}$ act in the natural way on these two terms. $\diamondsuit$

Final remarks: the more abstract theory of buildings. The geometric material emphasized in the section so far has focused on fairly explicit properties of buildings in Lie-type groups. This is the kind of information usually required for problems involving geometric applications of those simple groups.

But the reader should be aware that there is a great deal more material in Tits's general abstract theory [**Tit74**] of buildings; which we will only very briefly mention at this point.

For example, Tits defines buildings as abstract chamber complexes, via axioms based on the complex Σ for a Coxeter group W. See e.g. the discussion in Section 4.1 of [**AB08**] (and compare [**Ron89**, Sec 3.1]).

That general theory has also been applied, in various significant ways; again see [**AB08**] for some of those directions.

For our finite-simple-group purposes here, perhaps the most crucial result is Tits's characterization of finite buildings of rank at least 3; see e.g. [**AB08**, Ch 9] and [**Ron89**, Ch 8]. Very roughly:

THEOREM 7.2.9. *A finite (that is, having "spherical" apartment Σ) thick building of rank at least 3 comes from a Lie-type group over a finite field $\mathbb{F}_q$.*

As we mentioned early in Section 4.2, this relies on Tits's topological result 4.3 in [**Ron89**] that:

(7.2.10) Finite thick buildings of rank ≥ 3 are simply connected.

We recall that for a geometry equipped with a diagram, e.g. in the sense of Remark 7.2.8, the *rank* of the geometry is the number of nodes in the diagram.

We conclude that our approach to finite buildings earlier in this section, namely via parabolic structures in explicit Lie-type groups, is in fact suitably general—in view of Theorem 7.2.9.

7.3. Geometries for sporadic groups

For sporadic geometries, we will mainly follow the viewpoint of "Option S" in the book [**Smi11**]; cf. also Chapter 6 of [**BS08a**].

The success of Tits's theory of buildings ($\sim$ 1965) led to much further use of geometry in group theory. Particularly influential was the viewpoint in Tits's "local approach" paper [**Tit81**]; which roughly emphasized the diagram-inductive property of Remark 7.2.8, over the original approach.

One popular direction was the search for suitably-analogous geometries for *sporadic* groups; this analysis was pioneered by Buekenhout, notably in [**Bue79**]. We provide some context:

We note first that the rank-2 subgeometries in buildings are generalized polygons; very roughly:

REMARK 7.3.1 (Generalized polygons). A generalized n-gon has point-line circuits of length $2n$, but not less. ♢

For example, the projective-plane geometry (namely of type A_2) is a generalized triangle; cf. the pictures in Example 7.1.2. And the polar-space geometry of type C_2 is one example of a generalized quadrangle; cf. the pictures for $Sp_4(2)$ in Exercise 7.2.4. Similarly the building of type G_2 gives a generalized hexagon.

Hence the search for new geometries *other* than buildings suggested the idea of using rank-2 geometries other than generalized polygons.

Buekenhout primarily used the *circle* geometry on a set S: where points and lines are replaced by elements, and element-pairs, from S. He was able to give "diagram geometries", in the spirit of 7.2.8, for many of the sporadic groups. And his work inspired a great deal of further research. The diagram-inductive feature of these sporadic geometries gives partial analogues of various geometric properties of Lie-type groups which we saw in the previous Section 7.2; such as type for simplices, and structure of stabilizers resembling parabolics. However, primarily because of Tits's result 7.2.9 of buildings as coming only from Lie-type groups, we cannot expect any close analogues of defining structures like apartments.

p-local geometries. Nevertheless, the analogy of simplex-stabilizers with parabolic subgroups in Lie-type groups can be further developed: especially if we seek geometries in which simplex stabilizers are p-local subgroups. This is not usually the case in Buekenhout's geometries; but it does hold in the 2-*local geometries* introduced by Ronan-Smith [**RS80**].

For fuller details, see e.g. [**Smi11**, Sec 2.3] (or [**BS08a**, Ch 6]). In this section, we'll just extract a few high points:

The 2-local geometry for M_{24}. Historically the first 2-local geometry discovered was for the Mathieu group M_{24}. The observations arose when Ronan and Smith combined their geometric and group-theoretic viewpoints; mainly based on standard facts such as the following (see e.g. [**Con71**, p 225]):

EXAMPLE 7.3.2 (The 2-local geometry for M_{24}). The group $G = M_{24}$ preserves a Steiner system $\mathcal{S}(5, 8, 24)$: this includes a collection of 759 special 8-subsets (called *octads*) of the 24 permuted points—where any 5-subset of the 24 points lies in exactly one octad. The structure also includes the 8^3-partitions of the 24 points via octads, called *trios*; and the 4^6-partitions where any pair gives an octad, called *sextets*. One can form a simplicial complex Δ on these objects as vertices, using the obvious containment as adjacency; this defines the *2-local geometry for M_{24}*. $\diamondsuit$

Group-theoretically, it turns out that the vertex-stabilizers[2] $P_{\hat{O}}$, $P_{\hat{T}}$, $P_{\hat{S}}$ (of an octad, trio, sextet) are in fact 2-local subgroups, with the structures:

$$2^4 : L_4(2), \quad 2^6 : \big(L_2(2) \times L_3(2) \big), \quad 2^6 : 3Sp_4(2) .$$

One can naively observe that these groups "look like" parabolics P_J, in a Lie-type group over $\mathbb{F}_2$—those have a Levi decomposition $U_J L_J$, in which a 2-group U_J is extended by a Lie-type group L_J over $\mathbb{F}_2$.

Ronan and Smith then observed that the Dynkin diagrams for these local subgroups can be combined [**Smi11**, p97] as sub-diagrams of a larger "Dynkin-like" diagram—in a way consistent with the diagram-inductive spirit of 7.2.8:

$$\overset{O}{\circ}\!=\!\overset{T}{\circ} - \overset{S}{\circ} - \square$$

This picture is reminiscent of the Dynkin diagram of type C_4; except that the right-hand node has been replaced by a new symbol $\square$—roughly indicating we should *not* here expect a vertex, or a corresponding stabilizer-subgroup "$P_{\hat{\square}}$".

To see the consistency with the diagram-inductive feature, note that the subdiagram $\circ - \circ - \square$ for an octad O really does express the geometry of the link of the vertex O: namely there are 15 trios, and 35 sextets, adjacent to O; and their geometry is that of the 15 projective points, and 35 projective lines (but *not* the 15 planes) of the projective 3-space for $P_{\hat{O}}/O_2(P_{\hat{O}}) \cong L_4(2)$. This is called a *truncation* of that projective 3-space; and gives an example of a rank-2 subgeometry which is not a generalized polygon.

Similarly, the subdiagram $\circ \quad \circ - \square$ for a trio T describes its link as the join:

$$(3 \text{ octads in } T) * (7 \text{ sextet-partitions refining } T) .$$

Geometries for other sporadics. This view of M_{24} led to the discovery of 2-local geometries for many other sporadic groups. Indeed in a somewhat formal sense, it can be applied to all the sporadic groups—in [**BS08a**, Ch 6]; but it must be admitted that some of those more formal cases have little real geometric content.

We mention just a few of the cases with substantial geometric structure:

For example, a one-node extension of the diagram for M_{24} leads to the diagram for Co_1; and then a further one-node extension leads to the diagram for the Monster M. These lie in a class sometimes called *tilde geometries* in the literature.

[2]We mention that in the literature, these are usually denoted more simply by P_O etc, corresponding to vertex type; here we have written $P_{\hat{O}}$ etc, for consistency with our conventions in 1.3.20(4), notably as in 1.3.21.

A similar series via one-node extensions proceeds from M_{22} to Co_2 to BM: These geometries use a non-polygon rank-2 geometry called *Petersen geometry*: with vertices given by the 10 vertices and 15 edges of the classical Petersen graph. See also various relevant papers of Ivanov and Shpectorov, e.g. [**IS94**].

For odd p, there are a more limited number of analogous p-local geometries of some interest and complexity.

Some other unusual geometries were discovered in the years fairly soon after 1980. These were typically made for single simple groups of various types, and in a fairly "sporadic" fashion. We'll be mentioning a few of those, as we continue into the applications-portion of this chapter.

Some applications of geometric methods

The geometric theory sketched in the chapter so far has been applied in quite a number of areas. In the remainder of the chapter, we can only choose a few of these to demonstrate.

7.4. Geometry in classification problems

The discovery of new geometries such as those in Section 7.3 suggested the possibility of further classification theorems within geometry itself. Ideally these might characterize collections of geometries, properly including the finite buildings.

Unfortunately, the characterization of buildings is so precise, that it is hard to relax any particular axiom—without letting in a vast and uncontrollable new set of examples. Nonetheless, there have been some partial results.

The class of Tits geometries. One interesting such extension class is usually called *Tits geometries*; this term is now used by many authors, to replace the original term "type M" in Tits's local-approach paper [**Tit81**].

Roughly, these require (cf. [**Smi11**, 2.2.34]) that all rank-2 residues—that is, links—should be generalized polygons as in 7.3.1. Thus a larger-rank geometry built up from these might be thought of as a "generalized polytope"—though this terminology does not actually seem to be used. Tits in fact defined his geometries using a symmetric "Cartan" matrix M; where an (i, j)-entry of value k determines the rank-2 residue of type i, j as a generalized k-gon, in the diagram-inductive spirit of 7.1.6. But his definition did *not* require apartment-type properties, as in his original theory of abstract buildings. (Because of the restriction of rank-2 residues to generalized polygons, the context of Tits geometries in fact excludes most of the geometries for sporadics indicated in Section 7.3.)

We will indicate some unusual non-building geometries which are included among Tits geometries; and even though the groups involved may not be sporadic, we can consider the complexes thus arising roughly as "sporadic geometries":

One such example in fact corresponds to one of the usual finite Dynkin diagrams:

EXAMPLE 7.4.1 (Neumaier's C_3 geometry for A_7). For fuller details see for example [**Smi11**, 2.3.7]; this geometry has the C_3-diagram:

$$\overset{P}{\circ} - \overset{L}{\circ} = \overset{\pi}{\circ}$$

Here P indicates the 7 permuted *points*; L indicates all $\binom{7}{3}(= 35)$ subsets of size 3, as *lines*; and π indicates *planes*: an A_7-orbit (two choices are possible) of size 15—where each plane has all 7 points, plus a subset of 7 lines, together forming a projective plane over $\mathbb{F}_2$ as in Example 7.2.1. ◇

Finite geometries with affine diagrams. There are also various finite non-building Tits geometries, with *affine* diagrams; these were unexpected for *finite* groups—since they correspond to infinite *affine* Weyl groups $\tilde{W}$: Namely each finite Weyl group W has an infinite extension $\tilde{W}$: which is "affine" in the sense that it has an infinite normal subgroup given by the $\mathbb{Z}$-lattice for the root system of W, extended by the finite group W itself. The inifinte group $\tilde{W}$ has affine diagram of type $\tilde{X}$, where X is the type of W. Here $\tilde{X}$ is obtained as a certain one-node extension of X; see e.g. [**GLS98**, p 12] for the list of these extended diagrams. For example, there are geometries for Suz and for $U_4(3)$ with diagram $\tilde{C}_2$ ($\circ = \circ = \circ$), and geometries for Ly and for $G_2(3)$ with diagram $\tilde{G}_2$ ($\circ - \circ \equiv \circ$); see e.g. subsections 3.4, 3.5, and 3.12 in [**RSY90**].

Indeed there was a period in the 1980s when quite a number of such finite geometries with affine diagrams were discovered; see e.g. [**Smi11**, 9.3.9] for an overview of this area.

One avenue for possibly explaining this phenomenon came from studying a finite geometry as the quotient of an infinite affine building, acted on by a corresponding infinite Lie-type group. For the work of Bruhat-Tits showed that an infinite group—of finite type W, but defined over an infinite local field (e.g. $\mathbb{Q}_p$)—has a discrete subgroup which acts on a building of the corresponding infinite affine type $\tilde{W}$—but defined instead over the finite residue field (e.g. $\mathbb{F}_p$).

So, should the finite $\tilde{X}$-examples arises as quotients in this way? This is certainly suggested by certain results of Tits; one important such result, in fact underlying the simple-connectivity result (7.2.10)—see e.g. 7.9 in [**Ron89**]—states roughly that:

THEOREM 7.4.2. *The universal cover of most Tits-geometries having rank* ≥ 3 *are buildings.*

Here "most" refers to a complication when the diagram involves subdiagrams of the types A_3, C_3, or H_3 (dodecahedron); then an extra hypothesis is needed, namely that the covers of such subdiagrams are buildings. (That this is required should be clear from the Neumaier C_3-geometry in Example 7.4.1 above: for that geometry is equal to its own universal cover—without being a building.) A more directly suggestive result of Tits (cf. [**Ron89**, 10.25] states roughly that:

THEOREM 7.4.3. *An affine building of rank* ≥ 4 *arises via a local field, as in the Bruhat-Tits construction.*

This result applies to many of the finite geometries with affine diagrams mentioned above. And indeed Kantor [**Smi11**, 9.3.10] was able to determine, for such cases, exactly which such infinite Lie-type groups and local fields provided the relevant universal cover.

There is further literature in this direction; e.g. [**KLT87**]. The discussion in [**Smi11**, pp 290ff] provides a rough overview.

Some other directions. We briefly mention some other geometric classification results:

Timmesfeld [**Tim83**] and others used related group-theoretic hypotheses, to produce some classifications of finite groups with subgroups defining a Tits-geometry and diagram. During the process of proofs, various subcases (mainly for the small field $\mathbb{F}_2$) led to the discovery of some previously-unknown sporadic geometries. Many of these were later explained via Kantor's affine coverings indicated above.

Onofrei [**Ono11**] extended that group-theoretic approach to the modern topological context of fusion systems.

7.5. Geometry in representation theory

We should perhaps first recall from Section 5.2 that *algebraic* geometry is basic for the Deligne-Lusztig theory of ordinary representations of Lie-type groups. But we won't be pursuing that direction in this chapter, and instead continue our focus on projective geometry.

In overview, we will examine some application-areas for simple groups: first exploiting the "model-case" geometry of the Lie-type building; and thereafter, mentioning some suitable analogues for sporadic geometries.

We first recall some complexes coming from standard posets of p-subgroups:

We saw at Definition 3.3.11 the poset $\mathcal{B}_p(G)$ of p-radical subgroups—namely with the property that $1 < X = O_p(N_G(X))$. Indeed for G of Lie type in characteristic p, we recall from Theorem 5.4.2 that $\mathcal{B}_p(G)$ consists exactly of the unipotent radicals of parabolics. Since the poset of parabolics gives one view of the building Δ (cf. Remark 7.2.5), the poset of unipotent radicals (where inclusions are reversed: i.e., $P_J < P_K$ implies $U_J > U_K$) gives another such view.

Furthermore for general G, we have a standard equivalence (e.g. [**Smi11**, 4.3.4]) with other posets, introduced earlier at (2.4.2) and (2.5.1):

THEOREM 7.5.1. *$\mathcal{B}_p(G)$ is homotopy-equivalent to the poset $\mathcal{S}_p(G)$ (of all nontrivial p-subgroups) and to the poset $\mathcal{A}_p(G)$ (of nontrivial elementary abelian p-subgroups).*

Sometimes sporadic geometries, such as those in Section 7.3, are equivalent to these complexes.

One reason for mentioning these p-subgroup posets here, is that various useful results (especially from the topological literature) are stated for them; some of these will appear as the section proceeds.

The generalized Steinberg module and projectivity. For example: We had mentioned at earlier (2.4.4) the Brown-Quillen result for the reduced Lefschetz module $\tilde{L}$ that:

$\tilde{L}(\mathcal{S}_p(G))$ is a projective—called the generalized Steinberg module.

We will see that this construction does indeed generalize one standard construction of the usual Lie-type Steinberg module 5.2.8:

Constructing the usual Steinberg module. For in the case where G is of Lie type in characteristic p, by means of the equivalence in (2.4.4) of $\mathcal{S}_p(G)$ with $\mathcal{B}_p(G)$, which in turn gives the building Δ by (2.4.4), the module is equivalent to $\tilde{L}(\Delta)$—which is the construction of the Steinberg module given by the Solomon-Tits Theorem [**Smi11**, 3.4.15]. We state this latter result in the form:

THEOREM 7.5.2 (Solomon-Tits Theorem). *For Lie-type G in characteristic p, with building Δ, the alternating sum $\tilde{L}(\Delta)$ gives the Steinberg module.*

We give a quick sketch of the proof; since the underlying "Solomon-Tits argument" has been very influential:

We saw in the geometric properties of Δ discussed after (7.2.7) that a chamber "B" is on $q^{|\Phi^+|}$ apartments Σ; and indeed via (7.2.7) itself that every chamber appears in one of these apartments. Now each such sphere Σ gives, via the alternating sum over its chambers at the various distances, a top-dimensional cycle—and so potentially appears in $\tilde{L}(\Delta)$. And indeed, we will show that these cycles give a basis for that space.

In more detail: Each such Σ determines a sphere in top dimension, and so has 1-dimensional image in the top homology $\tilde{H}_{|\Pi|-1}(\Delta)$. Furthermore they are linearly independent, since each involves a maximal-distance chamber not in any of the others. So since they are $q^{|\Phi^+|}$ in number, they span a subspace of the right dimension for the Steinberg module. Indeed one can use Steinberg's original definition of the Steinberg module to recognize it as this subspace.

So it remains to show that any remaining homology vanishes, in the alternating sum defining $\tilde{L}$. To see this, we form Δ^-, by removing the $q^{|\Phi^+|}$ chambers at maximal distance from B; and we will show that Δ^- is contractible (and hence makes no contribution to homology). For this, we use the "gate" property of buildings:

Each panel has a *unique* chamber closest to B.

Thus the panels π farthest from B lie on just one chamber c—since in Δ^-, we had already removed all remaining chambers on π, which lie at the maximal distance. It is standard in topology that this condition means that we can homotopically "collapse" that chamber c: from π, down to the rest of the boundary ∂c. The effect of these removals is to collapse Δ^- down to Δ^{--}, in which all the chambers at the *two* largest distances from B have now been removed. However: we can iterate this collapsing-argument, for the panels of Δ^{--} at maximal distance; and so on—eventually contracting down to the single chamber B, and so to a point. □

EXERCISE 7.5.3. Use the $L_3(2)$-graph of Exercise 7.2.2, say with initial chamber "B" given by the edge $(a \subset \langle a, b \rangle)$, to verify the contractibility of Δ^-—as in the Solomon-Tits argument just sketched. Then the 8 apartments on B give, essentially via their 8 corresponding chambers at maximal distance from B, an $\mathbb{F}_2$-basis for the 8-dimensional Steinberg-module.

Similarly explore the $Sp_4(2)$-graph of Exercise 7.2.4, to give a construction of the 16-dimensional Steinberg module. ◇

Of course it is a feature of the actual Lie-type Steinberg module 5.2.8 that it is not just projective, but also irreducible.

For more general G, we cannot expect that the generalized Steinberg module $\tilde{L}(\mathcal{S}_p(G))$ should also be irreducible (as well as projective). But there are at least some restrictions on the irreducibles I, such that the projective cover $P(I)$ appears in $\tilde{L}(\mathcal{S}_p(G))$: typically $\dim I \geq |G|_p$—for details see [**Smi11**, p 213].

Some other results related to projective modules. Soon influential work of Webb extended the Brown-Quillen projectivity result (2.4.4) to further more general G-complexes Δ; by first extracting from their proof a sufficient condition (e.g. 4.3.4 in [**Smi11**]):

(7.5.4) If Δ^P is contractible $\forall$ p-groups $P > 1$ then $\tilde{L}(\Delta)$ is projective.

We mention that this condition is often close to showing that Δ is homotopy-equivalent with $\mathcal{S}_p(G)$—see e.g. [**Smi11**, 4.4.12]; but the projectivity result also holds for many non-equivalent Δ.

Projectivity was verified via (7.5.4) for many sporadic G and their p-local geometries Δ, in Ryba-Smith-Yoshiara [**RSY90**]—including for example:

- the 2-local geometry for M_{24} in Example 7.3.2, which turns out to be equivalent to $\mathcal{S}_2(M_{24})$; and also
- the C_3-geometry for A_7 in Example 7.4.1, which is *not* equivalent to $\mathcal{S}_2(A_7)$.

For some geometries Δ which fail the projectivity condition in (7.5.4), it turns out that $\tilde{L}(\Delta)$ satisfies the weaker condition of *relative* projectivity, with respect to fairly small p-subgroups P; by contrast, $P = 1$ would give full projectivity. There is some literature in this direction; see especially the recent fairly general treatment of Maginnis-Onofrei [**MO09**]. Chapter 6 of [**Smi11**] also surveys further literature on projectivity.

Finally: for a general finite group G, the generalized Steinberg module, and the underlying *Steinberg complex* (an associated chain complex developed by Webb), are used in topological contexts; see e.g. Grodal [**Gro02**].

Irreducible modules and coefficient systems. Again we start with the model-case of G of Lie type in characteristic p; consider G acting on some module V in that natural characteristic p. Earlier we had mentioned Theorem 5.2.11, namely that for V irreducible, and $P_J = U_J L_J$ a parabolic:

$$V^{U_J} \text{ is also irreducible under } L_J.$$

Application at a single parabolic P_J. When 5.2.11 above is applied for a single parabolic P_J, the result has various uses; especially in the modular representation theory of Lie-type groups and algebraic groups. See for example Kleshchev [**Kle97**] on branching rules for GL_n and S_n—namely the decomposition rules, when irreducibles are restricted to subgroups. But we also mention some "adjacent" areas of application:

For maximal subgroups: see e.g. Liebeck-Saxl-Seitz [**LSS87**], for application to the inclusion-setup of irreducible $X < Y$ in classical G on natural V, which we had mentioned at (6.2.5).

For p-compact groups (an analogue of compact Lie groups): see e.g. the work of Andersen-Grodal-Møller-Viruel [**AGMV08**], for an application to the Steinberg module V, in order to analyze its restriction to elementary abelian p-subgroups.

Application at all J: coefficient systems. Now consider applying Theorem 5.2.11 for all proper subsets $J \subset \Pi$ simultaneously; we get:

REMARK 7.5.5 (The irreducible presheaf on V). The maps $P_J \mapsto V^{U_J}$ define a *coefficient system* (or *presheaf*) on the building Δ. This is irreducible in the sense of coefficient systems, by applying Theorem 5.2.11 at each J. ◇

The presheaf-viewpoint has also had various applications; for background and further development, see [**Smi11**, Ch 10]. Here we mention just a few directions:

Ronan-Smith [**RS85**] observed that there is a 1:1 correspondence between irreducible modules V, and irreducible presheaves $\{V^{U_J} : J \subseteq \Pi\}$. This for example gives an approach to constructing the irreducibles, via the homology of presheaves.

A different application is to *embedding* a geometry Δ in a vector space V: Here the idea is that the "points and lines" of Δ should be consistently mapped to projective points and lines, in the projective space of V; or in the language of linear dimensions, to 1-subspace and 2-subspaces in V. There is a considerable geometric literature on such embeddings. For example, in the case of Lie-type G, embeddings of generalized hexagons are considered by Cooperstein in [**Coo01**]. For sporadic G, consider the 2-local geometry Δ for M_{24} in Example 7.3.2: one can check that the subgroups $P_{\hat{O}}$, $P_{\hat{T}}$, $P_{\hat{S}}$ there fix subspaces of dimensions 1,2,4—in the 11-dimensional irreducible Golay-code module V over $\mathbb{F}_2$. It follows that for points and lines given by octads and trios, the geometry Δ is embedded in V.

EXERCISE 7.5.6. Show that the C_3-geometry for A_7 in Example 7.4.1 is not embeddable—in any vector space V over $\mathbb{F}_2$.

Hint: Recall that lines are given by *all* 3-subsets of the 7 points. Thus the relation among the three nonzero vectors of a 2-space over $\mathbb{F}_2$ shows that all 3-set sums are 0. Finally show that this forces all point-vectors to be 0. ◇

For various further results on embeddings, see e.g. [**Smi11**, pp320ff].

We mention that the coefficient-system viewpoint is also used in Grodal's approach [**Gro02**] to higher limits in topology.

7.6. Geometry applied for local decompositions

For fuller detail on this topic, see e.g. [**Smi11**, Ch 3] or [**BS08a**, Ch 5].

Decompositions of group cohomology. Webb observed [**Smi11**, 7.2.5] that in the above projective situation, applying Ext*-functors and Frobenius reciprocity gives a decomposition of group cohomology, in terms of cohomology of the stabilizers:

$$\text{(7.6.1)} \qquad \text{Under (7.5.4)}, \ H^*(G)_p \ = \ \bigoplus_{\sigma \in \Delta/G} (-1)^{\dim \sigma} \ H^*(G_\sigma)_p \ .$$

This formula (7.6.1) has sometimes been used for the explicit computation of cohomology; for example, Adem-Maginnis-Milgram [**AMM91**] used the 2-local geometry for M_{12} as Δ.

But probably the major influence of (7.6.1) was to stimulate, around the early 1990s, a research direction in algebraic topology—which "explained" the cohomology decomposition, in terms of an underlying decomposition of the p-completed classifying space $BG_p^\wedge$ of G. The topological decompositions are phrased in the language of *homotopy colimits*; and they are typically indexed via standard subgroup posets like $\mathcal{S}_p(G)$, so that they apply to any finite group G. Such work of Jackowski, McClure, Oliver, Dwyer, Grodal and others is summarized (from a fairly group-theoretic viewpoint) in Chapter 5 of [**BS08a**]. We mention that in addition to the usual p-subgroup posets like $\mathcal{S}_p(G)$ mentioned above, variants using the p-centric subgroups (which we had indicated in the discussion leading up to Lemma 3.6.4) also play a prominent role in this area.

The above topological approach was applied methodically to sporadic simple groups G, in Benson-Smith [**BS08a**, Ch 7]: For each such G, they first show that some suitable 2-local geometry Δ for G is homotopy-equivalent to one of the standard p-subgroup posets such as $\mathcal{S}_p(G)$. Then applying the above results to the latter, they obtain a decomposition of the cohomology (and indeed of the classifying space) for G—in terms of that of the simplex stabilizers G_σ for $\sigma \in \Delta$.

Geometric decompositions for the Alperin Conjecture. We also mention some geometric approaches to the Alperin Weight Conjecture, which we had introduced earlier as Conjecture 5.4.3. For further detail on this topic, see e.g. [**Smi11**, Ch 13].

We saw in the discussion before 5.4.3 that the sum of the Alperin-weights in the AWC is indexed in effect by the p-radical poset $\mathcal{B}_p(G)$ of 3.3.11. Influential work of Knörr and Robinson [**KR89**] produces an equivalent statement of the conjecture—now indexed by an equivalent complex Δ (see e.g. [**Smi11**, 4.6.2]): namely chains of p-subgroups, each of which is normal in the last term of the chain.

In fact, they consider the AWC partitioned over the various p-blocks B of the group algebra. For B of defect 0 in the sense of 5.0.2, the unique projective irreducible in B automatically supplies the requirement of their alternative version of the conjecture. Thus they reduce to blocks B of positive defect: $\operatorname{def}(B) > 0$. Their version of the AWC requires vanishing of:

$$\sum_{c \in \Delta/G} (-1)^{\dim c} \, |\mathrm{Irr}(B_c)| \stackrel{?}{=} 0; \tag{7.6.2}$$

where B_c is a block of the stabilizer G_c, which lifts to B in the standard Brauer correspondence.

This alternating-sum formula should seem reminiscent of the definition of a reduced Lefschetz module, such as that appearing in (7.5.4). And indeed they show that (7.6.2) is in fact the degree-term of the fixed-points $\tilde{L}(B)^G$, in their *Lefschetz conjugation module* $\tilde{L}(B)$: given by the alternating sum of the induced modules $\mathrm{Ind}_{G_c}^G(B_c)$—where G acts by *conjugation*, rather than the more usual right-multiplication convention for module action. And paralleling (7.5.4), they show in fact that $\tilde{L}(B)$ is virtually projective.

This suggests stating a version of the AWC in terms of the *module* cohomology of the block B: namely that $H^*(G, B)$ should decompose via the terms $H^*(G_c, B_c)$ at the stabilizers G_c. In fact this does hold for positive dimension $H^{>0}$; but the statement for degree-0 is perhaps just as hard as the AWC itself. For some recent promising developments related to this approach, see the talk at URL:

`www.math.uic.edu/~smiths/talkl.pdf`

CHAPTER 8

Some fusion techniques for classification problems

In this chapter, we deviate somewhat from our main theme of applications of the CFSG; and instead consider some influential early results related to 2-fusion. These techniques quickly came to be regarded as fundamental; and they have been very frequently applied—not only throughout the CFSG, but in many other classification-type problems over the subsequent years.

In the first few sections of the chapter, we will overview some of those applications; and suggest some similar possible uses in other types of problems. In the final sections of the chapter, we discuss the possibility of extending the 2-local results to suitable analogues for odd primes p.

To begin, recall that at (3.5.1) we had introduced the notion of 2-fusion: that is, for T a Sylow 2-subgroup of a finite group G, we study the pattern of G-conjugacies among elements of T. We will be examining some fundamental early results on this topic. A good source for much of this material is [**GLS96**, Sec 15–17], from which we will quote frequently.

8.1. Glauberman's Z^*-theorem

This result is often applied very early on in a classification problem: to show that there must be *some* fusion, namely conjugacy among involutions in T; which provides the foundation on which the main argument can then be built.

The Z^*-Theorem. Glauberman's result [**ALSS11**, B.2.1] deals essentially with an involution weakly closed in T (indeed in $C_T(z)$); it can be simply stated for general H in the form:

THEOREM 8.1.1 (Glauberman Z^*-Theorem). *If an involution z of H commutes with no distinct conjugate of itself, then z lies in $Z^*(H)$—the preimage in H of $Z\big(H/O_{2'}(H)\big)$.*

So if H is simple, then z is *conjugate to some other $z^h = t \in C_T(z)$.*

The result is probably used most often in the form of the second statement. Note that initially, by simplicity we can quote the contrapositive of the initial statement—giving us just $z^h \in C_H(z)$; but we may as well choose z *extremal*, namely so that $C_T(z)$ in Sylow in $C_H(z)$—and then apply conjugacy of Sylows of $C_H(z)$.

We mention that Glauberman's proof in [**Gla66**] used 2-modular representation theory, and later analysis reduced this to ordinary representation theory. The recent approach of Waldecker [**Wal13**] instead emphasizes local group theory; but roughly assumes knowledge of possibilities for involution centralizers using the CFSG.

In the remainder of the section, we will be discussing some applications, of types which have proved fairly significant.

As context, we might ask something like: In a problem which reduces to a simple counterexample G, how does the above information, namely having $z \neq z^g \in C_T(z)$, actually help?

Of course, the answer depends on the hypotheses of the particular problem. Here is one of my favorite such situations:

An application to large extraspecial subgroups. Choosing this topic will allow us to say a little more about a situation which we had only briefly summarized in our CFSG outline, namely the treatment of the $GF(2)$-type branch (3) in the Trichotomy Theorem 2.2.8:

Large extraspecial 2-subgroups, in the context of $GF(2)$ type. Note that in the Lie-type groups defined over the smallest field $\mathbb{F}_2$, the root subgroups are of order only 2. This leads to certain restrictions on the structure of unipotent radicals, especially in involution centralizers. In particular, in the Lie types with Dynkin diagrams having only single bonds (namely A, D, E), below we abstract certain features of the centralizer of an involution z generating a root-group; these are visible in particular in $GL_n(2)$, in our earlier Even Case Example 2.0.6:

DEFINITION 8.1.2 ($GF(2)$ type). We say that G is of *GF(2) type* if for an involution z, with centralizer $M := C_G(z)$, we have $F^*(M) = O_2(M) =: Q$; this condition says Q is *large* in G; and furthermore $Q := O_2(M)$ is of *symplectic type*—this means that every elementary abelian subgroup which is characteristic in Q must in fact be cyclic. ♢

The symplectic-type condition arises here, because the only characteristic abelian subgroup of Q so arising in the Lie-type examples is a root group—which is cyclic since that root group just has prime order 2. In fact in those Lie-type cases, the group Q is even *extraspecial*: namely we have $Q' = \Phi(Q) = Z(Q) = \langle z \rangle$ of order 2. Thus those examples in fact satisfy the slightly sharper condition of:

DEFINITION 8.1.3 (large extraspecial 2-subgroup). This is the subcase of Definition 8.1.2 in which Q is extraspecial. ♢

EXERCISE 8.1.4 (More large-extraspecial cases). Check Q is large-extraspecial, in suitable Lie-type groups over the small field $\mathbb{F}_2$—beyond linear $GL_n(2)$ in 2.0.6.

Hint: Here "suitable" means that the Dynkin diagram should be a single-bond type—A,D,E; i.e. the classical linear, unitary, and orthogonal types. The exceptional cases E_6, E_7, E_8—and in fact also G_2—also work; but those require more detailed root-system knowledge. The same holds for the single-bond twisted types ${}^2D_n$, ${}^3D_4$, ${}^2E_6$.

Type A_3, for the group $L_4(2)$, is explored in Remark B.2.1: see the discussion of the unipotent radical U_2 there. More generally, in the indicated Lie types, z will generate a long-root subgroup U_α for the highest root α; $C_G(z)$ will be, for the appropriate $J \subseteq \Pi$, the parabolic[1] P_J which stabilizes that long-root group; and Q will be the unipotent radical U_J of P_J. Now U_J is defined as in Remark 1.3.20(4) via roots which are not combinations from J; and in these cases for J, we check that these roots consist of α, together with various pairs $\gamma, \alpha - \gamma$, having the property

[1]Indeed a maximal parabolic, except in type A_n.

that no other pair among this set sums to a root in Φ^+. Then (1.3.13) leads to the conclusion that Q is extraspecial. Some details for the cases of G_2 and E_6 appear in appendix Remark B.4.3. ◇

In fact, Aschbacher in [**Asc76**] (cf. [**ALSS11**, 7.1.1]) reduced the $GF(2)$ type problem to the large-extraspecial problem; by showing that the groups which have Q of symplectic type, but not extraspecial, are;

$$L_2(2^m \pm 1),\ M_{11},\ L_3(3),\ U_3(3), \text{ and } HS.$$

Beyond the indicated Lie-type groups over $\mathbb{F}_2$, the large-extraspecial situation was notorious for also containing the majority of the sporadic groups. We mention one example: This will reflect the connection of the $GF(2)$ type condition with the Klinger-Mason analysis of characteristic $\{2, p\}$ type—an analysis which we had mentioned briefly in our discussion leading up to the Trichotomy Theorem 2.2.8. Namely the Harada-Norton group HN has characteristic $\{2, 5\}$ type; with $2^{1+8}(A_5 \times A_5)2$ for its involution-centralizer structure—where $Q := 2^{1+8}$ gives $GF(2)$ type—and indeed large extraspecial. (See e.g. Franchi-Mainardis-Solomon [**FMS08**].)

Before proceeding with our discussion of the large-extraspecial situation, we digress briefly to indicate the generalization of $GF(2)$ type to $GF(2^n)$ type, which we had mentioned only briefly, after the Trichotomy Theorem 2.2.8; here, the analogue of the root group $\langle z \rangle$ is a root group B—which can now be of any order 2^n:

DEFINITION 8.1.5 ($GF(2^n)$ type). We say G is of $GF(2^n)$ *type* if we have the following generalization of the $GF(2)$-type condition of Definition 8.1.2: We have a 2-subgroup Q, again with the "large" restriction that $M := N_G(Q)$ satisfies $F^*(M) = O_2(M)$; but now Q is "semi-symplectic": roughly, this means that $Q' = \Phi(Q) = Z(Q) = B$—which arises from the strong local condition that a maximal normal elementary abelian 2-subgroup B of M is a TI-set[2] in G. ◇

Applying the Z^*-Theorem in large extraspecial subgroups. For fuller reference on the following material, see e.g. [**ALSS11**, Ch 7; esp 7.0.5].

We assume the large-extraspecial situation: Thus we have G simple, with involution z, such that: setting $M := C_G(z)$, and $Q := O_2(M)$, Q is extraspecial (that is, $Q' = \Phi(Q) = Z(Q) = \langle z \rangle$ is of order 2), and large in the above sense that $F^*(M) = O_2(M) = Q$.

On applying the Z^*-Theorem 8.1.1, we get some further $a := z^g \in M$. How can we use this? Here is a rough sketch:

First, we may take $a \in Q$: For Aschbacher showed (e.g. [**ALSS11**, 7.2.3]) that otherwise, G is either a unitary group over $\mathbb{F}_2$, or Co_2.

Now set $\tilde{Q} := Q/\langle z \rangle$—since Q is extraspecial, this quotient is elementary abelian; and we can study the action of $\overline{M} := M/Q$ on $\tilde{Q}$, particularly with respect to the $\overline{M}$-conjugates of $\tilde{a}$. So we set $Q_a := Q^g$; it is advantageous that most of this must lie in M. Indeed the group $A := Q \cap Q_a$ is elementary abelian (since $A' \leq \langle z \rangle \cap \langle a \rangle = 1$); further $L := Q(Q_a \cap M)$ has quotient $\overline{L}$ which is elementary abelian, and normal in $C_{\overline{M}}(\tilde{a})$. This provides a great deal of structure, which can be exploited in various ways.

[2]Recall this means: having trivial intersection with its distinct conjugates

EXERCISE 8.1.6. Exhibit these structures, in the case $G = L_5(2)$, where we have $C_G(z) \cong 2^{1+6}L_3(2)$. ♢

As a very rapid summary of the subsequent solution of the large-extraspecial problem: Timmesfeld (e.g. [**ALSS11**, 7.3.1]) determined a list of cases for $\overline{M}$; and then Smith and others (cf. the summary at [**ALSS11**, 7.0.1]) verified that these cases lead to the expected sporadic and Lie-type groups G.

An application to the Sylow 2-subgroup of $U_3(4)$. In our discussion leading up to the Dichotomy Theorem 2.0.9, we had briefly mentioned Lyons among the authors who handled the "small" subcase $m_2(G) \leq 2$ of the Odd Case. More precisely, Lyons in [**Lyo72**] characterized $U_3(4)$ by its Sylow 2-group T. This is in effect an internal-hypothesis recognition theorem, in the language of the introductory section of Chapter 4.

The proof begins with a discussion of fusion of involutions: The only involutions of T are those of $Z(T)$ of rank 2; denote them by z_1, z_2, z_3. Lyons in [**Lyo72**, Lm 1] applies the Z^*-Theorem in some unknown simple G with Sylow T—to see that z_1 must be conjugate to another involution of T. Now the only other choices for an involution are z_2, z_3; so we may as well assume that z_1 is conjugate to z_2. But we may equally well apply the Z^*-theorem to z_3, to see that it is conjugate to z_1 or z_2. Hence all 3 involutions z_i of T are in fact conjugate in G.

Here is one significant structural consequence of that conjugacy: We recall Burnside's Fusion Theorem (cf. [**GLS96**, 16.2]):

THEOREM 8.1.7 (Burnside's Fusion Theorem). *For W weakly closed in T with respect to G, G-fusion in $Z(W)$ is induced in $N_G(W)$.*

Applying this in the case $W = T$, the above conjugacy inside $Z(T)$ must be induced by $N_G(T)$. So $N_G(T)$ contains a 3-element; and hence induces the full permutation group S_3 on the three involutions z_i. Thus we have found that one 2-local subgroup is as in the target group $U_3(4)$, inside our still-unknown G.

Of course, considerable further work remains for the characterization, and we won't here describe those technical details; as before, we are primarily emphasizing that the fusion information is what initially gets the proof off the ground.

An application to semi-dihedral and wreathed Sylow groups. We continue to expand, as above, on our brief mention before Theorem 2.0.9 of the Small Odd Subcase $m_2(G) \leq 2$. For full details and definitions, see e.g. [**ALSS11**, Sec 1.4].

In the work of Alperin-Brauer-Gorenstein [**ABG73**], their intermediate analysis includes treatment of certain non-simple groups H, containing a Sylow 2-subgroup T which is semidihedral or wreathed; these H are there called "Q-groups", and I won't reproduce that technical definition here. Inside such groups, they apply the Z^*-Theorem in the positive direction, rather than the contrapositive as in the cases just above: that is, they use the first sentence in our statement of Theorem 8.1.1:

Namely for such a Sylow group T, the center $Z(T)$ is cyclic. And then using (ii) in [**ABG73**, Props 1.1,1.2], subgroups of $Z(T)$ are weakly closed in T (with respect to such groups H). They show at their result 3.1 that this leads to:

(8.1.8) $$H = O_{2'}(H)C_H\big(Z(T)\big).$$

We quickly sketch the deduction: By induction, we may as well assume $O_{2'}(H) = 1$; so in effect, we really need to show then that $Z(T) \leq Z(H)$. If this fails, we may assume by way of contradiction that $Z_0 := Z(T) \cap Z(H)$ is proper in $Z(T)$. So we may take some $Z \leq Z(T)$ with Z_0 of index 2 in Z; hence in $\overline{H} := H/Z_0$, we see $\overline{Z}$ is generated by an involution $\overline{z}$. From the subgroup weak-closure property (ii) indicated above, we get the analogue for the element $\overline{z}$: namely $\overline{z}$ is weakly closed in $\overline{T}$. (In particular using Sylow's Theorem, $\overline{z}$ cannot commute with any distinct conjugate.) So the "forward" direction of the Z^*-theorem forces $\overline{Z} \leq Z(\overline{H})$. But then standard arguments lead to $Z \leq Z(H)$—contrary to the definition of the proper subgroup Z_0 of Z as $Z(T) \cap Z(H)$. □

And now as before, the structural information (8.1.8) produced by the initial fusion analysis then sets up their main argument characterizing the groups H; and eventually leads to the simple G with such a Sylow group T.

8.2. The Thompson Transfer Theorem

In contrast to the Glauberman Z^*-Theorem, which is typically applied at an early stage in arguments, the Thompson Transfer Theorem is more usually applied toward the end of arguments: to show that a "shadow" configuration is *not* simple—or more precisely, that it is not perfect. But there is still an analogy with the Z^*-Theorem, in that the contrapositive form *can* be used to force the existence of some involution-fusion in a simple group G: this time, into a subgroup of index 2 in T.

We state the most elementary form of the result (for example [**ALSS11**, B.2.9] or [**GLS96**, 15.16]), since this case is the situation most common in applications:

THEOREM 8.2.1 (Thompson Transfer Theorem). *If an involution t of a finite group H has no H-conjugate within a subgroup T_0 of index 2 in a Sylow group T, then $t \notin O^2(H)$. (In which case $H > O^2(H)$, and in particular H is not simple.)*

For example, notice this holds in the case of a transposition t in the non-perfect group $H = S_n$.

We mention that the proof is comparatively elementary: just compute directly the homological "transfer"[3] homomorphism—in effect, the natural map of H/H' into T/T'. See also "control of transfer", e.g. [**ALSS11**, p 271].

This result was widely applied in the CFSG; and a number of such applications are indicated throughout the exposition of [**ALSS11**]; see e.g. the Index in that work. As before, below we have selected some representative applications:

Some applications to quasithin groups. There are more than 30 such quasithin applications to choose from. Here is a fairly typical one:

Eliminating a shadow related to $U_4(3)$. We start with the observation that the quasithin group $U_4(3)$ has a 2-local subgroup $2^4 : A_6$; but not $2^4 : S_6$.

So how might we eliminate, in some unknown quasithin simple G, that very similar but slightly larger 2-local configuration $L := 2^4 : S_6$? This 2-local subgroup *does* occur in the non-simple "shadow" $H := U_4(3)\langle t \rangle$—where t induces an orthogonal reflection (denoted 2_2 in the Atlas [**CCN**$^+$**85**]) on $\Omega_6^-(3) \cong U_4(3)$. So ideally,

[3]See the supplementary notes in appendix Section A.2.

we should use Thompson Transfer to force our unknown G to be non-perfect, as is the case for H. And indeed this is implemented at [**AS04b**, 13.5.16]; in overview:

Some preliminary results show much of the local structure of G is just as it *should* be in H: For $z \in O_2(L)$, set $M := C_G(z)$ and $Q := O_2(M)$; then we find that $Q \cong 2^{1+4}\langle t \rangle$, with $M = 2^{1+4}(3^2.4)\langle t \rangle$. Also $Z(Q) = \langle z, t \rangle$, and $N_G(Q) = M$; while for T Sylow in H, z is conjugate to all the involutions in $T_0 := T \cap L = 2^{1+4}.4$—indeed, these involutions even lie in 2^{1+4}.

EXERCISE 8.2.2. Check some of these facts explicitly, in the indicated non-simple group $H = U_4(3)\langle t \rangle$.

Hint: Much of this can be obtained from the Atlas [**CCN⁺85**, p 52]. The conjugacy of z with some $a \in 2^{1+4}$ follows as in the discussion of "z, a" in the large-extraspecial applications in earlier Section 8.1. ◇

Furthermore, we get that Q is weakly closed in M, and hence in T, with respect to G. So by the Burnside Fusion Theorem 8.1.7, G-fusion in $Z(Q)$ is induced in $N_G(Q) = M$. Now consider the above involution $t \in T \setminus T_0$; we saw there that $t, z \in Z(Q)$. Clearly $M = C_G(z)$ cannot conjugate z to t. Hence by the fusion-control in the Burnside statement, t cannot even be G-conjugate to z; and so cannot be G-conjugate to to any involution $a \in T_0$, since we saw that these *are* conjugate to z. Consequently we may apply the Thompson Transfer Theorem 8.2.1, to obtain $t \notin O^2(G)$—so that G is not simple. □

We mention also that a subcase of this shadow-configuration H, in which t is not an involution, but instead of order given by a 2-power at least 4, is also handled around [**AS04b**, 13.5.16]—using a more general version of Thompson Transfer.

Eliminating shadows in the quasithin $C(G,T)$-Theorem. Important applications of Thompson Transfer also arose in the proof of 2.1.1 of [**AS04b**]: the analogue for quasithin groups of the Global $C(G,T)$-Theorem 3.3.8: recall this deals with the case where T lies in a unique maximal 2-local subgroup M of G.

For example, we wish to eliminate the shadow of non-simple $H = L_3(2^n)\langle x \rangle$, where x denotes a graph automorphism. We sketch the procedure followed in Section 2.4 of [**AS04b**]:

We have a local P resembling a parabolic of $L_3(2^n)$, given by $L_2(2^n)$ acting on its natural module $N := O_2(P)$. So set $R := NN^x$; then $R\langle x \rangle$ essentially gives T, and $N_G(T)$ gives the unique maximal 2-local M over T. Much further work then leads to the fusion result [**AS04b**, 2.4.21.2]:

$$\text{For } i \text{ any involution of } R, \text{ we get } i^G \cap T \subseteq R.$$

EXERCISE 8.2.3. Check this holds, in explicit non-simple $H = L_3(2^n)\langle x \rangle$. For example when $n = 1$, the group R is dihedral of order 8; and its involutions lie inside its two 4-subgroups, which are the unipotent radicals $U_{\hat{1}}$ and $U_{\hat{2}}$ as in 1.3.4—which appear as N and N^x above. ◇

Thus x is not G-fused to any $i \in R$. And here R is of index 2 in T; so using Thompson Transfer 8.2.1, we get $x \notin O^2(G)$ at [**AS04b**, 2.4.22.2]. □

We mention also a related shadow: described as for H above—but where x instead involves not a graph- but a field-automorphism. This is also eliminated by a Thompson Transfer argument, after [**AS04b**, 2.4.24]. (We note that the wording "after 2.4.24" above corrects the inadvertent mis-statement "in 2.4.24" during the discussion at [**ALSS11**, p 99].)

An application to connectivity of the graph on 4-groups. In our discussion of the Dichotomy Theorem 2.0.9, we saw that the proof actually involved a trichotomy: where two cases were determined by whether or not connectivity holds, for the graph on the 2-groups of rank 3, with an edge determined by a common rank-2 subgroup. It's essentially equivalent to consider the *dual* graph: on rank-2 subgroups as vertices, with edges determined by a common overgroup of rank 3.

Some technical details related to connectivity in this latter viewpoint are visible in the more extended discussion in the outline volume [**ALSS11**]. For example, the discussion after 1.5.1 and the result B.4.10 there in fact make use of the technical result [**ALSS11**, B.4.9]; and we now sketch some details of applying Thompson Transfer in B.4.9:

That result assumes that $G = O^2(G)$ and $m_2(G) \geq 3$; and shows that if V is a 4-subgroup of T satisfying $m_2\big(C_T(V)\big) = 2$ ("isolated " in T), then we can find some conjugate $V^g \leq T$ which satisfies the stronger rank condition $m_2\big(C_T(V^g)\big) \geq 3$—and in particular, is not isolated.

The proof proceeds roughly as follows: Take a 4-group A normal in T. Using $m_2(G) \geq 3$, we get that $T_0 := C_T(A)$ is of index 2 in T, with $m_2(T_0) \geq 3$. We then show that such a V is of the form $\langle z, v\rangle$, where z is the *unique* involution in $R := C_{T_0}(v)$. We also get $V < C_T(V) = \langle v\rangle \times R$; and it follows that z is a square in R. Since we are assuming that $G = O^2(G)$, the Thompson Transfer Theorem 8.2.1 gives us some conjugate some $v^g \in T_0$; indeed we may assume that this conjugate is extremal: namely $C_T(v)^g$ is Sylow in $C_G(v)$, so that we may take $C_T(v)^g \leq T$. Then z^g is a square in R^g, and $R^g = C_{T_0}(v)^g \leq T$ by extremality; so that $z^g \in T_0$, since T_0 has index 2 in T. We get $\langle z^g, v^g\rangle = V^g \leq T_0 = C_T(A)$. We conclude that $m_2\big(C_T(V^g)\big) \geq 3$, as desired—either via V^gA if $V^g \neq A$; or if $V^g = A$, via $C_T(V^g) = C_T(A) = T_0$ having rank ≥ 3 at the start of the proof. □

8.3. The Bender-Suzuki Strongly Embedded Theorem

Recall we had stated this result earlier as Theorem 2.0.17, during our sketch of the proof of the Dichotomy Theorem 2.0.9. In particular, the simple groups that occur in the conclusion are the rank-1 Lie-type groups $L_2(2^n)$, $U_3(2^n)$, $Sz(2^{\text{odd}})$. And in Remark 1.3.20(6), this value of the rank means that there is a unique maximal 2-local subgroup M over a Sylow 2-group T.

We will use the relation of this subgroup M to the 2-fusion, to view the Theorem within the fusion-context of this chapter.

The strongly embedded condition and control of fusion. A recurring theme in the group theory literature has been the notion of "control of fusion": meaning results showing that G-fusion is induced by one or more local subgroups.

In this direction, the most fundamental result is the Alperin Fusion Theorem 3.5.3; recall this asserts that fusion in T can be accomplished by a *sequence* of "local fusions"—each of these being a conjugation in $N_G(T_i)$, for suitable subgroups T_i of T.

Strong control of fusion. In this section, however, we focus on the restricted situation, where fusion is controlled by a *single* local subgroup. Notice this single-subgroup control is in the spirit of the Burnside Fusion Theorem 8.1.7. We now recall the relevant standard terminology:

DEFINITION 8.3.1 (strong control of fusion). We say a subgroup $M < G$ *strongly controls* p-fusion in G, if any G-conjugacy in T is induced in M. That is, if we are given $A, B \subseteq T$ with $A^g = B$, then we may write $g = cm$, with $c \in C_G(A)$ and $m \in M$. ♢

The relationship with the strongly-embedded condition. Our earlier statement of the Strongly Embedded Theorem 2.0.17 used the fairly standard version (2.0.16) of the strongly embedded condition, namely:

For $g \in G \setminus M$, $M \cap M^g$ has odd order.

But we observed there that the condition in fact arises in the particular situation that was relevant at that point in our discussion, namely: The graph on four-groups, via containment in rank-3 groups, is disconnected. In fact the derivation of strongly embedded proceeds via reduction from there to a form defined via a different graph:

The commuting graph on involutions is disconnected.

For notice in this latter situation that:

(8.3.2) strong embedding variant: If $1 < S \leq T$, then $N_G(S) \leq M < G$.

Here M is the stabilizer of a connected component. And this connectivity-related variant of strongly embedded is most directly relevant to our earlier discussion of the Dichotomy Theorem 2.0.9.

But we now proceed a little farther: We see that for the subgroups T_i of a conjugation family in the Alperin Fusion Theorem 3.5.3, the terms $n_i \in N_G(T_i)$ in a local conjugation must in fact lie in M; hence the same holds for the product of the sequence, giving the overall conjugation. Since furthermore any $C_G(A)$ in Definition 8.3.1 lies in M, we conclude that M strongly controls fusion.

We mention that various other equivalent versions of the strongly embedded condition are used in the literature; see for example [**ALSS11**, B.4.7] for some of these further variants.

EXERCISE 8.3.3. Exhibit the strongly embedded subgroup M, in the rank-1 conclusion groups such as $L_2(2^n)$ above; and check that it satisfies some of the basic variant versions of strongly embedded indicated above. ♢

As before, in the remainder of the section, we exhibit some representative applications of the Strongly Embedded Theorem 2.0.17.

An application related to connectivity. As already suggested above, probably the most typical applications of strong embedding are related to connectivity of the 2-structure graphs in the many contexts where signalizer functors are relevant (notably the Dichotomy Theorem 2.0.9). And indeed various kinds of connectivity arise; mostly based on the general notion of component-stabilizers lying in proper subgroup $M < G$. Relationships among a number of such relevant graphs are discussed e.g. around [**ALSS11**, p 36].

For example: In Aschbacher's Proper 2-Generated Core Theorem (for example [**ALSS11**, 1.5.10]), the role of the M containing component stabilizers is played by the *2-generated core* $\Gamma_{2,T}(G)$: this is generated by the normalizers $N_G(S)$, for subgroups S of rank ≥ 2 in T. Of course this is a weakening of (8.3.2), which considers S of rank ≥ 1, i.e. all S. Aschbacher's result shows, under the assumption $m_2(G) \geq 3$, that if $\Gamma_{2,T}(G) \leq M < G$, then J_1 is the only "new" group to arise—beyond the "old" Bender groups of the usual Strongly Embedded Theorem 2.0.17. We sketch how this is accomplished:

Aschbacher at [**Asc74**, 3.7] reduces to the situation where $C_G(z) \leq M$ for all the *2-central* involutions z of G: For otherwise, this is where the new conclusion-group J_1 arises—using his earlier result 2.5,[4] which in turn applies an earlier classification result of Gorenstein and Walter.

And now, if x is another involution of T, and we have $C_G(x) \not\leq M$, then by his 3.8, we can conclude that $C_M(x)$ is strongly embedded in M. This result on $C_M(x)$ gives the sufficient local-condition in his [**Asc74**, Thm 2], to force M to be strongly embedded in G. And so at this final point in the proof, the old conclusion-groups now arise: namely the rank-1 Lie type groups in the Strongly Embedded Theorem 2.0.17 □

An application to quasithin groups. We recall from our discussion of the Small Even Subcase of the CFSG in Section 2.2 that when Aschbacher and Smith analyzed quasithin groups, they worked not under characteristic 2-type, but instead under the weaker notion of even characteristic in Definition 2.2.3: where at least the 2-central involutions z satisfy $F^*\big(C_G(z)\big) = O_2\big(C_G(z)\big)$.

We now expand a little on that earlier discussion, on the work of Chapter 16 in [**AS04b**]: which extends the Aschbacher-Smith classification of quasithin groups of even characteristic, to the still-weaker hypothesis of even type used in the GLS revisionism program [**GLS94**, p 36]. In the latter situation, the above centralizers $C_G(z)$ might instead have some components L of characteristic 2. And there the corresponding result [**ALSS11**, Sec 3.12] shows that the only new group to arise under even type, beyond the previous conclusions under even characteristic, is the Janko group J_1. (This is similar to the role played by J_1, in our discussion above of proceeding from strongly embedded to a proper 2-generated core.) We briefly summarize the deduction of this result:

The analysis of a new G, now with such a component L in some $C_G(z)$, leads in a fairly straightforward way to the situation where L is standard in G: and of course we saw after Definition 2.1.2 that this concept was crucial in the treatment of the Odd Case of the CFSG. But the subsequent arguments are made directly—that is, not quoting that earlier literature on standard form—roughly as follows:

For $K := C_G(L)$, we obtain at [**AS04b**, p 1183] a conjugate $R = K^g$ distinct from K, with $N_R(K)$ of even order: for otherwise, we would get the sufficient condition of [**AS04b**, I.8.2], for $N_G(L)$ to be strongly embedded. And of course, the

[4]Note that the references "3.5,2.1" just before [**Asc74**, 3.7] should be "3.3,2.5".

Strongly Embedded Theorem 2.0.17 gives the rank-1 Lie type groups in characteristic 2: in particular, these are old conclusion-groups, of even characteristic—and have no such component L, contrary to our assumption of a new-G involving L.

In the subsequent argument, most possibilities for L are eliminated. The only case remaining is for $L \cong L_2(4)$: and then the R above has order 2 (and indeed lies in L), so that $C_G(z) \cong \mathbb{Z}_2 \times L_2(4)$. Finally G is identified as the new-conclusion J_1, using the involution-centralizer recognition theorem of Janko [**AS04b**, I.4.9]. □

An application for Holt's theorem in permutation groups. One further variant of the condition of M strongly embedded—indeed the one referenced in the title of Bender's paper—is:

All involutions fix exactly one point of G/M.

Holt (see [**ALSS11**, B.2.1]), and independently F. Smith, extended the analysis to transitive groups satifying the weaker condition:

(8.3.4) (Holt:) *Some* 2-central involution fixes just one point of G/M.

In the extended result, the only new simple groups to arise, beyond the Bender groups in the strongly embedded situation, are S_n and A_n for odd n.

EXERCISE 8.3.5. Check Holt's 2-central condition in these "new" groups. ◇

In a moment, we will briefly summarize Holt's argument.

However, we mention that the reader wishing to directly consult Holt's paper [**Hol78**] may have some difficulty with the very terse exposition style which was common in that era. For this reason, I had provided to the class at the Venice Summer School 2015 some additional online notes on certain arguments in that paper—which now appear in appendix Section A.3.

Holt reduces at [**Hol78**, 4.1] to a certain more technical fusion condition: since failure would give Aschbacher's sufficient condition [**Asc73**] for G to have a strongly embedded subgroup. That is, the "old" conclusion groups, given by the Bender groups, arise here.

Then following further analysis of that technical fusion condition, the new conclusion groups S_n and A_n arise toward the bottom of [**Hol78**, p 182].

Holt's theorem was used at many points in the CFSG; for various applications, see e.g. the Index in [**ALSS11**]. It was typically used toward the end of proofs: when a group H has been constructed, which might conceivably be of odd index in the desired conclusion-group G; to force $H = G$. For G/H should here have Holt's hypothesis; but ideally the details of the problem being considered should be sufficiently specific to rule out Holt's Bender-group and alternating conclusions.

Subsection-appendix: a related result of Parker-Stroth. I thank Gernot Stroth, who (during the course of the lecture that became this chapter) pointed out a version of Holt's Theorem, using a variant-hypothesis, in Parker-Stroth [**PS14**]. Below I have transcribed a Web pdf file that Stroth kindly posted for the class:

This is a version of Holt's Theorem [**Hol78**], taken from Chris Parker and Gernot Stroth [**PS14**].

LEMMA 8.3.6. *Suppose K is a simple group, P is a proper subgroup of K, and r is a 2-central element of K. If $r^K \cap P = r^P$ and $C_K(r) \leq P$, then K is one of:*

$$PSL_2(2^a), PSU_3(2^a)(a \geq 2);\ {}^2B_2(2^a)(odd\ a \geq 3)\ ;\ Alt(n);$$

where in the first three cases P is a Borel subgroup of K, and in the last case we have $P \cong Alt(n-1)$.

PROOF. Set $\Omega := K/P$ and assume that $P < K$. The conditions $C_K(r) \leq P$ and $r^K \cap P = r^P$ together imply that r fixes a unique point of Ω. Let J be the set of involutions of K which fix exactly one point of Ω. Since r is a 2-central element of K, any 2-group which fixes at least 3 points when it acts on Ω commutes with an element of J. Hence Holt's criteria (*) from [**Hol78**] are satisfied. In addition, the simplicity of K yields that $K = \langle r^K \rangle = \langle J \rangle$. Thus [**Hol78**, Thm 1] implies that K is isomorphic to one of the following groups: $PSL_2(2^n), PSU_3(2^n)(n \geq 2)$; or ${}^2B_2(2^n)(n \geq 3$ and odd); or $Alt(\Omega)$; where in the first three classes of groups the stabiliser P is a Borel subgroup and in the latter case it is $Alt(\Omega \setminus \{P\})$. □

Analogous p-fusion results for odd primes p

The remaining sections of the chapter discuss the possibility of odd-p analogues of the results in the previous sections.

8.4. The Z_p^*-theorem for odd p

The odd-p analogue of the Z^*-Theorem states:

THEOREM 8.4.1 (Odd Z_p^* Theorem). *If an element z of odd order p in H commutes with no distinct conjugate, then $z \in Z_p^*(H)$—the preimage of $Z\big(H/O_{p'}(H)\big)$.*

Unlike Glauberman's situation for $p = 2$, for odd p no "elementary" proof of Theorem 8.4.1 is known. Instead, the CFSG is used in proving the result.

Experts were aware fairly early on that 8.4.1 could be obtained using the CFSG. For example, Artemovich published a proof in [**Art88a**]. And more recetnly:

A sketch of a CFSG-based proof. Guralnick and Robinson outlined a proof at [**GR93**, 4.1], from which we now extract a few features:

The hypothesis implies that z is central in a Sylow p-group P of H. We may assume that P is noncyclic: as otherwise the result follows using the Frobenius Normal p-Complement Theorem [**GLS96**, 16.10].

Then we reduce to z acting nontrivially on all components L, which we may assume are simple. Earlier Gross [**Gro82**] had studied the simple L in the CFSG-list 1.0.2, showing for p odd that p-central automorphisms are inner. For example, recall the description of $\mathrm{Out}(L)$ from Theorem 1.5.4: note e.g. that for a Lie-type group in characteristic p, nontrivial field automorphisms (including any of order p) do not centralize a Sylow p-group.

EXERCISE 8.4.2. Check that the analogue for $p = 2$ of Gross's conclusion fails. (Consider a transposition in S_6.) ◊

This inner-restriction leads in fact to $z \in L$; indeed with just a single component L providing the full group H. And earlier Gorenstein [**Gor82**, 4.250] had examined the list of L in the CFSG-list, showing in fact there there are only a few groups L that even satisfy the weaker condition, namely:

all L-fusion of z in P falls inside $\langle z \rangle$.

But in those cases, one sees that z *is* fused to a nontrivial power z^i; whereas our z is assumed to commute with no distinct conjugate. □

Some other directions. In the Guralnick-Robinson work [**GR93**], in fact the Z_p^*-theorem arises in their wider setup—namely, of generalizations of the Baer-Suzuki Theorem [**GLS96**, 15.5]. That result states that if x is a p-element such that $\langle x, x^h \rangle$ is a p-group for all $h \in H$, then $x \in O_p(H)$.

Robinson has also considered in [**Rob90**][**Rob09**] some approaches to proving the Z_p^*-theorem via p-block theory.

The reader may also wish to explore the discussions in Broué [**Bro83**], Rowley [**Row81**] and Toborg [**Tob16**].

8.5. Thompson-style transfer for odd p

By contrast with the situation odd Z^*-Theorem 8.4.1, there *is* an elementary proof of odd-p analogues of the Thompson Transfer Theorem 8.2.1: since the result for $p = 2$ is proved just by computing the transfer homomorphism. Such extensions are often called *Thompson-Lyons transfer*.

One fairly general such extension appears in [**GLS96**, 15.15]; instead we state the somewhat simpler special case in [**GLS96**, 15.17]:

THEOREM 8.5.1 (Thompson-Lyons Transfer). *Assume that Q has index p in a Sylow p-subgroup P of H; and that z of order p outside Q satisfies $z^H \cap P \subseteq zQ$. Then $z \notin O^p(H)$.*

We could in fact weaken the hypothesis to *extremal* $x \in z^H \cap P$; namely with $C_P(x)$ Sylow in $C_H(x)$.

Lynd [**Lyn14**] extends the Thompson-Lyons analysis to fusion systems.

My impression is that the case of odd p has not been as influential in applications as the original Thompson Transfer for $p = 2$.

8.6. Strongly p-embedded subgroups for odd p

There is a substantial literature on the odd-p analogue of the strongly embedded condition (2.0.16); we state the version analogous to (8.3.2):

DEFINITION 8.6.1 (strongly p-embedded). For odd p, and P denoting a Sylow p-subgroup of G, we say that M is *strongly p-embedded* in G, if for all $1 < X \leq P$, we have $N_G(X) \leq M < G$.

Here in contrast to the Strongly Embedded Theorem 2.0.17 at $p = 2$, for odd p there is no elementary treatment of the strongly p-embedded condition: instead, proofs depend on using the CFSG.

The Gorenstein-Lyons analysis of strongly p-embedded in the CFSG. We saw, during our discussion of the treatment of the pre-uniqueness case (2) of the Trichotomy Theorem 2.2.8, that the weaker condition of *almost* strongly p-embedded arises within the "Uniqueness" subcase of the Even Case. So a treatment of strongly p-embedded is required in that more general context.

Indeed for use in their inductive situations, Gorenstein and Lyons list at 24.1 of [**GL83**] (and see also [**GLS98**, 7.6.1]) the cases satisfying the strongly p-embedded condition, for the known simple groups G in the CFSG-list 1.0.2; and for P noncyclic, since the cyclic case is essentially uncontrollable. Hence when the CFSG was completed, their list became a result covering all simple G.[5] We note that the list contains the expected analogues of the Bender groups: that is, the rank-1 Lie type groups in characteristic p; but it also contains the alternating groups A_{2p}, and a few scattered cases for small primes p.

Of course their proof involves detailed examination of the p-local structure of the simple groups in the CFSG-list.

We mention that a similar analysis of strongly p-embedded configurations is also required, in the later GLS revisionism-approach to the CFSG—which we had briefly mentioned in Section 2.3 as new-approach (2) to the CFSG. For details of this later treatment, see e.g. [**GLS96**, Sec 17] and [**GLS99**, Ch 3].

Strong p-embedding also arises in new-approach (3) to the CFSG, namely the characteristic-p viewpoint of Meierfrankenfeld-Stellmacher-Stroth. See for example Parker-Stroth [**PS11**] for a discussion of that treatment.

Applications of strongly p-embedded in other directions. The condition of strong p-embedding arises naturally in group-theoretic situations other than the CFSG itself. Here is a sample:

In p-modular representation theory: Zhang [**Zha94**] deduces from the condition the existence of a p-block of defect 0, in the standard language of 5.0.2. And Robinson [**Rob11**] and others study the condition, in the context of endotrivial modules.

In p-local structure theory: Strongly p-embedded subgroups are especially important for *conjugation families*, in the context of the Alperin Fusion Theorem 3.5.3; In fact the members of those families can be taken to each possess a strongly p-embedded subgroup. An early discussion of these connections appears in Miyamoto [**Miy77**].

These lead in turn to various contexts in algebraic topology which depend on fusion, notably those related to group cohomology. We mention in particular: For saturated fusion systems, see e.g. Oliver-Ventura [**OV09**]; and for rings of group invariants, see e.g. Kemper [**Kem01**].

We also mention that Brown [**Bro00**] considers strong p-embedding, in relation to the probability of generating a group G (a context we had mentioned in Section 6.5).

---

[5] One point that often seems to be left implicit in the literature is the reduction from the general case, to the case where G is almost-simple: in fact this follows essentially using a Frattini Argument 3.0.3 on the Sylow p-subgroup of a component. I thank Jesper Grodal for interesting discussions on this background.

CHAPTER 9

Some applications close to finite group theory

I am grateful to many colleagues, who suggested quite a number of intriguing applications for this book—only a small fraction of which I've been able to discuss so far. And I also thank the referee for some additional suggestions.

In the remaining two chapters, I'll try to cover a wider array of those applications, than I've done in the earlier chapters; of course, this comes at the cost of treating that larger number much more briefly.

The present chapter will be devoted to applications in areas which are still reasonably close to standard group theory. As a general context for these fairly scattered applications, the discussion we gave in Section 6.4 of some major areas of applications—for the specific topic of maximal subgroups—is still largely applicable for the more general material from here on.

But since the first five chapters of the book focused largely on applications to group structure, we should probably precede the list in Section 6.4 with a slightly-expanded version of our introductory remarks before Section 1.4:

(0) *Internal structure for finite groups.* As we've seen in earlier chapters, the CFSG has had a fundamental impact throughout the structure theory of general finite groups. We'll now continue to consider various further applications in that direction; including (among many possible topics): general subgroup structure for Lie-type groups, in Section 9.3; solutions of special equations in groups, in Sections 9.4 and 10.5; and coverings by a union of subgroups or subsets, in Section 9.7, and also toward the end of Section 10.6.

We'll summarize some more distant application-areas at the start of Chapter 10.

9.1. Distance-transitive graphs

I thank Cheryl Praeger for assistance with this section.

For this very active area, a number of surveys are available: in fact we had already mentioned Praeger-Li-Niemeyer [**PLN97**], in Chapter 6 on maximal subgroups. Some others include: Brouwer-Cohen-Neumaier [**BCN89**], Ivanov [**Iva94**], Cohen (in Beineke-Wilson [**BW04**]), and van Bon [**vB07**]. In this section, we'll sample a few features of this now well-advanced project:

Definition and examples. A group G is *distance-transitive* on a connected graph Γ if for each i, G is transitive on pairs of vertices at distance i.

There are some standard classes of examples, including: vertices of hypercubes (within the larger class of Hamming graphs); Johnson graphs, Grassman graphs, odd graphs; and some examples from bilinear forms and codes. For definitions and discussion of these examples, see for example [**BCN89**], [**PLN97**, Sec 7.1], and [**vB07**, Sec 2].

EXERCISE 9.1.1. Check distance-transitivity for the usual 3-cube. (The longest distance is 3—and the automorphism group acts as S_4 on the 4 diagonals exhibiting this distance.) $\diamondsuit$

Ideally, of course, there should not be too many more examples—if the relevant classification problem is to be tractable.

Progress toward classification. There is a process for reducing to the case where G is primitive on the vertices. So we can then apply the actions-list in the Aschbacher-O'Nan-Scott Theorem, which we discussed in the form of the table in Remark 6.1.4. (And as we'd mentioned, that result depends on the CFSG, via the Schreier Conjecture 1.5.1.)

Which of the actions in Remark 6.1.4 should actually arise? The work of Praeger-Saxl-Yokoyama in [**PSY87**] showed that if G is primitive and distance transitive, and Γ is of diameter ≥ 3, then the primitive action has one of the following types:

- PA (product action)—with Γ a Hamming graph;
- HA (holomorph of abelian)—we usually refer to this case as "affine";
- AS (almost-simple).

Analysis of the HA and AS cases involves further use of the CFSG:

For the affine case HA, with $F^*\big(\ G/O_r(G)\ \big)$ simple, various authors (Liebeck-Praeger, van Bon, Ivanov, Saxl, Cohen ...) separately treated the usual alternating, sporadic, and Lie-type cases in the CFSG-list 1.0.2; with the final steps done by van Bon—see e.g. [**vB07**, Sec 5].

For the almost-simple case AS, again various authors (Ivanov, Saxl, Liebeck, Praeger, van Bon, Cohen ...) treated the corresponding alternating and sporadic cases; as well as the linear groups among the Lie-type cases. Partial results are available for some other Lie-type cases; for the current status, see for example Section 4 of [**vB07**].

Proofs involve detailed properties of the various simple groups; notably maximal subgroups—e.g. using results we indicated in Chapter 6. But the conclusion-graphs are comparatively rare; and proofs often require strong restrictions on various parameters, so that arguments are often highly computational in nature.

An application via the Sims Conjecture. The paper of Cameron-Praeger-Saxl-Seitz [**CPSS83**] uses the CFSG to establish, in their Theorem 1, a conjecture of Sims: namely that there is a function f, so that for a primitive permutation group G on a finite set, with d the length of any nontrivial G_α-orbit, we have for the stabilizer order (not index!) that $|G_\alpha| \leq f(d)$.

They in turn use this to establish their Theorem 2: that there are only finitely many connected distance-transitive finite graphs of any fixed valency greater than 2.

The proof of Theorem 1 applies the Aschbacher-O'Nan-Scott Theorem, using the cases as in Remark 6.1.4, to reduce to the subcase AS with a simple socle. Subsequent sections then consider the structure of the simple groups in the families in CFSG-list 1.0.2, to establish the needed bounds.

9.2. The proportion of p-singular elements

I thank Cheryl Praeger and Bill Kantor for suggesting this topic.

A surprising result on the proportion. Computational group theorists are interested in the efficiency of algorithms for various group-theoretic operations; in particular, we mention searching for p-singular elements (that is, of order divisible by p). Of course a random search can rely on the *proportion* of such elements in a group G. Isaacs-Kantor-Spaltenstein [**IKS95**] used the CFSG to establish:

THEOREM 9.2.1. *For p dividing the order of a permutation group G of degree n, the proportion of p-singular elements in G is at least $\frac{1}{n}$.*
Equality holds iff either $G = S_p$; or $n = p^a$ with G sharply 2-transitive.

EXERCISE 9.2.2. Check the indicated proportion in the case $G = S_p$.

Notice that the main bound $\frac{1}{n}$ depends on the degree n—but not on the prime p. This may not seem intuitively clear a priori!

The proof using maximal subgroups of S_n. Section 2 of [**IKS95**] reduces to the case where G is almost-simple. It would have been possible to obtain this by just quoting the structures-list in the O'Nan-Scott Theorem 6.1.1, and reducing to the almost-simple case (6) there. In fact the authors give an independent direct proof: though perhaps not surprisingly, the main logic sequence roughly follows the deduction of the actions-list in the Aschbacher-O'Nan-Scott Theorem in Remark 6.1.4—and their basic argument culminates in a reduction to the almost-simple case AS.

We had already discussed some deductions of this type—in appendix Section A.1, and in the applications-Section 6.5; so we will not here follow that reduction argument in Section 2 of [**IKS95**].

Instead, we'll sample a little of the argument in the later sections of that paper: since there we have $L := F^*(G)$ simple, those sections treat the various possibilities for simple L in the CFSG-list.

In fact Section 3 there indicates the comparatively easy calculations which treat the cases where L is alternating or sporadic; thus we are reduced to L of Lie type.

Section 10 handles the somewhat easier case where p is the characteristic of L: here p-elements are unipotent, so the theory of unipotent groups can be exploited.

After Section 10, p is *not* the characteristic prime of L. Consequently the p-elements lie in some torus—possibly non-split as in Example 5.2.2. In fact, tori are parametrized by conjugacy classes of the Weyl group W, as we had mentioned in (5.2.1), with p-elements realized as block-diagonal matrices. These typically commute with suitable unipotent elements, again allowing use of unipotent structure. Here, the calculations use the minimal permutation degree of L; the possibilities are tabulated in Section 4, using existing estimates from the literature.

Section 7 then makes the needed calculations (with some computer use) for exceptional L—one Lie type at a time; while Section 8 handles the classical groups in a more uniform way.

Some applications and extensions. Aside from the obvious use in searching for p-elements, Theorem 9.2.1 has also been applied to various other computational problems, including:

- random generation of G;
- recognition of G, e.g. as a classical group; and
- testing elements of G for containment in a subgroup H.

These and other applications can be looked up via MathSciNet, starting at [**IKS95**].

Niemeyer-Praeger [**NP10**] analyzed the context of the result, and further extended the methods.

9.3. Root subgroups of maximal tori in Lie-type groups

For a group G of Lie type, we first recall some notions leading up to *root subgroups* in 1.3.20(1):

We saw in Remark 1.3.7 that underlying a Lie-type group G is a simple Lie algebra $\mathcal{G}$ over $\mathbb{C}$; including a Cartan subalgebra $\mathcal{H}$, whose action decomposes $\mathcal{G}$ via root spaces, in terms of the root system Φ and Weyl group W. And a corresponding Cartan subgroup H determines root subgroups in G, as in Remark 1.3.20(1).

Next: Various subgroups X of G, notably those generated by individual root subgroups, are in fact H-invariant. And then of course the product XH is also a subgroup of G. Such groups, which we might call H-root groups (and indeed more general H-invariant subgroups, such as arbitrary overgroups of H) describe much of the interesting subgroup structure in G. For example, we saw in Section 6.2 that some natural maximal subgroups of G arise via maximal-rank-as-reductive root subsystems of Φ.

Seitz's theorem on overgroups of a maximal torus. In the above situation, H is in fact a split torus for G; we may also wish to consider similar notions of root groups, with respect to a nonsplit torus T in the language of Example 5.2.2. Such subgroups provide even more of the interesting subgroup structure for finite Lie-type groups G.

A basic work of Seitz [**Sei83**] extends the above notions of H-root groups to the case of nonsplit T. We won't here reproduce the more complicated statement of his main theorem at [**Sei83**, p 154]; but the spirit of that result is roughly that overgroups of $T_0 := T \cap O^{p'}(G)$ are generated essentially by what he calls *T-root groups*. To describe these T-root groups, we'll develop a bit more background.

First just for practice:

EXERCISE 9.3.1. Describe nonsplit tori in the case $L_3(4)$ of Example 5.2.2. $\diamondsuit$

Seitz's work proceeds mainly in the context of algebraic groups; but since subgroups of finite G might not display Lie structure, we mention that the CFSG was relevant in treating such further configurations. We also note that to avoid complications in groups over small fields of definition for G, say $\mathbb{F}_q$, where $q = p^a$ for a prime p, some of Seitz's results assume that q and p are roughly "not too small".

The construction of the T-root groups. We continue with our background development: We recall from Remark 1.3.23 that the finite Lie-type group G arises as the fixed points $\overline{G}^{\sigma}$, in an algebraic group $\overline{G}$ over the algebraic closure $\overline{\mathbb{F}_p}$, under an automorphism σ (combining a field automorphism possibly with a graph automorphism); with $T = G \cap \overline{T}$, for a maximal torus $\overline{T}$ of $\overline{G}$. Since $\overline{T}$ is a full maximal torus, it has $\overline{T}$-root groups in analogy with those for the split torus H of the finite group G.

Seitz then constructs the finite *T-root groups*: First he considers subgroups given by the intersection of finite G with the $\overline{T}$-root groups; and then he takes groups generated by σ-orbits of such intersections. From [**Sei83**, 3.1], each such group is either a p-group (e.g. when σ is just a field automorphism), or a possibly twisted Lie-type group over an extension field of $\mathbb{F}_q$ (e.g. when a graph automorphism is involved). And as we had noted earlier, these T-root groups are the building blocks, for the overgroups of T_0 in Seitz's main theorem.

EXERCISE 9.3.2. Give some examples of such subgroups in $L_3(4)$. These include the usual root groups; and orbit-determined groups like $U_3(2)$. ♢

Remarks on proof and applications. To sketch the proof: The T-root groups are used e.g. in (10.1) and (10.2) of [**Sei83**], to describe arbitrary overgroups Y of T_0 in G. Then, assuming that the main theorem about generation by T-root groups fails, Seitz reduces at (10.11) to the situation where $F^*(G)$ consists of at most 2 simple groups L. Now he can apply the CFSG-list 1.0.2 to determine possible L; and then examining the usual three cases (alternating, Lie-type, and sporadic L), he uses the known structure of L to obtain numerical contradictions.

The results in [**Sei83**] are used e.g. to describe more general subgroup structure in Lie-type groups; notably in the study of:

- maximal subgroups via the algebraic-groups approach,

which we described in the relevant subsections of Section 6.2. But there are applications in various other areas which make use of the CFSG; e.g.:

- generation and random walks (compare Section 6.5);
- fixed-point ratios in groups (cf. later Section 10.3); and
- logic: model theory related to algebraic groups—e.g. [**BB04**].

Some applications more briefly treated

The remaining sections of the chapter provide much less detailed discussions.

9.4. Frobenius' conjecture on solutions of $x^n = 1$

Frobenius (1895) showed that the number of solutions of $x^n = 1$ in G must be a multiple of n; and in the extremal case of exactly n, he conjectured:

CONJECTURE 9.4.1. *If exactly n elements satisfy $x^n = 1$ (for n which divides $|G|$), then these elements should in fact give a subgroup of G.*

EXERCISE 9.4.2. Explore this: e.g. for S_4 with $n = 8$; for A_4 with $n = 4$; etc. ♢

Various special cases had been established over the subsequent years; for example, the case of G solvable, by M. Hall in [**Hal76**, 9.4.1].

Later Zemlin [**Zem54**] reduced the problem to the case of G a simple group.

Finally Iiyori and Yamaki announced the completed proof in [**IY91**], having checked the various types of simple groups in a number of earlier papers. To give a quick idea of how they proceeded:

When p divides both n and $|G|/n$, their Lemma 1 shows that either p is odd, and G has a cyclic Sylow p-subgroup; or $p = 2$, and $m_2(G) \leq 2$. Their Lemma 2 eliminates the former case; and in the latter, they proceed by explicitly examining the simple groups arising in the Small Odd Subcase of the CFSG. We had indicated these groups in the applications-subsections following the Z^*-Theorem 8.1.1.

This reduces to the case where n and $|G|/n$ are coprime. Some sample arguments for various subcases are illustrated in [**IY91**]: e.g. for Lie-type G, they exploit the known structure of tori.

The result can be applied for example in the context of the *prime graph* of G, with vertices given by the primes p dividing $|G|$, and edges whenever G has an element of order pq. There is a corresponding literature on this graph; see e.g. [**Luc99**].

9.5. Subgroups of prime-power index in simple groups

For general finite groups H, there is no "co-Sylow" theory: that is, no theory of subgroups K having p-power *index* in H. (Though for solvable groups H, there is a well-known corresponding theory of Hall p'-subgroups—see e.g. [**ALSS11**, A.1.14].)

On the other hand, for simple G, the possible cases for subgroups which have p-power index are given in Guralnick [**Gur83**]. The method of proof had essentially appeared in Liebler-Yellen [**LY79**], though that paper only considered solvable subgroups K. (Cf. also Kantor in [**Kan85**] and Arad-Fisman in [**AF84**].)

The argument is a comparatively easy deduction, using the CFSG-list 1.0.2:

The cases where G is alternating or sporadic are fairly straightforward to deal with; so below we'll restrict attention to G of Lie type.

If G is defined over a field of characteristic p, then $G = KU$ for a full unipotent group U; and it follows that K is flag-transitive on the building of G. For this condition, the possibilities had been listed by Seitz [**Sei73**].

Otherwise G has characteristic other than p; so we may assume $K \geq U$, and then K is normal in a parabolic P. Now from the standard order formulas for G of Lie type, $|G : P|$ is rarely a prime power.

A corollary of the main result of [**Gur83**] is a new proof of the list of permutation groups of prime degree: that list had earlier been obtained as a consequence of the CFSG, by Feit [**Fei80**]—see 4.1 and 4.2 there.[1]

The main result has also been used in an impressively wide variety of further applications. I won't attempt to give details here; but a quick search on MathSciNet indicates areas such as:

- maximal subgroups;
- permutation groups;
- ordinary and modular character theory;
- group factorizations;
- profinite groups;
- number theory: Mersenne primes, Galois groups, group zeta functions;
- codes, association schemes, game theory;
- and even the Yang-Baxter equation in physics!

The result also describes permutation groups of prime-power degree.

It seems natural to also mention here the determination, again as an application of the CFSG, of the primitive groups of odd degree—in Liebeck-Saxl [**LS85**] and independently Kantor [**Kan87**]. Kantor also makes use of Aschbacher's Classical Involution Theorem; and applies his result to geometry, e.g. projective planes.

[1] Seemingly the values $p = 11, 23$ are only implicit in the statement of Feit's 4.2.

The sparsity of non-alternating primitive degrees. We close the section with a brief sketch of a general result on primitive degrees due to Cameron-Neumann-Teague [**CNT82**].

As background, recall that the "structured" primitive cases in (3)–(5) for S_n in the O'Nan-Scott Theorem 6.1.1—and indeed in the more detailed cases for primitivity in the Aschbacher-O'Nan-Scott Theorem in Remark 6.1.4—required rather special values for the index n. So we might expect roughly that for almost all n, the groups A_n and S_n are the only primitive groups of that degree.

The main result [**CNT82**, Theorem] makes this precise, in the following way: Let E denote the set of "exceptional" natural numbers—namely those n for which some group other than A_n or S_n has a primitive representation of degree n; and let $e(x)$ denote the number of members of E which are $\leq x$. Then the density of E is the limit of $\frac{e(x)}{x}$ for increasing x. The Theorem shows that $\frac{e(x)}{x} \sim \frac{2}{\log x}$; so from the limit we get density 0.

The proof uses the Aschbacher-O'Nan-Scott Theorem (though in effect arguments from Jordan's thesis already suffice), to express E as a union of four subsets, corresponding to possible socle-structures for a primitive group. Of these, the first three are easily shown to have density 0; leaving for analysis just the fourth set E_4, corresponding to the simple-socle case AS in 6.1.4.

So now the structures of the groups in the CFSG-list 1.0.2 are examined; the proof separates alternating groups from Lie-type groups (and the finite number of sporadic cases can be ignored because of the asymptotic context). Suitable calculations then give estimates establishing that E_4 also has density 0.

9.6. Application to 2-generation and module cohomology

For module cohomology $H^*(H, V)$, namely of a group H with coefficients in a module V, a number of results show (or else conjectures assert) that for suitable small dimensions n, $H^n(H, V)$ should be "not too large" in terms of V. Aschbacher and Guralnick in [**AG84**, Thm A] show:

(9.6.1) For H faithful on irreducible $V/\mathbb{F}_p$, we have $|H^1(H, V)| < |V|$.

We give a very quick overview of their proof:

They reduce to the case where H is simple; and they consider the cohomology group $H^1(H, V)$ using the standard interpretation via conjugacy of complements to V in the semidirect product HV.

Now their Theorem C allows them to express the generation of the product HV, in terms of cohomology $H^1(H, V)$ and the generation of H. And they get 2-generation of simple H using their [**AG84**, Thm B].

As we had mentioned at the start of Section 6.7, that result depends on the CFSG. We mention that their proofs for H alternating and Lie-type are fairly short; while the sporadic cases require a few more individual details.

Again these results are applied in many areas, such as:

- complements and module cohomology (e.g. $H^2(H, V)$);
- generation and presentations;
- permutation groups and Cayley graphs; and
- profinite groups.

9.7. Minimal nilpotent covers and solvability

Work of a number of authors has recently culminated in:

(9.7.1) A group with a minimal nilpotent covering is solvable.

Here a covering is via the union of a set of proper subgroups; it is minimal, if no subgroup can be removed; and nilpotent, if the subgroups are nilpotent.

To summarize the treatment:

Bryce-Serena [**BS08b**, 2.1] reduce to the case of $F^*(G)$ simple; and in the subcase where G is itself simple, they handle the alternating and sporadic cases, as well as several Lie-type cases.

Blyth-Fumagalli-Morigi [**BFM15**, Thm 2] complete the analysis, covering the usual 3 classes in the CFSG-list 1.0.2 for the simple group $F^*(G)$; here are a few features:

Their Lemma 2 easily reduces the Lie-type subcase to small rank. Then for any simple $F^*(G)$, their Proposition 6 eliminates the possibilities for outer automorphisms—using basic facts such as those in our discussion of these automorphisms in Section 1.5. So now G is simple; and they can finish by individually treating the only two small-rank Lie-type cases remaining after [**BS08b**] above.

9.8. Computing composition factors of permutation groups

I thank Bill Kantor for suggesting this topic.

An important task in computational group theory is the determination of composition factors of an input-group G. In particular, Luks [**Luk87**] gave a polynomial-time algorithm for finding composition factors, in the case where G in a permutation group. The proof of correctness of the algorithm uses the CFSG; here are a few features:

Some sub-algorithms of the main algorithm, in Sections 4–5 of [**Luk87**], are shown to involve primitive actions.

The possible actions are given by the Aschbacher-O'Nan-Scott Theorem; and the cases needed are summarized in [**Luk87**, 2.1]. Recall we described the actions via the table in Remark 6.1.4; and 6.1.4 uses the CFSG, via the Schreier Conjecture 1.5.1. Indeed at [**Luk87**, p 98], that dependence becomes visible in Luks' analysis: for arguments there reduce down to the almost-simple case AS; and at that point, they further require the Schreier Conjecture: namely solvability of the automorphism group of the relevant simple group.

These arguments are in effect applied to the output of the main algorithm, namely a section H of G which is a potential composition factor: in order to prove that H is indeed simple, and hence really is a composition factor.

Luks' algorithm was later substantially improved; see for example Babai-Luks-Seress [**BLS97**]—which still uses the Aschbacher-O'Nan-Scott Theorem.

Recently Babai in [**Bab16**] has shown that graph isomorphism can be solved in quasipolynomial time. The proof heavily uses the CFSG; and also earlier work of Luks [**Luk82**], showing that graph isomorphism for bounded valence is polynomial-reducible to other problems—which do explicitly involve groups.

CHAPTER 10

Some applications farther afield from finite groups

This final chapter contains a number of applications that are farther removed from finite group theory. Of course since they are beyond my own experience, I am grateful to the colleagues nearer to those other areas, who suggested them to me.

Before examining these individual specific topics, we indicate some more general contexts of applying finite groups in other areas of mathematics:

(1) *Geometry and topology*. Here the groups which occur naturally are mostly infinite, rather than finite. But often they satisfy weaker conditions, which still involve some suitable kind of finite behavior: Such conditions include for example finitely generated, and residually finite; these arise in the application we discuss in Section 10.1. Furthermore locally finite groups and generalizations appear in the application in Section 10.4.

(2) *Number theory and field theory*. Here groups arise in various ways: One standard area is given by Galois groups of field extensions; these are often but not always finite. Field theory also includes the study of *Brauer groups*; and we discuss an application to relative Brauer groups of suitable field extensions in Section 10.2.

(3) *Algebraic geometry*. This wide area of course has considerable overlap with the two areas just indicated. For example, many papers of Abhyankar on Galois groups invoke various applications of the CFSG. In Section 10.3, we examine an application to the monodromy groups of coverings of the Riemann sphere.

(4) *Computer science*. We had indicated one aspect, namely computational group theory, in our discussion of application areas in Section 6.4. But groups also arise more generally in computer science: One standard example is symmetry in network design—for example, expander graphs, which are the topic of the application we discuss in Section 10.6.

It would be possible to indicate many further research areas here. Indeed, as we have indicated explicitly in many of the sections in Chapters 9 and 10, such areas turn up in a MathSciNet search on the further papers which quote the application-papers discussed in those sections.

10.1. Polynomial subgroup-growth in finitely-generated groups

In this section, we focus on *subgroup growth* in a group G which is finitely generated and residually finite. (Recall that the latter means that the intersection of all finite-index subgroups is trivial.) The growth is measured by expressing the number of subgroups of index n in G, as a function $f(n)$. And one finiteness-type condition on this measure is given by *polynomial* subgroup-growth (PSG): when this function is polyomial in n—as opposed to say exponential in n.

A characterization of PSG. Lubotzky-Mann-Segal in [**LMS93**] completed a characterization of such groups:

THEOREM 10.1.1. *A finitely generated, residually finite G has (PSG) iff it has finite rank and is virtually solvable.*

Here *rank* r means that finite-index subgroups are r-generated; and *virtually* solvable means that some finite-index subgroup is solvable.

The paper [**LMS93**] builds on earlier work in Lubotzky-Mann [**LM91**] and Mann-Segal [**MS90**]. It also gives a good introduction to the literature.

The CFSG is used at [**LMS93**, pp367ff], to identify nonabelian chief factors of finite quotients of G; and similarly at [**MS90**, 4.1].

We indicate some features of the overall proof: The Lemma in [**LMS93**] (which is similar to an earlier argument of J.S. Wilson) shows that for N the centralizer of such chief factors, G/N must be a linear group in characteristic 0. By earlier results applying to this explicit situation, they can conclude using (PSG) that G/N is a finite extension of a solvable group X/N. Now any finite nonabelian chief factor of X would lie in a finite nonabelian chief factor of G—but such chief factors can't appear either in solvable X/N, or in N which centralizes them; so they lie *above* X/N. Consequently any finite chief factors of X must be solvable. And then an earlier result[1] leads to X itself being a finite extension of a solvable group, so that the same structure holds for G. □

Further developments. Theorem 10.1.1 has inspired much further work: for example, on analogous properties such as "polynomial index growth". Furthermore the result has been applied in many other areas, including:

- profinite groups;
- arithmetic groups and their zeta functions;
- crystallographic groups; and
- branch groups of trees.

A good survey article on the general area of growth in groups is Helfgott [**Hel15**]. That discussion is fairly explicit about uses of, as well as avoidance of, the CFSG.

10.2. Relative Brauer groups of field extensions

We turn to a topic in field theory; beginning with a rough sketch of some background material:

For a field K, the *Brauer group* $B(K)$ is given by the set of Morita-equivalence[2] classes of finite-rank central-simple algebras over K. These are relevant to the classification of division algebras, and to class field theory.

For a field extension L/K, the *relative* Brauer group $B(L/K)$ is the kernel of the natural map of $B(K) \to B(L)$.

We restrict attention to *global fields*: This term covers certain types of fields roughly arising via "one generator" constructions: namely algebraic number fields (i.e., finite extensions of $\mathbb{Q}$, which are of form $\mathbb{Q}(\alpha)$ by a standard result in Galois theory); function fields of algebraic curves (functions of the "one parameter" defining the curve); and finite extensions of the rational functions $\mathbb{F}_q(t)$. They have

[1] I suspect their mention of a result "in Section 3" is a misprint for "in Section 2"?

[2] That is, equivalence of module-categories for the indicated algebras.

the property that a extension L of K of global fields can similarly be realized in the form $K(\alpha)$.

Relative Brauer groups of global fields are infinite. The work of Fein-Kantor-Schacher in [**FKS81**, Cor 4] showed:

(10.2.1) For global fields $L > K$, $B(L/K)$ is infinite.

We'll now extract some features from the proof in [**FKS81**]:

This result (their Corollary 4) is deduced using their more general Theorem 2—which describes the p-part $B(L/K)_p$; and indeed more directly from Corollary 3 of Theorem 2, essentially showing that, for E the Galois closure of L/K,

$B(L/K)_p$ is finite iff p-elements of $\mathrm{Gal}(E/K)$ fix roots;

where the indicated roots are those of the minimal polynomial of some α, such that $K(\alpha)$ realizes the extension L.

Corollary 4 now follows—using their permutation-group theoretic Theorem 1, which states that:

(10.2.2) A transitive G has, for some p, a p-element fixing no points.

This result is independently interesting, for the theory of permutation groups.

We turn to some features of its proof. If (10.2.2) fails, we have:

(*) Each p dividing $|G|$ also divides $|G_\alpha|$.

Inductive arguments reduce to the case where G is simple (and in fact primitive); so we can examine the usual three types in the CFSG-list 1.0.2:

If $G = A_n$, using (*) leads to having G_α also transitive on the n-set for A_n; and a contradiction can be reached using (*) along with some known related number-theoretic estimates.

For G of Lie type, the arguments use (*) along with other standard structural features, such as: the subgroups of $L_2(q)$; subgroups generated by long-root elements—for this influential topic, see e.g. [**Kan79**][**LS94**]; Seitz's determination [**Sei73**] of flag-transitive subgroups; etc.

When G is sporadic, usually G_α is in the normalizer of some simple subgroup, and then it is fairly easy to check that (*) can't hold.

Further directions. The results in [**FKS81**] have inspired extensions in other parts of field theory; and have also been applied in a number of other research areas, including:

- "elusive" groups (meaning that no element of p fixes a point);
- orbital partitions and other topics in permutation groups;
- conjugacy class sizes in groups;
- solvability criteria for groups; and
- factorizations in graph theory.

We mention that Degrijse-Petrosyan [**DP13**] approach (10.2.1) instead via Bredon-Galois cohomology.

10.3. Monodromy groups of coverings of Riemann surfaces

Our next topic involves algebraic geometry; again we sketch some background:

A *Riemann surface* X is a 1-dimensional $\mathbb{C}$-manifold. Ordinarily we restrict attention to the case where X is connected and compact. A standard example is the *Riemann sphere*, namely the complex projective line. The *genus* of X (say g) is the number of "handles", viewed in $\mathbb{R}$-space; for example, the Riemann sphere has genus 0, a torus has genus 1; etc. A *branched covering* is given by a meromorphic function ϕ, which surjects X onto the Riemann sphere. Removing the finitely many branch-points leads to a covering in the usual topological sense—so that we can apply corresponding theory such as the fundamental group. And then lifting loops around those branch points maps the potentially huge fundamental group down to the finite *monodromy group* of the cover.

It is known that such monodromy groups often have composition factors which are cyclic, or alternating; what other simple composition factors could occur?

The Guralnick-Thompson Conjecture. Guralnick and Thompson [**GT90**] proposed the following:

CONJECTURE 10.3.1 (Guralnick-Thompson Conjecture). *For fixed g—but varying over all covers X, ϕ of that genus—only finitely many non-alternating non-abelian simple groups can arise as composition factors of a monodromy group.*

Various authors contributed to the proof of the Conjecture; and it was completed by Frohardt-Magaard in [**FM01**, Thm A]. We extract some features:

Note first that using the CFSG-list 1.0.2, there are only finitely many sporadics; so the "non-alternating" restriction in the Conjecture amounts to saying "only finitely many Lie-type simple groups".

Here is some general background to the work:

A standard variant of the Conjecture given in [**GT90**] reformulates the monodromy group in terms of the action of a permutation group H on a set Ω: it has r generators x_i; along with a relation $x_1 x_2 \cdots x_r = 1$ coming from the fundamental group; and the action satisfies:

$$\sum_i \left(\ |\Omega| - \#(\text{orbits of } x_i) \ \right) = 2(\ |\Omega| + g - 1 \ ) \ . \tag{10.3.2}$$

The corresponding strategy is then to try to eliminate composition factors above some particular size—by getting lower bounds on the left side of (10.3.2), sufficiently large to exceed the right-hand side. In [**Gur92**], Guralnick showed that such lower bounds could in fact be obtained via upper bounds on the *fixed-point ratio*:

$$\frac{|\mathrm{Fix}_\Omega(x_i)|}{|\Omega|} .$$

Within the above context, [**FM01**, Sec 1] reviews the earlier literature:

First Guralnick [**Gur92**, 5.1ff] reduced to showing that only finitely many *almost-simple subgroups* K can arise in the monodromy group. So as usual we apply the CFSG-list 1.0.2 to examine the cases for $G := F^*(K)$; recall we had already reduced to considering just Lie-type groups.

Liebeck-Saxl [**LS91**] then treated large Lie-type groups of exceptional types; so the problem was reduced to showing only finitely many *classical* G are possible.

Next Liebeck-Shalev [**LS99b**] treated classical G which have actions on Ω which are not of "subspace" type; this reduced to the case of Ω arising from the subspaces of natural module V for classical G.

Now the main work of Frohadt and Magaard in [**FM01**] can be quickly summarized:

Using fixed-point ratios for such V, obtained in an earlier work [**FM00**], they obtained in their Theorem B a group-theoretic result in the context of (10.3.2); leading to their main result Theorem A, the Guralnick-Thompson Conjecture 10.3.1.

Further developments. The proof of the Guralnick-Thompson Conjecture in [**FM00**] has inspired a number of further refinements; and has led to applications in various directions, such as orbits on tuples of group-elements closed under braid operations, and non-simple abelian varieties.

The particular technique of fixed-point ratios has also been applied in some other areas; for example, Keller's [**Kel05**] route to the $k(GV)$ problem: this is an alternative to e.g. the Robinson-Thompson route mentioned earlier, in (2) of the introductory remarks in Section 6.4.

See also Magaard-Waldecker [**MW15b**][**MW15a**], for some related applications of the CFSG.

Some exotic applications more briefly treated

These last few applications seemed unusual and possibly even surprising to me; see if you agree.

10.4. Locally finite simple groups and Moufang loops

See [**Hal06**] for fuller definitions and references of material only informally sketched in this section.

Recall that *locally* finite means that all finitely-generated substructures are finite. For example, we mention for reference below the standard fact (e.g. 1.2 in [**Hal06**]) that the fields which are locally finite but not finite are the infinite subfields K of the algebraic closure $\overline{\mathbb{F}_p}$ of a prime field $\mathbb{F}_p$.

Locally finite simple groups. A natural wider context for Hall's result on Moufang loops, indicated later in this section, is provided by the locally finite simple groups. It's probably hopeless to expect a full classification; but there is progress in classifying these groups under reasonable restrictions.

The usual standard example of an infinite simple group which is locally finite, but not finite, is the alternating group A_Ω on an infinite set Ω. Another class of examples is provided by the simple Lie-type groups $G(K)$, where K is a locally finite field as above.

Here the $G(K)$, but not the A_Ω, are *linear*: that is, they are embedded in a suitable $GL_n(K)$, for some finite dimension n.

Indeed the $G(K)$ were characterized as the only locally finite simple groups which are linear—in a series of papers by Belyaev, Borovik, Hartley, Shute, and Thomas. (The original proof depended on the CFSG; but later work of Larsen and Pink removed that dependence.)

Now the groups A_Ω do satisfy the weaker property of being *finitary-linear*: that is, they can be represented linearly on an infinite-dimensional space V over K, with every element g having finite-dimensional commutator $[V, g]$. Indeed further examples of locally finite simple groups arise from the finitary subgroups of the usual classical simple groups (linear, unitary, symplectic, orthogonal) on V.

Correspondingly Theorem 1.1 in Hall [**Hal06**], shows that the only finitary examples are those listed above. His proof uses the indicated linear-subcase result. Hall's paper quotes, as 5.1 and 5.2 there, an earlier result of Hall-Liebeck-Seitz, along with its extension by Guralnick-Saxl, on bounds for the minimal number of generators from a conjugacy class. Their proofs use the CFSG—including some detailed properties of representations, maximal subgroups, and Bruhat structure in Lie-type groups. But Hall actually only needs his weaker 5.3 and 5.4; the finitary condition gives some geometric leverage, which might lead to CFSG-free proofs of these results.

Hall's result reduces the treatment of locally finite simple groups to the non-finitary case. Here a description by Meierfrankenfeld in [**Mei95**] should allow for further analysis, if not necessarily full classification.

Simple Moufang loops. One generalization of a group is provided by a *loop*—which has the group axioms, except perhaps associativity. A *Moufang* loop then adds a weak version of the associativity axiom, namely:

$$(ax)(ya) = a\big((xy)a\big) .$$

Moufang loops arise in various contexts, notably projective geometry.

In the absence of any concept like a normal subloop, the appropriate analogue of simplicity is to say that a Moufang loop is *simple*, if any surjective homomorphism must in fact be an isomorphism. (Hence any contemplated full treatment of locally finite simple Moufang loops would have to include a classification of locally finite simple groups, which may be out of reach.)

A result of J. Hall. Hall's [**Hal07**, Cor 1.3] has an intriguing statement:

THEOREM 10.4.1. *A simple locally finite Moufang loop which is uncountable must in fact be a group (i.e. associative).*

First some remarks, on various details in the statement:

(1) We cannot replace "uncountable" above with finite or countable; as the example of the Paige loop below shows.

(2) Associativity is checkable locally, indeed on triples; but seemingly an uncountable checking is crucial for the result here?

(3) Even simplicity is "fairly" local; indeed Zaleskii had noted early on that a locally-finite simple group need *not* be a limit of finitely-generated simple subgroups. (E.g. infinite-dimensional finitary symplectic cases, as noted above, are (in odd characteristic) limits only of quasisimple groups.)

The background to this result involves a standard non-group example, namely the *Paige* loop: it is given by $\mathrm{PSOct}(F)$, i.e. the norm-1 split octonions (mod ± 1) over a field F.

Some features of the proof. The story starts in the finite case; Liebeck in [**Lie87b**] had shown that a finite simple Moufang loop is:

either a finite simple group, or $\mathrm{PSOct}(F)$ for a finite field F.

He uses techniques of Glauberman and Doro related to triality; and applies the CFSG to determine the relevant simple G with $S_3 \leq \mathrm{Out}(G)$.

Hall [**Hal07**, Thm 1.2] extended this to the locally finite case: namely he showed that a locally finite simple Moufang loop is:

a locally finite simple group, or $\mathrm{PSOct}(F)$ for a locally finite field F.

And now his Corollary 1.3, namely Theorem 10.4.1 above, follows: Recall we had mentioned that a locally finite field is a subfield of countable $\overline{\mathbb{F}_p}$ for some p; and the octonions are of dimension 8 over F—so that the second case above is countable. Thus uncountable locally finite simple Moufang loops must in fact be groups.

We mention that Hall's application above is one of various later applications of Liebeck's result; primarily in further developments in loop theory.

10.5. Waring's problem for simple groups

For background on this area, see e.g. [**Tie14**, Sec 3.3], and the survey [**BGK14**].

We first recall the classical theorem of Lagrange: that any positive integer can be written as the sum of at most 4 squares. For powers x^k higher than x^2, this was generalized to the classical *Waring problem*; solved by Hilbert in 1909:

THEOREM 10.5.1. *For any k, there is a value $g(k)$, such that any positive integer is the sum of at most $g(k)$ k-th powers.*

Within a group G—and we will restrict attention to simple G—we can ask the analogous question about finding $g(k)$: for representing elements now as a product of k-th powers in G.

Indeed we can go on to ask about $g(w)$—where we ask about powers of w, where w is some "word" in appropriate variables, more general than just the single variable x. For example, we can express the Ore Conjecture 5.6.3 as the statement that $g(w) = 1$, for $w := [x, y] = x^{-1}y^{-1}xy$.

There is an increasing literature on questions of this type.

One strong result is that of Larsen-Shalev-Tiep [**LST11**]:

THEOREM 10.5.2. *For any w, there is value N_w such that: for $|G| > N_w$ we have $g(w) = 2$.*

This says for example that in large enough simple G, any element is a product of 2 squares; or 2 cubes; etc.

Of course the proof assumes the CFSG-list 1.0.2; and uses detailed structure of the groups in that list.

10.6. Expander graphs and approximate groups

The notion of an *expander* graph arose in computer science; for any reasonable subset S of vertices, and its neighbors ∂S, the ratio $\frac{|\partial S|}{|S|}$ should exceed some $\epsilon > 0$. This is roughly so that the distribution of information should generally expand, as paths in the graph lengthen; rather than being isolated in some subset of vertices.

Furthermore for scalability in possible construction, such graphs should come in increasing families—for a fixed value of ϵ, so that the expansion does not decay with the increase.

And indeed such infinite increasing families, with fixed ϵ, have been built from Cayley graphs of discrete subgroups in infinite Lie groups; see e.g. Lubotzky [**Lub94**] for background.

Similar constructions have been applied to families of finite Lie-type groups; typically fixing a single Lie type (and possibly characteristic), and increasing the field-size. Of course this assumes properties of the groups in the CFSG-list 1.0.2.

Indeed Kassabov-Lubotzky-Nikolov [**KLN06**] show that for Lie types other than the Suzuki groups $Sz(q)$ (i.e. twisted type ${}^2C_2(q)$), expander families can always be constructed in this way.

That remaining Suzuki-case was subsequently also shown to provide expanders, in work done by Breuilard-Green-Tao [**BGT11**]; they make use of Tao's notion of a *k-approximate group*. Roughly: this is a subset A of G, closed under inverses, with products $A \cdot A$ covered by k translates, that is, by k cosets gA. For an introduction to approximate groups and their literature, see e.g Green [**Gre12**] and Breuillard [**Bre14**].

A few more uses of simple groups with approximate groups appear in the later paper Breuillard-Green-Tao [**BGT12**]. There are also further developments in the expander literature on constructions via simple groups.

Appendix

This Appendix contains some supplementary notes to the text. Most of them were composed during the course of the original Venice Summer School 2015 lectures, in response to student questions; and were posted as pdf files on the Web at that time.

APPENDIX A

Some supplementary notes to the text

The material in this Appendix should provide some further background, on certain of the features introduced in the original text.

A.1. Notes for 6.1.1: Deducing the structures-list for S_n

The fairly technical details in this appendix-section are not really required for the main flow of the chapter. I present the material only because I had found it necessary to clarify for myself certain points in the literature which had seemed unclear. Ideally these details will also be useful at least for the reader who wants to pursue this area more fully.

Unfortunately, in the literature it seems difficult to find more explicit treatment of the further details—such as inclusions among action-types in the Aschbacher O'Nan-Scott Theorem in 6.1.4—which underlie any deduction, such as the one we indicated after that result, of the structures-list in the O'Nan-Scott Theorem 6.1.1.

For example, Scott at [**Sco80**, p 329] gives just a 5-line sketch for actions-list; and Aschbacher's later correction in [**AS85**, App], giving the final form of the actions-list (A) (i.e. 6.1.4, in our treatment), makes just a bare statement of the inclusion (TW $<$ (3))—which is crucial for seeing that the original statement of the structures-list (S) in 6.1.1 is unaffected by the correction to (A).

Some further details on containments within the types in (A) can be found in various early sources; and they are more methodically treated in Praeger [**Pra90**]. For example: it is fairly standard that the twisted wreath product TW is properly contained in the holomorph of its socle—leading to action-type HS or HC, and hence to structure-type (3). We mention that the containments (HS,HC $\leq$ PA) are in fact detailed in Kovacs [**Kov85**, Sec 3]. Finally the containments (HS $\leq$ SD), and (CD $\leq$ PA), are treated in [**Pra90**, 3.4,3.9].

Properness ($<$) of various containments is usually left implicit: but typically the reader can use the action and the socle-structure to see non-isomorphism of the smaller term with the larger.

Further discussion relevant to these issues appears in [**PS**].

We mention that Wilson [**Wil09**, 2.6.2] does *not* quote the actions-list (A) of 6.1.4, in deducing the structures-list (S) of 6.1.1; so in particular, he gives a CFSG-free proof of the O'Nan-Scott Theorem 6.1.1. We will give a quick overview:

For a proper subgroup H of S_n, distinct from A_n and primitive on the n points, he subdivides cases via partial information on socle-shapes and actions—rather than via the fuller information which would be given if he quoted (A). And then working in the explicit group S_n rather than with abstract actions—in effect, via determining the containments among his subcases—he reduces those subcases for primitive H down to just (3)–(6) of Theorem 6.1.1.

His main logic-sequence is roughly parallel to the proof in [**LPS88**], particularly the inclusions among the 8 types, which we gave in columns 2 and 3 of Remark 6.1.4. Namely: He first reduces to a non-abelian socle—otherwise structure (4) arises. Then, he reduces to H having a *unique* minimal normal subgroup N; for if there were at least two such N, then H would lie in a group having their product as its as unique minimal normal subgroup. This argument parallels the inclusions (HC,HS $<$ (3),(5)) from 6.1.4. The subsequent reduction is to N having at least two components (say isomorphic to T); otherwise structure (6) arises. Then he reduces to the case where N_α surjects on T; otherwise structure (3) arises—this is parallel to (TW $<$ PA). Finally he observes that a compound-diagonal is proper in structure (3), parallel to (HC $<$ CD $<$ PA); while a single diagonal gives structure (5), corresponding to (HS $<$ SD).

Thus Wilson's argument in effect shows in an *implicit* way that four of the primitive-types in column 2 of 6.1.4 are non-maximal; so the remaining four give the candidate-maximal structures (3)–(6) of the O'Nan-Scott Theorem 6.1.1.

A.2. Notes for 8.2.1: The cohomological view of the transfer map

In describing the proof of the Thompson Transfer Theorem 8.2.1, I only briefly mentioned the *transfer* map—of a group G into a subgroup H; and I also indicated that the map could be viewed from a homological perspective. Some of the students requested more detail on this connection—so these notes were provided (online, at that time), as a least the beginning of an answer.

This connection seems to be mentioned fairly frequently in the literature, especially the topological literature; but typically without much detail—especially detail in a format suitable for those with a mainly group-theoretic background. However, a fairly explicit reference is Exercise III.9.2 in Brown [**Bro94**]; later in the section, I'll try to expand a little on that view. Another possible reference is [**Wei69**, p 116].

The classical transfer map in group theory. The *transfer* map is sometimes denoted by V, from the German name *Verlagerung*. For a subgroup $H \leq G$, it is typically viewed as:

the group homomorphism $V : G/G' \to H/H'$ described below.

In many classic group-theory texts (see e.g. [**CR90**, 13.11]), the definition of V is given by a computational formula, as follows:

Assume that the cosets of H in G are given by Hg_i $(i = 1 \cdots n)$. Then corresponding to the action of $g \in G$ permuting these cosets, for each i, we get unique values i' and $h_i \in H$ such that:

$$g_i g = h_i g_{i'}$$

Using these, we define:

$$V(g) := \Pi_{i=1}^{n}\ h_i = \Pi_{i=1}^{n}\ (g_i g g_{i'}^{-1}) \in H; \tag{A.2.1}$$

where the g on the left is read modulo G', and the values on the right are read modulo H'.

Of course the standard references check that this map is well-defined—that is, independent of choice of coset representatives; and that it does indeed give a homomorphism.

Some other texts (e.g. [**Asc00**, Sec 37]) give a more general version, first defining a homomorphism $\alpha : G \to A$ to some abelian group A, and then the transfer V is recovered from the special case $A := H/H'$.

The viewpoint of group homology H_1. The definition (A.2.1) above makes no mention of homology. And as mentioned above, the literature can sometimes be rather terse about the connection: For example, [**Bro94**, Exer III.9.2] has the reader derive the formula in (A.2.1) from the abstract definition of the group homology $H_1(G)$ with coefficients in $\mathbb{Z}$. And [**CR90**, p 336] indicates just that such a derivation "can be shown" (but using the language of Tate cohomology $\hat{H}^{-2}$).

One common shorthand version of the connection is via the statements that: there is a natural "restriction" of $H_1(G)$ to $H_1(H)$; where one standard interpretation of $H_1(G)$ is the quotient G/G' (and similarly for H)—giving the desired map from G/G' to H/H'.

However, for those whose background is group theory (with perhaps less homological algebra), it may be difficult to extract these statements from the fuller generality of definitions given in many algebraic-topology texts. So below I'll try to give a somewhat more leisurely expansion.

For purely formal reasons in homological algebra, the group-inclusion given by $i : H \to G$ induces correponding maps $i_* : H_n(G) \to H_n(H)$ in group homology; see e.g. [**Bro94**, II.3] for general definitions of $H_*(G)$. We'll want to see that for dimension $n = 1$, the induced map i_* is in fact just the transfer map V above.

The transfer map i_* is often called "restriction" from G to H; though it does not correspond to a simple-minded restriction—for example, restricting the action on a G-module to H.

For our dimension $n = 1$, the first key point is to see that $H_1(G)$ can be naturally identified, as claimed above, with the group quotient G/G': To this end, see e.g. the displayed equation at the end of [**Bro94**, II.3]: The map from the chain space C_1 to $C_0 = 0$ is necessarily the zero map; so that the space C_1 of chains is just the space Z_1 of cycles—and furthermore a copy of G gives a spanning set for this space. On the other hand, the space B_1 of boundaries is the image of the map shown there from chains C_2; and that boundary map is just the additive form of the group commutator $hg(gh)^{-1} = hgh^{-1}g^{-1}$, so that the image B_1 is spanned by G'. Thus we get an identification of $H_1 = Z_1/B_1$ with G/G'. (See also 8.54 in [**CR90**] for a more formal discussion.)

At this point, we have expressed i_* as a homomorphism from G/G' into H/H'. So the remaining key point is to show that i_* "has to be" the map computed by the formula (A.2.1) for the group-theoretic transfer V.

And now, in a rough summary of my understanding of a standard way of proceeding: One can use the duality of H_1 with H^1 in e.g. [**Bro94**, VI.7.1]; this expresses the *non*-canonical isomorphism of a finite abelian group with its dual—which we can apply to each of G/G' and H/H'. Now $H^1(G)$ is defined as $\mathrm{Hom}(G, \mathbb{Z})$; and we can restrict such homomorphisms, defined on G, down to H

in the intuitive way—that is, the restriction $i^* : H^1(G) \to H^1(H)$ is straightforward. Then if we compose on each side with the non-canonical isomorphisms of H_1 with H^1 (for G and H, respectively), we obtain a map from G/G' into H/H'. Because of the non-canonical isomorphisms used, in effect one uses a dual-basis calculation of i_*, carried through that composition, to get the formula (A.2.1) for V. (Again, I'm not sure I can point to a source in the literature for fuller details.)

A.3. Notes for (8.3.4): Some details of proofs in Holt's paper

These notes are not really needed for the treatment in the main text.

Instead, they are intended as possible assistance to the more interested reader, who may wish to examine arguments in Holt's original paper [**Hol78**]—which determines transitive groups satisfying (8.3.4), a condition somewhat weaker than strong embedding.

Holt's paper is written in the terse style of exposition which was fairly common around the late 1970s . And I found that in researching the material that became this book, at the much later date of 2015, I sometimes had difficulties in recovering arguments at various points there. So I have provided these notes, indicating some places where for my own understanding I had to expand the original exposition; I hope they may save potential readers from having to repeat some of that effort. And I thank Derek Holt for reading these notes, and providing some improvements.

In the notes in this section, I will often refer to Holt's original numbering of results; but for consistency with the numbering used throughout the present book, I will typically also assign them LaTeX label-numbers of the form A.3.x—for convenient reference within this section. Of course I will also be assigning such numbers to the additional remarks that I have added.

Some preliminary remarks on Section 1 of [Hol78]. First we will discuss some issues which arise in Holt's introductory Section 1; including a number of additional lemmas which may help streamline the later exposition.

In Holt's paper, G is assumed to be a permutation group on a set Ω. We write the group action via a superscript: as in α^g, for $\alpha \in \Omega$ and $g \in G$.

The term "permutation group" by definition includes *faithful* action on Ω. For several later references, it will be convenient to assign a number to this admittedly very basic fact:

(A.3.1) (faithful:) $\ker_G(\Omega) = 1$. (Implicitly $G > 1$; so that $|\Omega| \geq 2$.)

We mention that G is not initially assumed to be transitive on Ω; instead this emerges at A.3.7.

Next we recall that Holt's analysis is motivated by the condition (8.3.4) that a 2-central involution fixes a unique point. Thus Holt begins with [**Hol78**, p 165]:

DEFINITION A.3.2. Following Holt, let J denote the set of involutions in G that fix exactly one point of Ω. For $\alpha \in \Omega$, set $J_\alpha := J \cap G_\alpha$.

And note, for this section only: For a subgroup $H \leq G$, we write $J(H)$ for the analogous subset within H. (Elsewhere in this book, $J(-)$ denotes the Thompson subgroup 3.1.1.) ◊

So we will mainly be concerned with groups G for which $J \neq \emptyset$. And indeed this condition will hold for G, from Holt's Section 2 onward.

Some consequences when $J \neq \emptyset$. At this point, Holt's exposition in Section 1 proceeds to a more general condition he calls (*)—which does not, at least explicitly, assume that $J \neq \emptyset$. We'll give that condition (*) in a moment, at A.3.5.

But first we "digress"—to add a number of useful properties, of the later situation from Section 2 on, where we do explicitly know that $J \neq \emptyset$:

LEMMA A.3.3. *Assume that* $J \neq \emptyset$*: say* $z \in J_\beta$*, for* $\beta \in \Omega$*. Then:*
(1) (unique fixed point:) $Fix_\Omega(z) = \{\beta\}$*.*
(2) $C_G(z) \leq G_\beta$*.*
(3) For $\gamma \neq \beta$*, we have* $z \notin G_\gamma$*; indeed* $J_\beta \cap J_\gamma = \emptyset$*.*
(4) $|\Omega|$ *is odd (and hence* ≥ 3*).*
(5) β^G *is the unique* G*-orbit on* Ω *of odd size.*
(6) (inheritance:) Assume that $z \in H \leq G$*, with* $H \not\leq G_\beta$*; and further that* $\beta^H \subseteq \Delta = \Delta^H \subseteq \Omega$*. Then in* $\overline{H} := H^\Delta$*, we have* $\overline{z} \in J(\overline{H})_\beta$*, and* $|\Delta|$ *is odd* ≥ 3*.*

PROOF. Note that (1) is just a re-phrasing of the basic unique-fixed-point condition (8.3.4), which gives J in Definition A.3.2. Next (1) implies (2), since $C_G(z)$ permutes $\mathrm{Fix}(z)$; and also (1) immediately implies (3). Furthermore as z has order 2, by (1) all z-orbits on Ω, other than β, have size 2—so that $|\Omega|$ is odd; and there must be at least one such nontrivial orbit, by faithfulness in (A.3.1)—giving $|\Omega| \geq 3$, completing (4). Indeed we see by oddness in (4) that G must have an orbit of odd size on Ω: and again as z has order 2, it must fix a point on any such orbit; so that (1) gives the uniqueness of that orbit, as required for (5).

Finally consider the situation in (6): in particular, recall that the standard notation H^Δ means the quotient $H/\mathrm{ker}_H(\Delta)$; and in particular $\overline{H}$ is faithful on Δ. Since $H \not\leq G_\beta$ by hypothesis, we have $|\beta^H| \geq 2$. So by (1), z must have a nontrivial orbit on β^H and hence on Δ. Then $\overline{z} \neq 1$, and so $\overline{z}$ is also an involution in $\overline{H}$; and further $|\Delta| \geq 3$. Using (1) again, we conclude that $\overline{z} \in J(\overline{H})_\beta$. Finally $|\Delta|$ is odd, by (4) applied in $\overline{H}$. □

The relationship with Holt's condition (*). We now return from our digression above on $J \neq \emptyset$, essentially to Holt's exposition at [**Hol78**, p 165]. However before stating his condition (*), it seems important to first discuss one background detail:

Namely, Holt does not—at least explicitly, as far as I can see—assume $|\Omega| \geq 3$. Of course for $|\Omega| = 2$, the only faithful permutation group is S_2, and there $J = \emptyset$; so it's clear that Holt is implicitly making the indicated assumption. So for clarity, we do make that detail explicit:

HYPOTHESIS A.3.4. $|\Omega| \geq 3$.

And indeed, we will continue to indicate that assumption explicitly, wherever it seems needed in the remainder of the notes in this section.

With that issue ideally clarified, we return to Holt's initial exposition; stating his alternative more technical condition (*), under which he does his main work:

HYPOTHESIS A.3.5 (Holt's condition (*)). *Whenever a 2-subgroup* $S \leq G_\alpha$ *satisfies* $|Fix(S)| \geq 3$*, there is* $t \in J_\alpha \cap C_G(S)$*.*

Notice that the condition "$|\mathrm{Fix(S)}| \geq 3$", in order to be non-vacuous when $S = 1$, implicitly requires $|\Omega| \geq 3$—which we have made explicit as Hypothesis A.3.4.

And again we digress briefly, to indicate the technical need for that hypothesis: If $|\Omega| = 2$ were to be allowed, then *no* 2-group S (including the case $S = 1$) can satisfy $|\mathrm{Fix(S)}| \geq 3$. And then Hypothesis A.3.5 by itself actually does hold—though only vacuously. However, it is easily checked that $G = S_2$ does not satisfy the conclusion of Holt's main theorem: either in the form at [**Hol78**, p 165]; or in the form of his Theorem 1. (In the latter case, we'd have "H" trivial—since $J = \emptyset$ in this case of S_2.)

So the admittedly-technical observation in the previous paragraph emphasizes that Holt was implicitly assuming Hypothesis A.3.4. And it motivates our policy of continuing to emphasize it explicitly, in the remainder of the notes in this section.

We also briefly continue our present digression from Holt's exposition, in another direction: it will be convenient to interpolate here an explicit statement of an inheritance-situation for Hypothesis A.3.5, in the spirit of our earlier A.3.3(6); this seems to be used implicitly later in Holt's proof:

LEMMA A.3.6. *Assume that Hypothesis* A.3.5 *holds. Take any* $S \leq G_\alpha$ *with at least 3 fixed points, and any* $t \in J_\alpha \cap C_G(S)$ *(these* t *exist in view of Hypothesis* A.3.5*). Then in* $\overline{C} := C_G(S)^{Fix(S)}$*, we have* $\overline{t} \in J(\overline{C})_\alpha$*, and* $|Fix(S)|$ *is odd.*

PROOF. Set $C := C_G(S)$. Since $\mathrm{Fix}(S)$ has size ≥ 3 by hypothesis, it contains some further point $\beta \neq \alpha$. Then also $S \leq G_\beta$; and we may apply Hypothesis A.3.5 also to β, to similarly get some $u \in J_\beta \cap C$. Now $u \notin G_\alpha$ by A.3.3(3), so $C \not\leq G_\alpha$. And as $\alpha \in \mathrm{Fix}(S)$, we see $\alpha^C \subseteq \mathrm{Fix}(S)$. So now we get $\overline{t} \in J(\overline{C})_\alpha$ and $|\mathrm{Fix}(S)|$ is odd, by A.3.3(6) applied with $t, \alpha, C, \mathrm{Fix}(S)$ in the roles of "z, β, H, Δ". □

We now return from our digression(s), to Holt's exposition at [**Hol78**, p 165].

At this point, Holt states that the case $S = 1$ in Hypothesis A.3.5 implies that G is transitive on Ω. To me this seemed to require some details. This point partly motivated some of the extra material that I added above; so I'll re-state the transitivity remark in the form:

LEMMA A.3.7. *Assume Hypotheses* A.3.4 *and* A.3.5*. Then* $J_\alpha \neq \emptyset, \forall \alpha \in \Omega$*; and* G *is transitive on* Ω.

PROOF. For the trivial 2-group $S = 1$, we have $\mathrm{Fix}(1) = \Omega$ and $C_G(1) = G$. Now as discussed above: here it is necessary to have $|\Omega| \geq 3$: for we had noted earlier that $J = \emptyset$ when $|\Omega| = 2$. But since we have added Hypothesis A.3.4, we see $\mathrm{Fix}(1) = \Omega$ has size ≥ 3; and then Hypothesis A.3.5, applied for *any* α and this choice of $S = 1$, gives $J_\alpha \neq \emptyset$, as required for the first statement.

For the final statement of transitivity: We see α is in the unique G-orbit of odd length using A.3.3(5). Since this applies to each $\alpha \in \Omega$, that unique odd-length orbit is all of Ω. □

Holt's next sentence states that transitive groups with the unique-fixed-point condition (8.3.4) in fact satisfy Hypothesis A.3.5. Again this seemed to me to require details; so having developed some tools above, I'll re-state that assertion as:

LEMMA A.3.8. *Any transitive* G *with* (8.3.4) *satisfies* A.3.4 *and* A.3.5.

PROOF. From the hypothesis (8.3.4), we have an involution $z \in Z(T)$ for some $T \in \mathrm{Syl}_2(G)$, with $z \in J_\beta$ for some β. In particular $J \neq \emptyset$, so via A.3.3(4), we get Hypothesis A.3.4 that $|\Omega| \geq 3$. For reasons explained earlier, it seems important to state this explicitly.

Furthermore $T \leq C_G(z)$, and $C_G(z) \leq G_\beta$ using A.3.3(2); so we have:

$$|G_\beta|_2 = |G|_2.$$

In particular, $\beta \in \mathrm{Fix}(T) \subseteq \mathrm{Fix}(z)$, while $\mathrm{Fix}(z) = \{\beta\}$ using the re-phrasing in A.3.3(1) of the unique-fixed-point condition (8.3.4). So we conclude that:

$$\mathrm{Fix}(T) = \mathrm{Fix}(z) = \{\beta\}.$$

With these facts in hand, we can start to verify the requirements for Hypothesis A.3.5: We take any $\alpha \in \Omega$, and any 2-group $S \leq G_\alpha$ with $|\mathrm{Fix}(S)| \geq 3$; and we must show $J_\alpha \cap C_G(S) \neq \emptyset$.

So embed $S \leq R \in \mathrm{Syl}_2(G_\alpha)$. Then using $|G_\beta|_2 = |G|_2$, and the hypothesis that G is transitive on Ω, we see that $R \in \mathrm{Syl}_2(G)$; so $R = T^g$ for some $g \in G$. We must have $z^g \in J$, since $z \in J$; and $z^g \in Z(R) \leq C_G(S)$. Now using the equality $\mathrm{Fix}(T) = \mathrm{Fix}(z) = \{\beta\}$ above, we have:

$$\alpha \in \mathrm{Fix}(G_\alpha) \subseteq \mathrm{Fix}(R) = \mathrm{Fix}(z^g) = \{\beta^g\}.$$

Thus $\alpha = \beta^g$, and so $z^g \in G_\alpha$. Combining this with earlier remarks, we now have $z^g \in J_\alpha \cap C_G(S)$, as required for Hypothesis A.3.5 □

We recall that (8.3.4) is the form of Holt's hypothesis that is used in most applications of his result; in particular, for those suggested in toward the end of our discussion in that area of Section 8.3.

Conjugacy of involutions in dihedral groups. The material above concludes our discussion of parts of the exposition of Holt in [**Hol78**, Sec 1].

Before continuing with Holt's further exposition, we explicitly state some standard facts about conjugacy of involutions in dihedral groups—these are in fact applied in later sections of [**Hol78**]:

LEMMA A.3.9. *Consider a dihedral group $D = \langle d, e\rangle$, generated by distinct involutions d, e. Set $x := de \neq 1$, to denote a generator of the cyclic subgroup of index 2 in D. Write $a \sim_D b$ when a, b are D-conjugate (or just $a \sim b$ when D is understood).*

(1) If $|x|$ is odd (hence ≥ 3), all involutions in D are conjugate; and in particular, d and e are interchanged by an involution in D.

(2) If $|x|$ is even, the unique involution of cyclic $\langle x\rangle$ is central in D; and we have $d \sim dx^2 \sim dx^4 \sim \cdots$, and $e \sim ex^2 \sim ex^4 \sim \cdots$—accounting for all the other involutions of D.

Notes for Section 2 of Holt. Next we will discuss some aspects of the more general exposition in [**Hol78**, Sec 2]; where Holt establishes some basic properties under his main hypothesis.

Preliminary observations. For the remainder of the notes in this section:

We assume Hypotheses A.3.4 and A.3.5.

We already indicated why we have added A.3.4, namely $|\Omega| \geq 3$, to Holt's original hypothesis.

In view of Lemma A.3.7, this means we can start out with the properties:

(A.3.10) G is transitive on Ω, and $J_\alpha \neq \emptyset$ for each $\alpha \in \Omega$.

We mention that the transitivity here is presumably the reason for the statement that "m is ... independent of α", in the second sentence of Section 2 of Holt. Furthermore $J_\alpha \neq \emptyset$ is seemingly required for the statement that $m_{\alpha\beta} \neq 0$ in 2.1 of [**Hol78**]. But in any case, from $J \neq \emptyset$ in (A.3.10):

We may apply the various properties in A.3.3.

In particular, we recall that A.3.3(4) gives us:

$$|\Omega| \text{ is odd (and } \geq 3).$$

Remarks on the proof of [**Hol78**, 2.1]. Here we expand on the first few lines of the proof of Holt's (2.1)—showing that certain involutions $\bar{t}, \bar{u}$ are interchanged by an involution in $\langle \bar{t}, \bar{u} \rangle$. In particular, this allows us demonstrate another use of A.3.3(6); as well as the use of the dihedral-conjugacy results above.

For any distinct $\alpha, \beta \in \Omega$, by (A.3.10) we can find $t \in J_\alpha$ and $u \in J_\beta$. We now set $D := \langle t, u \rangle$, and $Y := \langle tu \rangle$. We recall in this situation that:

$$Y \text{ is cyclic of index 2 in dihedral } D.$$

Set $\Delta := \alpha^D$, and $\overline{D} := D^\Delta = D/\ker_D(\Delta)$. Our initial goal is to show that $\overline{Y}$ has odd order ≥ 3.

Using A.3.3(2)(3) we have:

$$\text{(A.3.11)} \qquad u \notin G_\alpha, \text{ so } u \notin C_G(t).$$

In particular $[t, u] \neq 1$, so that $|tu| > 2$; and so we've already shown at least that:

$$\text{(A.3.12)} \qquad Y \text{ has order } \geq 3.$$

The oddness of $|\overline{Y}|$ will require a little more work.

Since $u \notin G_\alpha$ by (A.3.11), we see $D \not\leq G_\alpha$. So by the inheritance property A.3.3(6):

$$\bar{t} \in J(\overline{D})_\alpha, \text{ and } |\Delta| \text{ is odd } \geq 3.$$

So u of order 2 must fix some point of Δ—which by the usual unique-fixed-point condition A.3.3(1) can only be β; and in particular, we have $\beta \in \Delta$. And now we can argue symmetrically, interchanging t and u: Arguing as for (A.3.11), $z \notin G_\beta$ by A.3.3(3); so we may apply A.3.3(6) to u, to get that also $\bar{u} \in J(\overline{D})_\beta$. The arguments for (A.3.12), made now in $\overline{D}$, go through to show $|\overline{Y}| \geq 3$.

Further $\Delta = \alpha^{\overline{D}}$ has size $|\overline{D} : \overline{D}_\alpha|$. And since $\overline{Y}$ has index 2 in $\overline{D}$, while we got $\bar{t} \in \overline{D}_\alpha$ in the previous paragraph, we see that Δ in fact has size $|\overline{Y} : \overline{Y}_\alpha|$. But $\overline{Y}_\alpha$ is normal in $\overline{D}$, as Y is cyclic of index 2 in D, and $\overline{D}$ is transitive on Δ; so $\overline{Y}_\alpha$ must in fact fix all the points of Δ—and we get $\overline{Y}_\alpha = 1$, since $\overline{D}$ is faithful on Δ by construction. Thus we conclude that $|\Delta|$, which is odd (and ≥ 3) by the previous paragraph, is equal to $|\overline{Y}|$ (which we also saw there is ≥ 3). This achieves our initial goal.

In particular, $\overline{D}$ now has twice odd order: $2|\overline{Y}|$. And so by A.3.9(1), all involutions of $\overline{D}$ are conjugate; with $\bar{t}$ and $\bar{u}$ interchanged by an involution of $\overline{D}$. This completes our expansion of the first few lines of the proof of [**Hol78**, 2.1].

Remarks on [**Hol78**, 2.2]. We also comment on Holt's subsequent result, on an inheritance-property for his main hypothesis:

LEMMA A.3.13. *Assume that $S \leq G$ is a 2-group which satisfies $|Fix(S)| > 1$. Then $C_G(S)^{Fix(S)}$ satisfies Hypotheses* A.3.4 *and* A.3.5.

PROOF. We will provide some details—since the result is crucial for the later application of induction (for example, via [**Hol78**, 2.5]).

Recall that our overall hypotheses on G gave $J \neq \emptyset$ in (A.3.10); and this implies by A.3.3(4) that $|\Omega|$ is odd and ≥ 3. Since nontrivial orbits on Δ of the 2-group S have size given by a power of 2, we see that $|\text{Fix}(S)|$ is odd; so our present hypothesis that $|\text{Fix}(S)| > 1$ in fact guarantees that $|\text{Fix}(S)| \geq 3$. In particular, we have established Hypothesis A.3.4 for $\text{Fix}(S)$. So it remains to establish Hypothesis A.3.5, for $C_G(S)^{\text{Fix}(S)}$.

Set $C := C_G(S)$, $\Delta := \text{Fix}(S)$, and $\overline{C} := C^{\Delta} = C/\ker_C(\Delta)$. Let α denote any point of Δ; in particular $S \leq G_\alpha$. Then from Hypothesis A.3.5 in G, since we obtained $|\text{Fix}(S)| \geq 3$ in the previous paragraph, we get some $t \in J_\alpha \cap C$; and indeed from A.3.6, we even get that $\bar{t} \in J(\overline{C})_\alpha$.

But this isn't quite the same as concluding Hypothesis A.3.5 for $\overline{C}$. For this, we must consider any 2-group $\overline{T} \leq \overline{C}_\alpha$ with $|\text{Fix}_\Delta(\overline{T})| \geq 3$, and we must show that $J(\overline{C})_\alpha \cap C_{\overline{C}}(\overline{T}) \neq \emptyset$.

So let T denote the preimage in C of $\overline{T}$, so that $S \leq T$; and take U to be any Sylow 2-subgroup of T, so that we have $\overline{U} = \overline{T}$. Since S is central in C and hence in T, while U is Sylow in T, we have $S \leq U$. Now $\alpha \in \text{Fix}_\Delta(\overline{T})$ of size ≥ 3; and also $\text{Fix}_\Delta(\overline{T}) = \text{Fix}_\Delta(\overline{U}) \subseteq \text{Fix}_\Omega(U)$, so the latter has size ≥ 3 as well.

Thus we can apply Hypothesis A.3.5 in G, with U in the role of "S", to α. This provides us with some $z \in J_\alpha \cap C_G(U)$. Indeed as $S \leq U$, we have $C_G(U) = C_C(U)$, so $z \in J_\alpha \cap C$; and our earlier argument via A.3.6 shows that $\bar{z} \in J(\overline{C})_\alpha$. But also we have $\overline{C_C(U)} \leq C_{\overline{C}}(\overline{U}) = C_{\overline{C}}(\overline{T})$. Thus we have shown that $\bar{z} \in J(\overline{C})_\alpha \cap C_{\overline{C}}(\overline{T})$—as required for Hypothesis A.3.5 in $\overline{C}$. □

It is also convenient to explicitly state here one consequence of A.3.13:

LEMMA A.3.14. *For S in* A.3.13, *$C_G(S)$ is transitive on Fix(S) of odd size ≥ 3.*

PROOF. We saw early in the previous proof that $|\text{Fix}(S)|$ is odd ≥ 3. So A.3.14 gives us the hypotheses, inside $\overline{C}$, for A.3.7—establishing the indicated transitivity. (This is essentially the subcase $\overline{T} = 1$ in the previous proof.) □

We now skip over the remaining preliminary results in [**Hol78**, Sec 2].

Some remarks on Section 3 of Holt. Section 3 of [**Hol78**] begins the main proof: Namely Holt continues to assume Hypothesis A.3.5 (and implicitly also A.3.4 that $|\Omega| \geq 3$, as we have discussed). And he now further assumes that G is a counterexample to his Theorem 1, and aims to produce a contradiction. His Section 3 develops a number of general properties of this more specific situation.

In particular, the preliminary results of his Section 2 continue to be available: notably transitivity and $J_\alpha \neq \emptyset$ in (A.3.10). By the latter, we can continue to use the properties in A.3.3; especially $|\Omega|$ odd, via A.3.3(4).

The proof uses induction on $|G|$; so that faithful $G = G^\Omega$ is chosen as a counterexample of minimal order. We will record Holt's fundamental remark on induction, in his first paragraph, in the following form; and also we will indicate an explicit proof:

LEMMA A.3.15. *If $|Fix(S)| > 1$ for a 2-group $S > 1$, then $C_G(S)^{Fix(S)}$ satisfies Theorem 1.*

PROOF. We saw in A.3.13 that $\overline{C} := C_G(S)^{\mathrm{Fix}(S)}$ has Hypotheses A.3.4 and A.3.5. This part of the argument works even if we have $S = 1$.

To apply induction on $|G|$, we will need $C_G(S) < G$. So now we apply the hypothesis $S > 1$: If we had $C_G(S) = G$, then G would permute $\mathrm{Fix}(S)$; which would then by transitivity of G in (A.3.10) be all of Ω. But now faithfulness of G (recall (A.3.1)) would force $S = 1$, contrary to hypothesis.

Thus by induction, $\overline{C}$ satisfies Theorem 1. □

Notice we could re-phrase the argument on faithfulness in the previous proof as:

In A.3.15, $S \cap Z(G) = 1$; so that $C_G(S) < G$.

We now skip over the remaining results in [**Hol78**, Sec 3].

Notes on Section 4 of Holt. We had introduced Holt's Theorem in Section 8.3, as an application (indeed, extension) of the Strongly Embedded Theorem 2.0.17.

In fact the groups in Theorem 2.0.17 arise as conclusion-groups for Holt during the proof of his [**Hol78**, 4.1]—which we state below as Lemma A.3.16. Consequently in this subsection, we will primarily explore Holt's treatment of that result.

We observe that the condition in A.3.16 does *not* hold in the Bender groups, which are conclusion-groups in Theorem 2.0.17 and consequently also in his Theorem 1. Thus A.3.16 is true only in the context of his overall proof of Theorem 1 by contradiction: namely when the Bender groups do arise, they will contradict his assumption at the start of Section 3 that G is a counterexample to Theorem 1. However, if we choose to view the overall logic in a more "forward" direction, we are actually seeing the Bender groups arising as conclusions; and hence *reducing* the proof thereafter to the situation in A.3.16.

LEMMA A.3.16. *There exist distinct $a, b \in J_\alpha$, with $[a,b] = 1$ and $ab \notin J$.*

We will expand on Holt's proof of this lemma. The proof proceeds by contradiction. One aspect of assuming that A.3.16 fails is:

(A.3.17) For any distinct $a, b \in J_\alpha$ with $[a,b] = 1$, also $ab \in J_\alpha$.

Along the way, we will want to show that the elements of J are G-conjugate. We approach that condition via the following more technical statement:

(A.3.18) Given $t \in J_\alpha$, all $u \in J_\gamma$ (for $\gamma \neq \alpha$) are G-conjugate to t.

PROOF. We set up just as in our earlier remarks on the proof of 2.1 in [**Hol78**]: Set $D := \langle t, u\rangle$, $y := tu$ and $Y := \langle y \rangle$. We saw at (A.3.12) that $u \notin G_\alpha$, and Y has order ≥ 3. And using A.3.9(1) (as we did for $\overline{Y}$ at the end of those remarks), we see that if Y has odd order, then we have the conjugacy $t \sim_D u$, as required.

This reduces us to even $|Y| \geq 4$. Let v denote the unique involution in Y, which is central in D. Here we get $v \in C_G(t,u) \leq G_\alpha \cap G_\gamma$ using A.3.3(2). So setting $\Delta := \mathrm{Fix}(v)$, we have $|\Delta| > 1$.

Also set $C := C_G(v)$ and $\overline{C} := C^\Delta$. By A.3.15 with $\langle v \rangle$ taken in the role of "S", we see $\overline{C}$ satisfies Theorem 1. Here Holt's earlier lemma [**Hol78**, 2.5] shows that:

all elements of $J(\overline{C})$ are $\overline{C}$-conjugate.

We will use this fact, to pull back to suitable conjugacy in G.

For note that we have $t, u \in D \leq C$; and by the inheritance property A.3.6, we have $\bar{t} \in J(\bar{C})_\alpha$ and $\bar{u} \in J(\bar{C})_\gamma$. Hence $\bar{t} \sim_{\bar{C}} \bar{u}$ by the previous paragraph. So there is an element $w \in C$, with $\bar{w} = \bar{t}$ and $w \sim_C u$; in particular w is also an involution. Thus it will now suffice to show that $w \sim_G t$. We are done if $w = t$; so we may assume that $wt \neq 1$. Note since $\bar{w} = \bar{t}$ that also $w \in G_\alpha$.

At this point, Holt indicates "by assumption $|wt|$ is odd". Presumably he means that this follows by assuming failure of A.3.16—probably via something like (A.3.17). This doesn't seem to me to be immediate, so I'll supply some details:

Set $E := \langle w, t \rangle \leq G_\alpha$, $x := wt$, and $X := \langle x \rangle$.

Assume by way of contradiction that $|x|$ is even. Let z denote the unique involution of X, which is central in $E = \langle w, t \rangle$. Since $w \sim_C u$ with $u \in J(C)$, we also have $w \in J(C) \subseteq J$; so as we saw $w \in G_\alpha$, in fact $w \in J_\alpha$. Next since $\bar{w} = \bar{t}$, and w is an involution, we have $\bar{w}\bar{t} = \overline{(wt)} = 1$; and hence we see that $x = wt \in \ker_C(\Delta)$. So we get $z \in X = \langle x \rangle \leq \ker_C(\Delta)$. And then as $|\Delta| > 1$, we conclude from the unique-fixed-point condition A.3.3(1) that $z \notin J$. On the other hand, since $|x|$ is even, we can apply A.3.9(2): Since $tz \notin X$, it must be E-conjugate to either w or t. But in either case, we get $tz \in J_\alpha$, via that conjugacy under $E \leq G_\alpha$; and we also have $[t, tz] = 1$, since $z \in Z(E)$. We can now finally apply our contradiction-hypothesis (A.3.17) to conclude that also $t(tz) = z \in J_\alpha$; but this contradicts our observation above that $z \notin J$.

This contradiction establishes that $|x|$ is odd (and hence ≥ 3). So again arguing as at the outset with A.3.9(1), we have $w \sim_E t$. And as observed earlier, this suffices to complete the proof of (A.3.18). □

In particular (A.3.18) shows that all members of J_γ are G-conjugate; so by transitivity (A.3.10) this conjugacy also holds for all other J_α. And since (A.3.18) also shows that members of J_γ are fused into some other J_α, transitivity now establishes:

(A.3.19) All elements of J are G-conjugate.

We will now see how strongly embedded subgroups arise, as we complete the proof of Lemma A.3.16:

Let Γ denote the commuting graph of J. Here (A.3.11) shows that connected components lie within the separate J_α; so in particular, Γ is disconnected. Further we have $G = \langle J \rangle$ and $O(G) = 1$, by earlier 3.1 in [**Hol78**]; this essentially just applies induction on $|G|$.

These properties, together with our contradiction-hypothesis (A.3.17), corresponding to the failure of Lemma A.3.16, give the hypotheses for 1.2 in [**Hol78**]: this is the sufficient condition of Aschbacher [**Asc73**] for G to contain a strongly embedded subgroup, which we had mentioned in our briefer discussion of Holt's Theorem in Section 8.3.

Thus by the Strongly Embedded Theorem 2.0.17—given as 1.1 in [**Hol78**]—we can conclude that $F^*(G)$ is a Bender group: that is, a Lie-type group of rank 1 in characteristic 2.

This contradicts Holt's assumption that G is a counterexample to Theorem 1; and so completes the proof of Lemma A.3.16. □

But, as we noted before the proof: The more natural contradiction to observe here is that the Bender groups do *not* satisfy the condition of A.3.16—for in them, the role of J_α is played by the involutions in a root group; they commute, but they are also all conjugate, so that products remain in J. Thus really the Bender groups in the conclusion of Holt's Theorem 1 (they are described more explicitly in his Theorem 2) are actually *arising* here; and hence the remainder of the proof is *reduced* to the situation that holds in A.3.16.

We conclude the section with a final remark on the remainder of Holt's Section 4, where he completes the proof of Theorem 1. He proceeds with a pair t, u chosen as in Lemma A.3.16: namely $t, u \in J_\alpha$ with $[t, u] = 1$ and $tu \notin J$. This will lead, with considerable further work, to the alternating and symmetric conclusion-groups in Theorem 1 (again the groups are more clearly visible in Theorem 2); they arise notably toward the bottom of [**Hol78**, p 182].

APPENDIX B

Further remarks on certain exercises

B.1. Some exercises from Chapter 1

REMARK B.1.1 (More on Exercise 1.3.22: Some practice with root systems and parabolics). We first recall that Example 1.3.21 described the general root system of type A_{n-1}, for $SL_n(q)$; and in particular, positive roots are characterized as sums of simple roots which are *adjacent* in the ordering on Π. For fuller rerefence see e.g. [**Car89**, Sec 3.6(i)].

Subspace stabilizers in $L_4(2)$. Now we specialize to $n = 3$, for $G = L_4(2)$. We have simple roots given by $\alpha_1, \alpha_2, \alpha_3$ in Π; the remaining positive roots in Φ^+ are given by the pairs $\alpha_1+\alpha_2$ and $\alpha_2+\alpha_3$ (but *not* $\alpha_1+\alpha_3$!), and the triple $\alpha_1+\alpha_2+\alpha_3$.

First consider $k = 1$: namely the stabilizer $P_{\hat{1}}$ of a 1-space V_1 in the natural 4-dimensional module V. We use the description given in Remark 1.3.20(4): Here for $J = \hat{1} = \Pi \setminus \{\alpha_1\} = \{\alpha_2, \alpha_3\}$, we first compute the unipotent radical $U_{\hat{1}}$: The positive roots which are *not* linear combinations from J are α_1, $\alpha_1 + \alpha_2$, and $\alpha_1 + \alpha_2 + \alpha_3$. These generate the unipotent radical $U_{\hat{1}}$; notice that it is elementary abelian 2^3 using the property (1.3.13): for the sum of any two of these generating-roots is *not* in the root system Φ. We can in fact recognize $U_{\hat{1}}$ as the classical "point" transvection subgroup: For this we recall a standard general definition:

DEFINITION B.1.2 (transvection). An x of order p, acting on V in characteristic p, is a *transvection* if $C_V(x)$ is a hyperplane of V; or equivalently, if $[V, x]$ has dimension 1.

In our present case of $U_{\hat{1}}$, all the elements have the *same* commutator on V, given by the projective point V_1; in particular, they are trivial on both V_1 and V/V_1.

We turn now to the Levi complement $L_{\hat{1}}$: The positive roots that are J combinations are given by α_2, α_3, and $\alpha_2 + \alpha_3$. The root subgroups for these roots and their negatives generate $L_{\hat{1}} \cong L_3(2)$: which is trivial on V_1, but acts naturally on V/V_1.

Next consider $k = 3$: namely the stabilizer $P_{\hat{3}}$ of a 3-space V_3. Here since a graph automorphism of G interchanges α_1 and α_3, while fixing α_2, we can obtain the structure of $P_{\hat{3}}$ by making these same interchanges, in the setup of the previous paragraph. In particular, $U_{\hat{3}}$ is the "hyperplane" transvection subgroup: whose elements are transvections, all centralizing the hyperplane V_3, and whose commutators on V given by various 1-subspaces within V_3. This group is trivial on V_3 and V/V_3; while $L_{\hat{3}} \cong L_3(2)$ is natural on V_3, but trivial on V/V_3.

Finally consider $k = 2$, namely the stabilizer $P_{\hat{2}}$ of a 2-subspace V_2. This time we work with the subset $J = \hat{2} = \Pi \setminus \{\alpha_2\} = \{\alpha_1, \alpha_3\}$. The positive roots which are not J-combinations are $\alpha_2, \alpha_1 + \alpha_2, \alpha_2 + \alpha_3$, and $\alpha_1 + \alpha_2 + \alpha_3$. These generate the unipotent radical $U_{\hat{2}}$, which is elementary abelian of rank 4. And the positive roots that are J-combinations are α_1 and α_3; the root subgroups for these roots and their negatives then generate the Levi complement $L_{\hat{2}} \cong L_2(2) \times L_2(2)$, whose factors act naturally on V_2 and V/V_2, respectively. □

Isotropic-subspace stabilizers in $Sp_4(2)$. Here we have root system of type C_2; for reference see e.g. [**Car89**, Sec 3.6(i)].[1] Then the simple roots in Π are α_1 (short) and α_2 (long); with further positive roots in Φ^+ given by $\alpha_1 + \alpha_2$ (short), along with $2\alpha_1 + \alpha_2$ (long).

For the natural 4-dimensional module V, as in Exercise 1.3.6 we take a hyperbolic basis given by orthogonal hyperbolic pairs v_1, v_4 and v_2, v_3—that is, such that $(v_1, v_4) = 1 = (v_2, v_3)$ in the form. Then the totally isotropic subspaces are represented by $V_1 := \langle v_1 \rangle$ and $V_2 := \langle v_1, v_2 \rangle$.

So for $k = 1$ and the stabilizer $P_{\hat{1}}$ of the 1-space V_1: With $J = \hat{1} = \{\alpha_2\}$, the positive roots which are not J-combinations are α_1, $\alpha_1 + \alpha_2$, and $2\alpha_1 + \alpha_2$. The corresponding root subgroups generate the unipotent radical $U_{\hat{1}}$: which turns out, in characteristic 2, to be elementary abelian 2^3—though this is not immediate just from (1.3.13) and the root system, Note that $U_{\hat{1}}$ acts trivially on the quotients of the series $0 < V_1 < V_1^{\perp} < V$. Next for the Levi complement $L_{\hat{1}}$: The only positive root which is a combination from J is just α_2 itself; and the root subgroups for $\pm\alpha_2$ generate $L_{\hat{1}} \cong L_2(2)$; this acts naturally on $V_1^{\perp}/V_1$; but trivially on V_1 and $V/V_1^{\perp}$. The parabolic $P_{\hat{1}}$ for $Sp_4(2)$ can be studied in the intersection $P_{\hat{1}} \cap P_{\hat{3}}$ of parabolics for $L_4(2)$ above—since it is invariant under a graph automorphism switching α_1 and α_3.

Finally for $k = 2$ and the stabilizer of the totally isotropic 2-subspace V_2: With $J = \hat{2} = \{\alpha_1\}$, the positive roots which are not J-combinations are given by α_2, $\alpha_1 + \alpha_2$, and $2\alpha_1 + \alpha_2$. Again these generate an elementary abelian unipotent radical $U_{\hat{2}}$, which is trivial on V_2 and V/V_2. For the Levi complement $L_{\hat{2}}$: The only positive root which is a J-combination is α_1; and the root subgroups for $\pm\alpha_1$ generate $L_{\hat{2}} \cong L_2(2)$, which acts naturally on V_2 and V/V_2. Again this group can be studied inside the parabolic $P_{\hat{2}}$ for $L_4(2)$ above; in particular, the $L_2(2)$ in the Levi complement in $Sp_4(2)$ is a suitable diagonal embedding in the two factors $L_2(2)$ of the Levi complement in $L_4(2)$. □

This concludes our further remarks on practice with parabolics. ◇

B.2. Some exercises from Chapter 4

REMARK B.2.1 (More on Exercise 4.0.2: The Thompson Order Formula). The case A_8 suggested in Exercise 4.0.2 is perhaps a little complicated for the beginner. So I'll record here the details which I gave in the corresponding problem-session and in a web pdf file, at the time of the original lectures.

Recall that for G with classes of involutions z, t, Thompson's formula in Theorem 4.0.1 reads:

$$|G| = a(z)|C_G(t)| + a(t)|C_G(z)|, \tag{B.2.2}$$

[1] Note that the diagram labeled B_2 on [**Car89**, p 46] would usually be denoted C_2, since the first simple root is short.

where $a(u)$ is the number of ordered pairs (x, y) from the classes z, t with $u \in \langle xy \rangle$. Of course since x, y are involutions, $D := \langle x, y \rangle$ is dihedral, with xy generating a cyclic normal subgroup W of index 2 in D.

Thus if we wish to compute $a(z)$, we must consider cases with $z \in W$. In such cases, z is the unique involution in cyclic W, so that:

D is a dihedral subgroup, of order divisible by 4, in $C_G(z)$.

Consequently the structure of $C_G(z)$ is part of the local information that we will use, to implement Thompson's formula.

Indeed in calculating $a(z)$, we may as well compute the entire $C_G(z)$-*orbit* of each pair (x, y)—rather than trying to write down individually all the pairs in that orbit. And that orbit-size is just the index in $C_G(z)$ of $C_{C_G(z)}(x, y) = C_{C_G(z)}(D)$:

$$|C_G(z) : C_{C_G(z)}(D)| \tag{B.2.3}$$

We'll use the small cases $G = S_4$ and S_5 in Exercise 4.0.2 for a quick warmup.

The case $G = S_4$. Take $z := (12)(34)$ and $t := (12)$ to represent the two classes. Then we have $C_G(z) \cong D_8$ is dihedral of order 8; while we see $C_G(t) \cong S_2 \times S_2$ is a 4-group.

We first consider cases for $a(z)$:

Here $C_G(z) \cong D_8$, so dihedral subgroups D of order divisible by 4 can have size either 4 or 8. But we can't have $D = \langle x, y \rangle$ a 4-group: for this would give $z = xy$—impossible as z, x are even permutations, but y is odd.

The remaining order-8 case in fact forces $D = C_G(z)$: And we see we can take x as either of the involutions of A_4 other than z, say $(13)(24)$; and y either of the involutions of $C_G(z)$ outside A_4, say (12). Here we've directly counted 4 such pairs; but we can also derive that count by the general orbit-method in (B.2.3): for we see that $C_{C_G(z)}(D) = C_D(D) = Z(D) = \langle z \rangle$ has index 4 in $C_G(z)$. Further as $D = C_G(z)$ here, this is the *only* orbit of x, y.

From these calculations, we conclude that $a(z) = 0 + 4 = 4$.

We turn to cases for $a(t)$:

Here $C_G(t)$ is a 4-group; so the only possible dihedral subgroup D is $C_G(t)$ itself. And as we saw for z above, D of order 4 forces $xy = t$. Now in the explicit situation of $D = C_G(t) = \langle (12) \rangle \times \langle (12)(34) \rangle$, we see that $x = z = (12)(34)$ and $y = (34)$ give the only possible pair. Thus $a(t) = 1$.

The main formula (B.2.2) now gives $|S_4| = 4 \cdot 4 + 1 \cdot 8 = 24$, as desired. □

The case $G = S_5$. Here we can re-use much of the above setup for S_4: For example, we can again take $z = (12)(34)$ and $t = (12)$. And still $C_G(z) \cong D_8$; but now $C_G(t) \cong S_2 \times S_3$ has order 12 rather than 4.

Again $a(z) = 4$: as $C_G(z)$ is unchanged, the S_4-arguments still apply.

So we turn to cases for $a(t)$:

This time $C_G(t) \cong D_{12}$; so dihedral subgroups D of order divisible by 4 can have size either 4—or a new possibility given by 12.

We get from S_4 the order-4 case, with $D = \langle (12) \rangle \times \langle (12)(34) \rangle = \langle t, z \rangle$; notice that we have $x = z$ and $y = tz = (34)$. But this time $D < C_G(t)$, so the pair is not uniquely determined. Indeed since $C_G(t) \cong D_{12}$, we see that $C_{C_G(t)}(D) = D$ is of index 3 in $C_G(t)$. Thus from (B.2.3) this orbit-length contributes of 3 pairs.

The case D of order 12 forces $D = C_G(t)$. And such a D does indeed arise—for example, from $x = z = (12)(34)$ and $y = (35)$. Furthermore here we find that $C_{C_G(t)}(D) = C_D(D) = Z(D) = \langle t \rangle$. This has index 6 in $C_G(t)$; so via (B.2.3) we get a contribution of 6 pairs. Furthermore we see this is the only such orbit from $C_G(t)$.

We conclude that $a(t) = 3 + 6 = 9$.

Now the main formula (B.2.2) gives $|S_5| = 4 \cdot 12 + 9 \cdot 8 = 120$, as desired. □

The case $G = A_8$. We turn to the main work of this Exercise. Notice that A_8 provides the first "realistic" example: in the sense of being the smallest *simple* group with exactly two classes of involutions.

I'll regard A_8 primarily as $L_4(2)$ (recall the isomorphisms in Exercise 1.5.5), for several reasons:

- I'm not skillful with permutations, and could probably never get the calculation right in that notation. However, I will make occasional comparisons with the viewpoint of A_8.
- I find the linear-group view more natural for the structures of $C_G(t)$ etc. (You might disagree!) For example, sometimes using (1.3.13) with knowledge of the root system helps to describe the action of involutions.
- Historical value: The structures are similar to those in various calculations via the Thompson Order Formula in the literature. In particular, we'll see features that were used in large-extraspecial theory (cf. Definition 8.1.3); recall this is a part of the $GF(2)$-type case, which we saw as branch (3) of the Trichotomy Theorem 2.2.8.

And of course I hope that you (the readers) will find future applications: both of the formula, and of the methods used in implementing it.

Let me expand briefly on the "realistic" aspect mentioned just above: I will be giving explicit representatives in $L_4(2)$ of the cases for pairs x, y etc. But observe that if you remove the explicitness from the examples, and just use the abstract group-theoretic structures of $C_G(z)$ etc: you should still be able to see "abstract" representatives of x, y, and make the same orbit-length calculations. That is: you can work just as you'd have to do, in a classification problem: where you would have local information about your group G; but still have to prove it was $L_4(2)$, or something else—starting with finding its order via the formula.

Some general setup in $L_4(2)$:

A less advantageous feature of my $L_4(2)$-convention (compared with permutations in A_8) is that 4×4 matrices are rather bulky to write out. So I'll adopt a terse notational abbreviation for matrices; leaving the reader to write them out more fully. It will be possible to give representatives of our needed pairs (x, y) using essentially only unipotent matrices; for example:

$$\begin{pmatrix} 1 & & & \\ a & 1 & & \\ b & c & 1 & \\ d & e & f & 1 \end{pmatrix}.$$

DEFINITION B.2.4. For the remainder of this Exercise, I'll use vertical bars to delimit the set of nonzero values in such a matrix; for example: $|a, e|$, to denote the with $a = e = 1$—but with $b = c = d = f = 0$. Sometimes we'll need the transposes of such matrices; so $a' \cdots f'$ will index the corresponding transposed-positions. The reader will probably still find it helpful to expand the occurrences of this notation back to matrix pictures. ◇

REMARK B.2.5 (The root-system viewpoint). Notice that we can regard $|a|$ as the generator of the root subgroup U_{α_1}; and $|b|$ as the generator of the root subgroup $U_{\alpha_1+\alpha_2}$; and so on. Indeed we had already interpreted matrices via root-structure, in Example 1.3.10—for the smaller group $L_3(2)$ and root system A_2,

And for the present larger root system of type A_3, the root-viewpoint will again be useful: especially, as we had mentioned above, using (1.3.13) to reduce some of our calculations of actions of involutions below. In fact we recorded many facts about the A_3 root system, in Remark B.1.1. ◇

The involution classes in $L_4(2)$:

The class of an involution u is determined by the dimension (1 or 2) of the commutator $[V, u]$ on the 4-dimensional natural module V.

The case of of commutator of dimension 1 (namely a transvection) gives the class of 2-central z—it can be represented by the matrix $|d|$ in the above convention. (In A_8 it has cycle-type 2^4.) Notice that $C_G(z)$ essentially appeared earlier as the Even Case Example 2.0.6. This centralizer can be recognized Lie-theoretically as the minimal parabolic P_2, in the conventions of Remark 1.3.20(4). Indeed its unipotent radical $Q := U_2$ is determined by the positive roots which are not combinations of $J = \{2\}$—these are given by α_1, $\alpha_1 + \alpha_2$, $\alpha_1 + \alpha_2 + \alpha_3$, $\alpha_2 + \alpha_3$, and α_3; these come from the matrices $|a|, |b|, |d|, |e|, |f|$. So we see that U_2 is obtainable as the product $U_{\hat{1}}U_{\hat{3}}$ of the transvection subgroups which we saw earlier, in the cases "$k = 1, 3$" of Remark B.1.1. This product is extraspecial of type 2_+^{1+4}—as we can see using (1.3.13), by checking which sums of these roots are also roots in Φ^+; cf. also the discussion in later Exercise 8.1.4. And we see that P_2 has its Levi complement $L_2 \cong L_2(2)$ generated by the matrices $|c|, |c'|$; corresponding to $\pm\alpha_2$, where α_2 is the only positive root which is a J-combination. In particular, we see $C_G(z)$ has order $2^6 \cdot 3 = 192$. Note that Q is "large" in G, as in Definition 8.1.3, because $F^*(C_G(z)) = O_2(C_G(z)) = Q$.

The case of commutator dimension 2 gives the other class t; it can be represented by $|b, e|$. (In A_8 it has cycle-type 2^2.) Here $C_G(t)$ gives most of the maximal parabolic subgroup $P_{\hat{2}}$ as in Example 1.3.21: It contains the full unipotent radical $U_{\hat{2}} \cong 2^4$, which we saw in Remark B.1.1; in the matrix notation used here, it is generated by $|b|, |c|, |d|, |e|$. And $C_G(t)$ is in fact the extension of $U_{\hat{2}}$ by a single $L_2(2)$-subgroup (a diagonal embedding across the two such $L_2(2)$-subgroups in the Levi complement $L_{\hat{2}}$) which is generated by the matrices $|a, f|$ and $|a', f'|$. So we see $C_G(t)$ has order $2^5 \cdot 3 = 96$.

We also record a further useful fact:

$$z^G \cap Q = (U_{\hat{1}} \cup U_{\hat{3}})^\#, \tag{B.2.6}$$

where $U_{\hat{1}} = \langle |a|, |b|, |d| \rangle$ and $U_{\hat{3}} = \langle |d|, |e|, |f| \rangle$. This is visible from matrix forms: for $x \in z^G \cap Q$ must have 1-dimensional commutator $[V, x]$. These groups are the standard point- and hyperplane-transvection subgroups, which we saw after Definition B.1.2; namely the unipotent radicals of the corresponding point- and hyperplane-stabilizers $P_{\hat{1}}, P_{\hat{3}}$ there.

Cases for $a(z)$:

Since A_8 has no elements of order 8 or 12 (see e.g. [**CCN$^+$85**, p 22]), any dihedral subgroup D of order divisible by 4, in $C_G(z)$ of order $2^5 \cdot 3$, has size given by 4, 8, or 12.

For $D \cong D_4$: Here as in earlier cases where D has order 4, we must have $xy = z$. Further we cannot have $x \in Q$: for (B.2.6) shows in that case[2] that also $xz \in z^G$, whereas we need $xz = y \in t^G$. So now looking for x, and hence also y, in $C_G(z) \setminus Q$: We find that $x = |c'|$ with $y = xz = |c'||d|$ gives such a D. We compute then that $C_{C_G(z)}(D) = C_{C_G(z)}(x)$ has order 2^4—either work directly with matrices; or use root-subgroup theory, say as follows: For $x = |c'| \in U_{-\alpha_2}$ centralizes in Q the root groups corresponding to α_1, α_3, and $\alpha_1 + \alpha_2 + \alpha_3$; and in the viewpoint of (1.3.13), adding $-\alpha_2$ to any of those three roots produces a result that is *not* in the root system of type A_3—which was described in Remark B.1.1. This subgroup has index $2^2 \cdot 3 = 12$ in $C_G(z)$, giving the orbit-size for such pairs via (B.2.3). And essentially since $\langle x \rangle$ is Sylow in $C_G(z)/Q$, we get that all such pairs must arise from this orbit.

For $D \cong D_8$: In this case, we have $z = [x, y] = (xy)^2$. Suppose first $x \notin Q$: then via Sylow considerations as above, we may as well calculate with $x = |c'|$. The group $C_Q(|c'|)$, via our root-group argument above, also in fact equals $[Q, |c'|]$. From this equality, it follows that $|c'|$ inverts no element of order 4 in Q; so this forces $y \in Q$. Then $[x, y] = z$ means that x centralizes y modulo z, and so y lies in $C_{Q/\langle z \rangle}(x)$; but we check that the preimage of this group is in fact centralized by x. We conclude that $z \in Q$. Suppose next we had $y \notin Q$; then a similar argument, since the action of y on Q is essentially that of $|c'|$, again produces a contradiction. So we also have $y \in Q$. And now the pair given by e.g. $x = |b|$, $y = |a, f|$ can be used. This time we compute (again the root-group viewpoint of (1.3.13) is useful) that $C_{C_G(z)}(D) = \langle |a|, |d| \rangle$, of order 2^2. This has index $2^4 \cdot 3 = 48$ in $C_G(z)$, giving the orbit-size of such pairs via (B.2.3). However: this time our example does *not* give the unique orbit of such pairs. Instead, another orbit is given essentially by the transpose about the anti-diagonal: namely instead we take $x = |e|$, with the same y. This new orbit is distinct, basically because the point- and hyperplane-transvection subgroups $U_{\hat{1}}, U_{\hat{3}}$ in Q are not in fact conjugate in G. And the new orbit works just as well as the original—basically because the transevection subgroups are conjugate by an *outer* automorphism, of graph type.[3] Thus the two transvection subgroups have the same group-theoretic behavior—and so each contributes an orbit of suitable pairs x, y, of length 48. As a result, we get an overall contribution of $48 + 48 = 96$ such pairs.

---

[2]This conjugacy of the three elements of $\langle x, z \rangle^{\#}$ is a standard feature of the large-extraspecial situation in the abstract; see e.g. [**Smi80**, 1.6(b)].

[3]This extra-orbit property, corresponding roughly to reducibility of $C_G(z)$ on $Q/\langle z \rangle$, is a very unusual feature, within the context of large-extraspecial theory, of the linear groups; cf. e.g. [**ALSS11**, 7.2.6].

For $D \cong D_{12}$: Since 3-elements of $C_G(z)$ are fixed-point-free on $Q/\langle z\rangle$, elements of Q do not invert them; so $x, y \in C_G(z) \setminus Q$. We can now mimic the D_4 case above: continue with the $x = |c'|$ as there; but replace the y there by the product $|c'||c||d|$. We saw $|c'||c|$ of order 3 is fixed-point-free on $Q/\langle z\rangle$; hence $C_{C_G(z)}(D)$ is just $\langle z\rangle$. This has index $2^5 \cdot 3 = 96$ in $C_G(z)$, for the orbit-contribution via (B.2.3). And again Sylow considerations show that this orbit exhausts the possibilities.

We conclude from the cases above that $a(z) = 12 + (48 + 48) + 96 = 204$.

Cases for $a(t)$:

The same argument as for $C_G(z)$ above show that any dihedral D of order divisible by 4, in $C_G(t)$ of order $2^4 \cdot 3$, has size 4, 8, or 12. And indeed order 8 is ruled out: the power map [**CCN$^+$85**, p 22] shows that elements of order 4 square to elements in z^G; whereas here we would need to have $(xy)^2 = t \notin z^G$.

For $D \cong D_4$: As before for D of order 4, we get $xy = t$. Here we can take $x = z = |d|$, and $y = |b, d, e|$. In this case we check (say using root subgroups in (1.3.13)) that $C_{C_G(t)}(D) = C_{C_G(t)}(y)$ is a Sylow subgroup of $C_G(t)$, of order 2^5—an extension of the subgroup $\langle |d|, |b|, |e|, |a||f|\rangle$ by $\langle |c|\rangle$. This has index 3 in $C_G(t)$, giving the orbit-contribution via (B.2.3). Again we check that this orbit exhausts the possibilities: for example, if we had $x, y \in C_G(t) \setminus U_{\hat{2}}$, both would have a 2-dimensional commutator on V, and so lie in t^G—whereas we need $x \in z^G$.

For $D \cong D_{12}$: Involutions of $C_G(t)$ inverting 3 elements, such as the involutions x and y, lie outside the subgroup of index 2 given by $U_{\hat{2}}$ extended by a 3-element. But we saw just above that these lie in t^G, whereas we need $x \in z^G$. Thus there are no candidates for pairs x, y in this case.

We conclude that $a(t) = 3$.

And now the main formula (B.2.2) gives:

$$|A_8| = 204 \cdot (2^5 \cdot 3) + 3 \cdot (2^6 \cdot 3) = (34 + 1) \cdot (2^6 \cdot 3^2) = 2^6 \cdot 3^2 \cdot 5 \cdot 7,$$

as desired. □

This concludes our further remarks on the Thompson Order Formula. ◇

REMARK B.2.7 (More Exercise 4.1.1: Some classical 3-transposition groups). We indicate some details for one particular group—under three different names:

The unitary group $U_4(2)$. See e.g. [**CCN$^+$85**, p 26] for more detailed reference. Take $G = U_4(2)$, with 2-central class z; we have $|z^G| = 45$. The class consists of transvections: that is, dim $[V, z] = 1$, where V is the 4-dimensional unitary module.

In fact $W := [V, z]$ is an isotropic point, in the polar space on V. (Cf. Example 7.0.1; this case is in fact a generalized quadrangle in the sense of 7.3.1.) Since transvections z correspond 1:1 with these points, we can do some of our group computations in the context of that geometry.

In the viewpoint of the rank-3 discussion leading up to Exercise 4.3.2, we will want to compute the suborbits: that is, the orbits of $C_G(z)$ on z^G. Here $C_G(z)$ is the analogue of the 1-space stabilizer for the full linear group in Example 1.3.4: namely, the parabolic $P_{\hat{1}}$ which stabilizes the 1-subspace W; it is the extension of the extraspecial group $U_{\hat{1}} \cong 2^{1+4}$ by $L_{\hat{1}} \cong 3^2{\cdot}2$. And we find that $C_G(z)$ is transitive on the 12 isotropic points in $W^\perp$ other than W, as well as on the remaining 32 points in $V \setminus W^\perp$. For example, let A be an isotropic point in $W^\perp \setminus W$. Then the stabilizer in $C_G(z)$ of the isotropic 2-space $\langle W, Z\rangle$ has structure $2 \cdot 2^2 \cdot 3 \cdot 2$, of index 12. Next for isotropic $X \notin W^\perp$, we find that the stabilizer in $C_G(z)$ of the unitary 2-space $\langle W, X\rangle$ is a Levi group $L_{\hat{1}} = 3^2 \cdot 2$, of index 32. Since this now accounts for all $1+12+32 = 45$ isotropic points, these must be the suborbits—and we see G has permutation rank 3 on z^G.

To complete the verification of the 3-transposition property (1.2.2), we check the orders $|xy|$ of products from z^G: For $t \in C_G(z) \setminus O_2\big(C_G(z)\big)$ we get $|zt| = 2$; and we get $|tu| = 3$ for a pair t, u of such conjugates in $C_G(z) \setminus O_2\big(C_G(z)\big)$, generating an S_3 in $L_{\hat{1}} \cong 3^2 \cdot 2$. So we do indeed have a class of 3-transpositions. □

The orthogonal group $\Omega_6^-(2)$. We now mention a relevant standard isomorphism: Because the diagram D_3 is the same as A_3, we see that $\Omega_6^-(q) \cong U_4(q)$. So here, we may also regard $U_4(2)$ as the orthogonal group $\Omega_6^-(2)$. And because an orthogonal transvection t does not lie in the simple group $\Omega_6^-(2)$, we in fact consider the almost-simple extension given by $G = O_6^-(2) = \Omega_6^-(2)\langle t\rangle$. We get calculations fairly similar to those for $U_4(2)$ above, so we present them with less detail:

We have $C_G(t) \cong Sp_4(2) \times \langle t\rangle$, so $|t^G| = 36$. Now $W := [V, t]$ is a non-singular point, and again these points are in correspondence with t^G. This time there are 15 non-singular points other than W in $W^\perp$, and 20 in $V \setminus W^\perp$; and again $C_G(t)$ is transitive on each set, so G has permutation rank 3 on t^G.

And then for orders $|xy|$: Taking $u = tv$ for a suitable involution v taken from $Sp_4(2)' \cong A_6$, we get $|tu| = 2$; and taking $t' \in C_G(t) \setminus Sp_4(2)$ generating with t an S_3, we get $|tt'| = 3$. So again we have 3-transpositions. □

The symplectic group $Sp_4(3)$ *and the orthogonal group* $\Omega_5(3)$. Finally we mention the standard isomorphism $U_4(2) \cong Sp_4(3)$ (cf. p261 in [**ALSS11**]). Here we can take r to be a symplectic reflection—which lies outside the simple group $Sp_4(3)$, so that again we consider the almost-simple extension given by $G = Sp_4(3)\langle r\rangle$. And once again summarizing rapidly:

Taking $W := [V, r]$, we get correspondence of such points with reflections r^G. Here we have characteristic $p = 3$, and $C_G(r)$ is the parabolic $P_{\hat{1}}$ stabilizing a 1-space—of structure $3^{1+2} : SL_2(3)$; and we get $|r^G| = 40$. There are 12 points other than W in $W^\perp$, and 27 in $W \setminus W^\perp$. Again $C_G(r)$ is transitive on these sets—using calculations with the parabolic $P_{\hat{1}}$ much like those for $U_4(2)$ above; so G has permutation rank 3 on r^G.

Then for $|xy|$: If s is a reflection with $S := [V, s] \leq W^\perp$, we get $|rs| = 2$. But if $S \not\leq W^\perp$, we find that $\langle r, s\rangle$ induces on the 2-space $W \oplus S$ the natural group $GL_2(3) \cong 2S_4$—so that $|rs| = 3$. And yet again we see we have a class of 3-transpositions.

To conclude, we mention that since the root systems of type B_2 and C_2 are exchanged by a suitable graph automorphism, it is standard that the finite groups $Sp_4(q)$ and $\Omega_5(q)$ are isomorphic. Thus $Sp_4(3)$ above is isomorphic to $\Omega_5(3)$—and essentially the above calculations make the 3-transposition verification for the almost-simple extension of the group $\Omega_5(3)$ by an orthgonal reflection. □

This finishes our further remarks on 3-transpositions. ◇

B.3. Some exercises from Chapter 5

REMARK B.3.1 (More on Exercise 5.1.2: Some irreducible degrees for S_n). We first recall the background for the hook-length formula:

For a partition λ, by convention we arrange the parts in decreasing order of size. The *Young diagram* is the arrangement of "boxes", in descending rows, of lengths given by the part-sizes. The *hook length* of a box is the number of boxes in its "hook"—starting from the bottom, and pivoting at the box to reach the end of the row. The hook-length formula (e.g. 20.1 in [**Jam78**]) states:

$$\dim(I_\lambda) = \frac{n!}{\Pi \text{ (hook lengths in } \lambda)} \tag{B.3.2}$$

Now to verify the remarks stated just before Exercise 5.1.2:

For the partition with a single part, the single box has hook length n. Thus the formula gives dimension $n!/n! = 1$—for the fully-expected dimension of the trivial module.

For the partition with parts $(n-1), 1$: the boxes in the top row have hook lengths given by $n, (n-2), (n-3), \cdots, 2, 1$; with just 1 in the second row. And correspondingly the formula gives $n!/\big(n \cdot (n-2) \cdot (n-3) \cdot 3 \cdot 2 \cdot 1\big) = (n-1)$, for the dimension of the natural irreducible. □

Next let's examine all the partitions λ, for the case $n = 4$; these are:

$$4 \text{ and } 3,1 \text{ and } 2,2 \text{ and } 2,1,1 \text{ and } 1,1,1,1.$$

Of course, the first two cases are covered by our remark above: giving dimensions 1 and 3.

For $\lambda = 2, 2$, the hook lengths in the two rows are $3, 2$ and $2, 1$. So here the formula gives the dimension $4!/(3 \cdot 2 \cdot 2 \cdot 1) = 2$. This is just the natural irreducible for the quotient $S_4/O_2(S_4) \cong S_3$.

Finally we observe that the remaining partitions are in the image of the duality-operation: in the Young diagram, this just transposes the boxes about the usual diagonal. Since this leaves the hook lengths invariant, we see—using our earlier computations for $3, 1$ and 4—that $2, 1, 1$ and $1, 1, 1, 1$ give dimensions 3 and 1.

The latter is the *sign* representation—namely trivial on the even permutations A_4, and with value -1 on the odd permutations $S_4 \setminus A_4$ □

This concludes our further remarks on irreducibles of S_n. ◇

REMARK B.3.3 (More on Exercise 5.2.5: Weight theory for some representations). Recall we wish to mimic the observations for $L_3(2)$ given in Example 5.2.4.

$Sp_4(2)$ on its natural module. The root system of type C_2 is described in Remark B.1.1: Positive roots are α_1, α_2, $\alpha_1 + \alpha_2$, and $2\alpha_1 + \alpha_2$; the first and third are short roots.

Check that the highest weight is λ_1, which is fundamental and hence dominant; it can be expressed in terms of roots as $\alpha_1 + \frac{1}{2}\alpha_2$. The remaining, lower weights are $\lambda_1 - \alpha_1$, $\lambda_1 - (\alpha_1 + \alpha_2)$, and $\lambda_1 - (2\alpha_1 + \alpha_2)$; exhibiting via (1.3.9) the action of negative-root subgroups.

$L_3(2)$ on its adjoint module. Since the weights on the adjoint module are by definition roots, the highest weight λ is necessarily the highest root: namely $\alpha_1 + \alpha_2$. Now in Example 5.2.4, we saw that $\lambda_1 = \frac{2}{3}\alpha_1 + \frac{1}{3}\alpha_2$. Symmetrically $\lambda_2 = \frac{1}{3}\alpha_1 + \frac{2}{3}\alpha_2$;

for example, using the action of a graph automorphism. And so $\lambda = \alpha_1 + \alpha_2$ can also be expressed as $\lambda_1 + \lambda_2$—which we see is dominant (though this time not fundamental).

The remaining, lower roots are $\lambda - \alpha_1 = \alpha_2$ and $\lambda - \alpha_2 = \alpha_1$, exhibiting via (1.3.9) the action of negative-root subgroups; as does $\lambda - (\alpha_1 + \alpha_2) = 0$; in fact we get a 2-dimensional space for the 0-weight—corresponding to the Cartan subalgebra $\mathcal{H}$ of dimension 2 in the adjoint module afforded by the underlying Lie algebra $\mathcal{G}$. And further action of negative-root subgroups then leads to the three root spaces for those negative roots. ♢

REMARK B.3.4 (More on Exercise 5.2.7: Some Steinberg tensor-product modules). We'll adopt some shorthand notation, using the dimension as an abbeviation for the module itself.

Irreducibles for $SL_2(4)$. The basic irreducibles, for $SL_2(2)$, are the trivial module of dimension 1, and the 2-dimensional natural module.

So the Steinberg tensor-product irreducibles from 5.2.6 are expressed as:

$$1 \otimes \sigma(1),\ 2 \otimes \sigma(1),\ 1 \otimes \sigma(2), \text{ and } 2 \otimes \sigma(2).$$

Now $\sigma(1)$ is still the trivial module. So these tensor-product modules are: the trivial module; the natural 2-dimensional $SL_2(4)$-module, and its algebraic conjugate under a field automorphism; and a 4-dimensional module—this is in fact the natural module for $\Omega_4^-(2) \cong L_2(4)$, and is the Steinberg module for $SL_2(4)$ in the sense of Definition 5.2.8. Since $L_2(4)$ has Lie rank $n = 1$, altogether we get $q^n = 4^1 = 4$ irreducibles.

Irreducibles of $SL_3(4)$. The basic irreducibles, for $SL_3(2)$, are the trivial module 1; the natural module 3, and its dual $\overline{3}$; and the adjoint (Steinberg) module of dimension 8. To see completeness of this list, recall we have seen that they afford all four of the 2-restricted dominant weights—these are given by $0, \lambda_1, \lambda_2, \lambda_1+\lambda_2$—in earlier Example 5.2.4 and Remark B.3.3.

So mimicking the work above for $SL_2(4)$, we get: the trivial module; the natural module 3 and its conjugate $\sigma(3)$, and their duals $\overline{3}$ and $\sigma(\overline{3})$; four 9-dimensional modules—namely the product of (3 or $\overline{3}$) with conjugates under σ; the adjoint module 8 and its conjugate $\sigma(8)$; four 24-dimensional modules—namely the product of (3 or $\overline{3}$) with $\sigma(8)$; and their σ-conjugates; and the 64-dimensional Steinberg module $8 \otimes \sigma(8)$. The rank of $SL_3(4)$ is $n = 2$; and indeed we have $q^n = 4^2 = 16$ irreducibles altogether. ♢

REMARK B.3.5 (More on Exercise 5.2.10: Some Weyl module dimensions for the case of $Sp_4(2)$). We first set the stage for Weyl's formula: Write $\delta := \sum_{i=1}^n \lambda_i$ for the sum of the fundamental weights; and let λ be some dominant weight. Then the formula (e.g. [**Hum78**, p 139]) is:

$$\dim(W_\lambda) = \prod_{\alpha \in \Phi^+} \frac{(\lambda + \delta, \alpha)}{(\delta, \alpha)} \tag{B.3.6}$$

The meaning of these terms is actually reasonably straightforward: Since δ is the sum of the fundamental weights, the denominator term at α is essentially the "content" of α: namely the sum of its coefficients as a combination of the simple roots α_i—but *weighted* by the squared-lengths of the α_i, since the λ_i are actually dual to the simple co-roots, rather than to the roots themselves. And then the corresponding numerator term simply adds to the denominator term the value of (λ, α)—where each λ_i in λ contributes the weighted α_i-content of α.

We now apply this to the root system of type C_2: Recall this was given in Remark B.1.1: The positive roots are α_1, α_2, $\alpha_1 + \alpha_2$, and $2\alpha_1 + \alpha_2$.

Here the 2nd and 4th roots are long—with squared-length 2. Thus the corresponding denominators term give the weighted coefficient-sum values:

$$1, 2, 3, 4.$$

First consider $\lambda = \lambda_1$: To the above, we add the α_1-contents $1, 0, 1, 2$ to get:

$$2, 2, 4, 6.$$

So from the Weyl formula (B.3.6) we conclude that $\dim(W_{\lambda_1}) = 4$.

Next consider $\lambda = \lambda_2$: This time we add the (weighted!) α_2-contents—these are $0, 2, 2, 2$—to the denominators, obtaining:

$$1, 4, 5, 6.$$

So from the formula we conclude that $\dim(W_{\lambda_2}) = 5$. ◇

REMARK B.3.7 (More on Exercise 5.4.1: Alperin-weights $(U_J, \lambda \cdot St_J)$ in Lie-type groups). For G given by either $L_4(2)$ or $Sp_4(2)$, the natural module V has high weight "λ" given by the first fundamental weight λ_1—as in Example 5.2.4 and Remark B.3.3. So the high-weight space "X" on V is given by the 1-dimensional subspace we usually call V_1. Now $N_G(X)$ is a parabolic by 5.2.3(2), and we see it is one we recognize: Namely we can take "J" as $\hat{1}$—in the case $k = 1$ of Remark B.1.1: Thus our "P_J" is $P_{\hat{1}}$, so that "U_J" is the unipotent radical $U_{\hat{1}} \cong 2^3$ as described there. Furthermore for the Levi complement, we have $K_{\hat{1}} \cong L_3(2)$ or $L_2(2)$, respectively—and this has Steinberg module "St_J" given by the 8-dimensional adjoint in Exercise 5.2.9, or the natural module for $L_2(2)$.

Next consider the 6-dimensional orthogonal module W, for $G = L_4(2)$ regarded as $\Omega_6^+(2)$: One calculates (much as for V above) that the highest weight "λ" for this module is the second fundamental weight λ_2. Using (1.3.9), as in Example 5.2.4 and Remark B.3.3, we find that the parabolic $N_G(X)$ for a highest-weight space "X" in W as in 5.2.3(2) is in fact the maximal parabolic $P_{\hat{2}}$. So here we take "J" as $\hat{2}$, in the case $k = 2$ in Remark B.1.1. Our "U_J" is the unipotent radical $U_{\hat{2}} \cong 2^4$ described there. Further for our "St_J": the Levi complement $K_{\hat{2}} \cong L_2(2) \times L_2(2)$ has its Steinberg module of dimension 4 given by the tensor product of the natural modules for the two factors. we could instead use the language for the orthogonal group: The space X is spanned by a singular vector, which has stabilizer given by $2^4\Omega_4^+(2)$; now St_J of dimension 4 above is just the natural module for the quotient $\Omega_4^+(2)$. ◇

B.4. Some exercises from Chapter 6

REMARK B.4.1 (More on Exercise 6.1.2: Maximals in $S_4, \cdots, S_8$). We look for cases (1)–(6) in the O'Nan-Scott Theorem 6.1.1, as candidates for possible maximality. Recall those cases are in addition to the obvious maximal A_n of S_n.

In more detail: We will in effect prove the completeness of the maximals-lists for these groups in the Atlas [**CCN$^+$85**]. We will quote that reference only for more basic information—such as group orders, and existence of certain subgroups. And often we can then see just directly that various subgroups are maximal. But when we say below that a subgroup "is maximal", it might be more appropriate to say that it "will be maximal": For once we have found all occurrences of (1)–(6) in Theorem 6.1.1—and further check that there are no containments among them—then we will know first that our candidates really are maximal; and second that our list is in fact complete.

We note first of all that case (5) is ruled out in all our groups: For it would require $k \geq 2$; whereas even for our largest value of $n = 8$, the order $|A_n|$ is not divisible by the squared-order of a nonabelian simple group.

Furthermore case (3) is ruled out in all our groups: Numerically it could arise with $j = 2$; and $k = 2, 3$, when $n = 4, 8$. But a standard result (e.g. 3.2 in [**Cam81**]) shows that when $k > 1$, $j > 2$ is required for primitivity.

We now turn to the individual groups:

For $n = 4$: Simple subgroups as in (6) are ruled out, as S_4 is solvable. Case (4) with $r = 2 = d$ does not arise: for $2^2L_2(2)$ is all of S_4, and so is not even proper. But case (2) with $j = 2 = k$ does arise—with maximal $S_2 \wr S_2 \cong D_8$ transitive but imprimitive via a 2, 2 partition. Finally case (1) with $j = 3$ gives maximal S_3; but $j = 2 = k$ in (1) does not arise, since $S_2 \times S_2$ is not maximal—via case (2) above. So our candidate-list is:

$$S_3,\ D_8\ ;$$

and since there are no inclusions among them, we conclude that they are maximal, and that the list is complete.

For $n = 5$: Again (6) is ruled out: this time because A_5 is the smallest simple group, so that its proper subgroups are solvable. And case (2) is ruled out, since 5 is prime. But case (4) arises with $r = 5$ and $d = 1$: and in fact $5 : GL_1(5)$ is the Sylow 5-normalizer in S_5, which is maximal. Finally case (1) arises, with $j = 4$ or 3, giving maximal S_4 and $S_3 \times S_2$. Our final list is:

$$S_4,\ S_3 \times S_2,\ 5 : GL_1(5)\ ;$$

and there are no inclusions, so the list is complete.

For $n = 6$: Case (4) is ruled out, since 6 is not a power r^d. Since $|A_6|$ is not divisible by the order of $L_3(2)$, which is the only simple group beyond A_5 of order smaller than that of A_6, case (6) can only arise with $F^*(H) = A_5$. And indeed there is a maximal subgroup of S_6 which is given by an S_5 that is transitive on the 6 points—which are visible as the cosets in S_5 of the subgroup $5 : GL_1(5)$ of the previous paragraph. Note that we get an isomorphic group S_5 from the subcase $j = 5$ of (1)—but this S_5 in (1) is *not* transitive. The subcase $j = 4$ of (1) also leads to a maximal subgroup, of form $S_4 \times S_2$. But the subcase $j = 3 = k$ of (1) with $S_3 \times S_3$ is not maximal—it is proper in maximal $S_3 \wr S_2$, from the subcase $j = 3$ of (2). Finally the subcase $j = 2$ of (2) arises—and here the group $S_2 \wr S_3$ happens to be isomorphic to $S_4 \times S_2$ in (1) above; but that group in (1) was intransitive, whereas this group in (2) is transitive. In fact an outer automorphism of S_6 interchanges these two subgroups; as well as the two subgroups S_5 above. So our final list is:

(intrans:) S_5, $S_4 \times S_2$; (imprim:) $S_2 \times S_3 \cong S_4 \times S_2$, $S_3 \wr S_2$; (prim:) S_5 ;

with no inclusions, as desired.

For $n = 7$: Since 7 is prime, some arguments are similar to $n = 5$ above: Namely we rule out case (2); and case (4), given by $7 : GL_1(7)$, is in fact the

Sylow 7-normalizer, which is maximal. All three subcases of (1) also arise, giving maximal S_6, $S_5 \times S_2$, and $S_4 \times S_3$. Finally case (6) does not arise: For by order-divisibility much as for $n = 6$ above, the only simple groups that might arise as $F^*(H)$ in (6) are A_5, $L_3(2)$, A_6, and $L_2(8)$; but in (6), we see H must be transitive on 7 points—and among these groups, only $L_3(2)$ has a subgroup of index 7. Now $L_3(2)$ does appear as a maximal subgroup of A_7; but an outer automorphism in $S_7 \setminus A_7$ interchanges two such subgroups, so they are not maximal in S_7. Thus our list for S_7 is:

$$S_6,\ S_5 \times S_2,\ S_4 \times S_3,\ 7 : GL_1(7)\ ;$$

with no inclusions, as desired.

For $n = 8$: First (4) does not arise: for $r = 2, d = 3$ gives $2^3L_3(2)$; and while this is maximal in A_8, two conjugacy classes of such subgroups are interchanged by elements of $S_8 \setminus A_8$, so they are not maximal in S_8. The subcases $j = 2, 4$ of (2) do arise—giving $S_2 \wr S_4$ and $S_4 \wr S_2$, which are maximal in S_8; so in particular, the subcase $j = 4$ of (1) does not arise, as we see $S_4 \times S_4$ is not maximal. The remaining subcases $j = 7, 6, 5$ of (1) give maximal subgroups S_7, $S_6 \times S_2$, and $S_5 \times S_3$. However, the groups A_7, A_6, A_5 do not appear in the transitive case (6), since those groups have no subgroup of index 8 (e.g. using our lists above). Furthermore in the only other simple groups of smaller order dividing $|A_8|$, namely $L_3(2)$, $L_2(8)$ and $U_3(3)$, the latter two also have no subgroup of index 8; so the only possibility for (6) has $F^*(H) = L_3(2)$. And indeed $S_8 \cong L_4(2) : 2$ has a subgroup $L_3(2) : 2$, which is transitive on the 8 points: for it is not contained in any of the intransitive subcases in (1) above—in particular this uses our maximals-list for $n = 7$ above. So our final list is:

$$S_7,\ S_6 \times S_2,\ S_5 \times S_3,\ S_2 \wr S_4,\ S_4 \wr S_2\ ,\ L_3(2) : 2\ ;$$

with no inclusions, as desired.

This concludes our further remarks on maximals of S_n. ◊

REMARK B.4.2 (More on Exercise 6.2.2: Maximals of $L_3(2)$ and $L_4(2)$). We proceed much as we did for S_n in earlier Remark B.4.1, again verifying the maximals-list in the Atlas [**CCN$^+$85**]: This time, we first determine all cases from Theorem 6.2.1; and again, on checking that there are no inclusions in that list, we know that the subgroups are maximal, and that the list is complete.

For $n = 3$: Since 3 is prime, cases (2)(2′)(3) of Theorem 6.2.1 are ruled out. Also (4) does not arise: for $r = 3, d = 1$ requires a group 3^{1+2}, whose order 3^3 does not divide that of $L_3(2)$. Thus we are reduced to cases (1) and (6).

Now quasisimple groups in (6) are ruled out: for A_5 is the only simple group smaller than $L_3(2)$, and 5 does not divide the order of $L_3(2)$. But in (6) we get solvable $F^*(H) = GL_1(8) \cong \mathbb{Z}_7$: indeed the Sylow 7-normalizer of structure $\mathbb{Z}_7 : \mathbb{Z}_3$ is maximal in $L_3(2)$. Notice this is the extension-field subcase refining case (6), namely C_3 in Aschbacher's terminology after Theorem 6.2.1. Finally (1) in the cases $j = 1$ or 2 yields subgroups $2^2L_2(2)$, which we recognize as the parabolics $P_{\hat{1}}$,$P_{\hat{2}}$ of Example 1.3.21; these are conjugate by a graph automorphism. Thus our resulting list is:

$$2^2L_2(2) \text{ (two classes)},\ \mathbb{Z}_7 : \mathbb{Z}_3\ ;$$

since there are no inclusions, our subgroups are maximal, and the list is complete.

For $n = 4$: We will compare with our results for $\mathring{S}_8 \cong L_4(2) : 2$ in earlier Remark B.4.1. Indeed, note first that $L_3(2) : 2$ and $S_2 \wr S_4 \cong 2^4 S_4$, which were maximal in S_8 in B.4.1, in fact intersect $A_8 \cong L_4(2)$ in subgroups $L_3(2)$ and $2^3 S_4$—which are not maximal in $L_4(2)$. So these groups from S_8 will not appear in our list here.

Case (4) of Theorem 6.2.1 does not arise: for $n = 4$ is not a power of $r \neq 2$. Furthermore the cases (2)(2′)(3) do not arise: For $j = 2 = k$ in (2′) yields the group $L_2(2) \times L_2(2) \cong S_3 \times S_3$; while these same values $j = 2 = k$ in (2)(3) give $L_2(2) \wr S_2 \cong S_3 \wr S_2$. Both these groups normalize a subgroup E_9, which is Sylow in $L_4(2)$. Indeed the latter is the full Sylow 3-normalizer—and hence contains the former; but it is not maximal, as it lies in an $S_6 \cong Sp_4(2)$ to be discussed below. So as for $n = 3$ above, again we are reduced to cases (1) and (6).

Case (1) with the values $j = 1$ or 3 yields subgroups $2^3 L_3(2)$—which we recognize as the parabolic subgroups $P_{\hat{1}}$ and $P_{\hat{3}}$ of earlier Remark B.1.1; recall these were maximal in A_8 but not in S_8, in B.4.1. Again these are conjugate via an outer automorphism. Further $j = 2$ in (1) yields the structure $2^4\big(L_2(2) \times L_2(2)\big) \cong S_4 \times S_4$, namely the maximal parabolic $P_{\hat{2}}$; this is the intersection of A_8 with the maximal subgroup $S_4 \wr S_2$ of S_8.

We turn finally to case (6): Here a subgroup H must have an irreducible representation of dimension 4 over $\mathbb{F}_2$; so the quickest way to finish might be to quote the modular character tables in the Modular Atlas [**JLPW81**]. However, we can actually present many of the details, using just the basics of modular representation theory that we indicated in earlier Chapter 5:

We observe first that $GL_1(16)$ is not maximal, as it lies in $GL_2(4)$; but in fact $GL_2(4) : 2$ is maximal in $L_4(2)$—again we have the extension-field refinement of (6), namely C_3 in Aschbacher's terminology after 6.2.1. This subgroup, of structure $(A_5 \times 3) : 2$, is the intersection of $A_8 \cong L_4(2)$ with the earlier maximal $S_5 \times S_3$ of S_8. We remark also that the subgroup A_5 here exhibits a 2-dimensional modular irreducible over $\mathbb{F}_2$, read over the subfield $\mathbb{F}_2$, from the list of modules of dimensions 1,2,2,4 for $SL_2(4)$ given in Remark B.3.4. A different subgroup A_5 exhibits the 4-dimensional Steinberg module for $\Omega_4^-(2)$—but this subgroup is not maximal, as it is contained in the subgroup $Sp_4(2)$ to be discussed below.

Note next that since our group $L_4(2)$ is over $\mathbb{F}_2$, we see that $GL_4(2)$ is the same as simple $PSL_4(2) = L_4(2)$; so instead of quasisimple preimages, we just have simple groups as the remaining possibities for $F^*(H)$ in (6). Now just as for S_8 in B.4.1, we get only A_5, A_6, A_7, $L_3(2)$, $L_2(8)$, and $U_3(3)$ as choices for $F^*(H)$ using order-divisibility.

In fact $F^*(H) = A_5$ occurs in a subgroup $H = S_5$—but this is not maximal: for from the representation theory of $SL_2(4)$ indicated just above, it either acts as $\Omega_4^-(2)$ lying in $Sp_4(2)$; or as $SL_2(4)$—lying in the maximal subgroup given by $K := GL_2(4) : 2 \cong (A_5 \times 3) : 2$ above, where $F^*(K)$ also contains $F(K)$ of order 3. Next, the cases A_6 and A_7 do lead to maximal subgroups S_6 and A_7 of $L_4(2)$: we can conclude that both are irreducible, since they do not lie in the reducible possibilities in (1) above. They are the intersections with $A_8 \cong L_4(2)$ of the earlier maximal subgroups S_7 and $S_6 \times S_2$ of S_8. Furthermore $S_6 \cong Sp_4(2)$ gives the classical-subgroup refinement of (6) called C_8 in Aschbacher's terminology after 6.2.1; and it gives a different view of the irreducibility of S_6.

It remains to eliminate the three further simple groups listed prior to the previous paragraph: First any subgroup $L_3(2)$ necessarily acts reducibly: e.g. we noted in earlier Remark B.3.4 that the 2-modular irreducibles of $L_3(2)$ have dimensions given by $1, 3, 3, 8$. Next a Steinberg tensor-product analysis of the irreducibles for $SL_2(8)$, similar to that for $SL_2(4)$ in B.3.4, shows that 4-dimensional irreducibles for $L_2(8)$ are defined over $\mathbb{F}_8$, but not over $\mathbb{F}_2$. Finally $U_3(3) \cong G_2(2)'$—and the smallest irreducible for the latter corresponds to the 6-dimensional quotient of the Cayley-algebra module of Example 5.2.13.

So our final list of candidates consists of:

$$2^3 L_3(2) \text{ (two classes)}, \ 2^4\big(L_2(2) \times L_2(2)\big), \ GL_2(4) : 2, \ S_6, \ A_7 \ ;$$

as there are no inclusions, we see they are maximal, and the list is complete. ◊

REMARK B.4.3 (More on Exercise 6.2.7: Maximal parabolics for G_2 and E_6). We can proceed much as in Remark B.1.1: using [**Car89**, Sec 3.6] to describe root systems, and (1.3.13) to describe the interaction of root subgroups.

For type G_2: The root system is even sketched on [**Car89**, p 46]. The simple roots are α_1 (which is short) and α_2 (long). The remaining positive roots are the short roots $\alpha_1 + \alpha_2$ and $2\alpha_1 + \alpha_2$, and long $3\alpha_1 + \alpha_2$ and $3\alpha_2 + 2\alpha_2$.

For $J = \{1\}$, we have Levi complement determined by root subgroups for $\pm\alpha_1$, with the structure $GL_2(q)$. And observe that the remaining positive roots satisfy the condition indicated in the Hint to Exercise 8.1.4: namely in addition to the highest root $3\alpha_2 + 2\alpha_2$, we have pairs given by α_2,$3\alpha_1 + \alpha_2$ and $\alpha_1 + \alpha_2$,$2\alpha_1 + \alpha_2$, which sum to the highest root—but no other pairs sum to a root. It follows using (1.3.13) that the unipotent radical U_1 has "semi-extraspecial" structure q^{1+4}; when $q = p$ is prime, this is extraspecial.

For $J = \{2\}$, similar calculations give $L_2 \cong GL_2(q)$ via $\pm\alpha_2$; and U_2 has structure given by composition factors q^{2+1+2}.

For type E_6: The description of the root system given at [**Car89**, p 49] is fairly complicated. Instead the representation of positive roots at pages 5–6 of [**AS76**], essentially as a sum of simple roots, is more convenient for our purposes here: so we will adopt the notation of that paper. In particular, the Dynkin diagram is numbered so that 1 appears at the end of a long branch of the E_6 diagram, and 2 at the end of the short branch.

We begin with $J = \hat{2}$: The roots not involving α_2 form a subdiagram of type A_5; and the corresponding $\pm$ root groups determine a Levi complement $L_{\hat{2}}$ of structure $GL_6(q)$. For the unipotent radical $U_{\hat{2}}$, note that there are 21 positive roots involving α_2; namely, remove from the first 26 listed in [**AS76**] those which are numbered $2, 3, 5, 8, 17$. Note that the highest root is numbered 26; and just as for $J = 1$ in type G_2 above, again we have the condition that the other 20 roots fall into pairs summing to α_{26}, with no other pairs summing to a root; so that $U_{\hat{2}}$ has structure q^{1+20}—and further $L_{\hat{2}}$ is irreducible on $U_{\hat{2}}/Z(U_{\hat{2}})$.

Next we consider $J = \hat{1}$: Here $L_{\hat{1}}$ with diagram of type D_5 has structure given by $H \cdot \Omega_{10}^+(q)$. As for $U_{\hat{1}}$: The relevant positive roots are the first 16 listed in [**AS76**]. The usual calculations with (1.3.13) give $U_{\hat{1}}$ elementary of order q^{16}; this is a "half-spin" irreducible for $L_{\hat{1}}$.

Similar calculations apply for the other maximal parabolics, which we now just quickly summarize: In view of the graph automorphism of the diagram, we only need to consider the cases $J = \hat{3}$ and $\hat{4}$. From the corresponding subdiagrams, the Levi complement structures are determined by H with $SL_2(q) \times SL_5(q)$ and $SL_3(q) \times SL_3(q) \times SL_2(q)$, respectively. And the unipotent radicals have composition factors q^{5+20} and q^{2+9+18}. (Cf. Section 4.10.4 in [**Wil09**].) $\diamondsuit$

REMARK B.4.4 (More on Exercise 6.6.2: Some p-exceptional orbit sizes). In Theorem 6.6.1(iii): For $c := 2^r - 1$ or $2^r - 2$, the irreducible permutation module V has dimension $n = c - 1$ or $c - 2$, respectively. Consider the first two nontrivial values $r = 2, 3$:

For $r = 2$: We have for $c = 3$ that S_3 is transitive on the 3 nonzero vectors in V of dimension 2; so this example is 2-exceptional. But the case $c = 2$ is too small to be meaningful, as V would be of dimension $c - 2 = 0$.

For $r = 3$: With $c = 6$, we see that $S_6 \cong Sp_4(2)$ is transitive on the 15 nonzero vectors of V of dimension 4. With $c = 7$, we have $A_7 < A_8 \cong \Omega_6^+(2)$ for V of dimension 6; and the 35 singular vectors have stabilizer $2^4\Omega_4^+(2)$ in A_8, while the 28 nonsingular vectors have stabilizer $Sp_4(2) \cong S_6$. These stabilizers intersect A_7 in $(A_4 \times 3) : 2$, or A_6,S_5, respectively; so A_7 is transitive on the 35 singular vectors; and breaks the 28 nonsingular vectors into orbits of size 7 and 21. So both these examples for $r = 3$ are 2-exceptional.

In Theorem 6.6.1(iv), we use the inclusion $SL_2(5) < SL_2(9) < SL_4(3)$. Here we see $SL_2(9)$ is transitive on the 80 nonzero vectors of 4-dimensional V; and a vector stabilizer is a 3-Sylow subgroup of order 9. Hence the stabilizer in $SL_2(5)$ is a 3-Sylow of order 3, and index 40: so that the group has 2 orbits of length 40—and this example is 3-exceptional. $\diamondsuit$

Bibliography

Note: The citations use bibtex with the "amsalpha" style—and hence the convention of amsabs.bst: Namely the alphabetization is done *first* with respect to automatically-generated identifier-labels, such as [AO16]; and only after that, with respect to author surnames.

Thus when looking up citations with multiple authors, expect the initial letter of each surname followed by the year—e.g. [AO16] for the 2016 paper of Aschbacher and Oliver. For single authors, expect the first three letters of the surname followed by the year, as in [Asc76] for the 1976 paper of Aschbacher.

[AB08] Peter Abramenko and Kenneth S. Brown, *Buildings*, Graduate Texts in Mathematics, vol. 248, Springer, New York, 2008. Theory and applications. MR2439729

[ABG73] J. L. Alperin, Richard Brauer, and Daniel Gorenstein, *Finite simple groups of* 2*-rank two*, Scripta Math. **29** (1973), no. 3-4, 191–214. Collection of articles dedicated to the memory of Abraham Adrian Albert. MR0401902

[AD02] Alejandro Adem and James F. Davis, *Topics in transformation groups*, Handbook of geometric topology, North-Holland, Amsterdam, 2002, pp. 1–54. MR1886667

[AF84] Zvi Arad and Elsa Fisman, *On finite factorizable groups*, J. Algebra **86** (1984), no. 2, 522–548, DOI 10.1016/0021-8693(84)90046-2. MR732264

[AG84] M. Aschbacher and R. Guralnick, *Some applications of the first cohomology group*, J. Algebra **90** (1984), no. 2, 446–460, DOI 10.1016/0021-8693(84)90183-2. MR760022

[AGL81] Michael Aschbacher, Daniel Gorenstein, and Richard Lyons, *The embedding of* 2*-locals in finite groups of characteristic* 2*-type. I*, Ann. of Math. (2) **114** (1981), no. 2, 335–404, DOI 10.2307/1971297. MR632843

[AGMV08] K. K. S. Andersen, J. Grodal, J. M. Møller, and A. Viruel, *The classification of p-compact groups for p odd*, Ann. of Math. (2) **167** (2008), no. 1, 95–210, DOI 10.4007/annals.2008.167.95. MR2373153

[AJS94] H. H. Andersen, J. C. Jantzen, and W. Soergel, *Representations of quantum groups at a pth root of unity and of semisimple groups in characteristic p: independence of p* (English, with English and French summaries), Astérisque **220** (1994), 321. MR1272539

[AK90] Michael Aschbacher and Peter B. Kleidman, *On a conjecture of Quillen and a lemma of Robinson*, Arch. Math. (Basel) **55** (1990), no. 3, 209–217, DOI 10.1007/BF01191159. MR1075043

[AKO11] Michael Aschbacher, Radha Kessar, and Bob Oliver, *Fusion systems in algebra and topology*, London Mathematical Society Lecture Note Series, vol. 391, Cambridge University Press, Cambridge, 2011. MR2848834

[Alp87] J. L. Alperin, *Weights for finite groups*, The Arcata Conference on Representations of Finite Groups (Arcata, Calif., 1986), Proc. Sympos. Pure Math., vol. 47, Amer. Math. Soc., Providence, RI, 1987, pp. 369–379. MR933373

[ALSS11] Michael Aschbacher, Richard Lyons, Stephen D. Smith, and Ronald Solomon, *The classification of finite simple groups*, Mathematical Surveys and Monographs, vol. 172, American Mathematical Society, Providence, RI, 2011. Groups of characteristic 2 type. MR2778190

[AM04] Alejandro Adem and R. James Milgram, *Cohomology of finite groups*, 2nd ed., Grundlehren der Mathematischen Wissenschaften [Fundamental Principles of Mathematical Sciences], vol. 309, Springer-Verlag, Berlin, 2004. MR2035696

[AMM91] Alejandro Adem, John Maginnis, and R. James Milgram, *The geometry and cohomology of the Mathieu group M_{12}*, J. Algebra **139** (1991), no. 1, 90–133, DOI 10.1016/0021-8693(91)90285-G. MR1106342

[AO16] Michael Aschbacher and Bob Oliver, *Fusion systems*, Bull. Amer. Math. Soc. (N.S.) **53** (2016), no. 4, 555–615, DOI 10.1090/bull/1538. MR3544261

[Art88a] O. D. Artemovich, *Isolated elements of prime order in finite groups* (Russian), Ukrain. Mat. Zh. **40** (1988), no. 3, 397–400, 408, DOI 10.1007/BF01061317; English transl., Ukrainian Math. J. **40** (1988), no. 3, 343–345 (1989). MR952132

[Art88b] E. Artin, *Geometric algebra*, Wiley Classics Library, John Wiley & Sons, Inc., New York, 1988. Reprint of the 1957 original; A Wiley-Interscience Publication. MR1009557

[AS76] Michael Aschbacher and Gary M. Seitz, *Involutions in Chevalley groups over fields of even order*, Nagoya Math. J. **63** (1976), 1–91. MR0422401

[AS85] M. Aschbacher and L. Scott, *Maximal subgroups of finite groups*, J. Algebra **92** (1985), no. 1, 44–80, DOI 10.1016/0021-8693(85)90145-0. MR772471

[AS93] Michael Aschbacher and Stephen D. Smith, *On Quillen's conjecture for the p-groups complex*, Ann. of Math. (2) **137** (1993), no. 3, 473–529, DOI 10.2307/2946530. MR1217346

[AS04a] Michael Aschbacher and Stephen D. Smith, *The classification of quasithin groups. I*, Mathematical Surveys and Monographs, vol. 111, American Mathematical Society, Providence, RI, 2004. Structure of strongly quasithin K-groups. MR2097623

[AS04b] Michael Aschbacher and Stephen D. Smith, *The classification of quasithin groups. II*, Mathematical Surveys and Monographs, vol. 112, American Mathematical Society, Providence, RI, 2004. Main theorems: the classification of simple QTKE-groups. MR2097624

[Asc73] Michael Aschbacher, *A condition for the existence of a strongly embedded subgroup*, Proc. Amer. Math. Soc. **38** (1973), 509–511, DOI 10.2307/2038941. MR0318308

[Asc74] Michael Aschbacher, *Finite groups with a proper 2-generated core*, Trans. Amer. Math. Soc. **197** (1974), 87–112, DOI 10.2307/1996929. MR0364427

[Asc76] Michael Aschbacher, *Finite groups in which the generalized Fitting group of the centralizer of some involution is symplectic but not extraspecial*, Comm. Algebra **4** (1976), no. 7, 595–616, DOI 10.1080/00927877608822122. MR0407144

[Asc77a] Michael Aschbacher, *A characterization of Chevalley groups over fields of odd order*, Ann. of Math. (2) **106** (1977), no. 2, 353–398, DOI 10.2307/1971100. MR0498828

[Asc77b] Michael Aschbacher, *A characterization of Chevalley groups over fields of odd order. II*, Ann. of Math. (2) **106** (1977), no. 3, 399–468, DOI 10.2307/1971063. MR0498829

[Asc80] Michael Aschbacher, *A characterization of Chevalley groups over fields of odd order. II*, Ann. of Math. (2) **106** (1977), no. 3, 399–468, DOI 10.2307/1971063. MR0498829

[Asc81a] Michael Aschbacher, *Finite groups of rank 3. I*, Invent. Math. **63** (1981), no. 3, 357–402, DOI 10.1007/BF01389061. MR620676

[Asc81b] Michael Aschbacher, *Weak closure in finite groups of even characteristic*, J. Algebra **70** (1981), no. 2, 561–627, DOI 10.1016/0021-8693(81)90236-2. MR623826

[Asc83a] Michael Aschbacher, *Finite groups of rank 3. II*, Invent. Math. **71** (1983), no. 1, 51–163, DOI 10.1007/BF01393339. MR688262

[Asc83b] Michael Aschbacher, *The uniqueness case for finite groups. I*, Ann. of Math. (2) **117** (1983), no. 2, 383–454, DOI 10.2307/2007081. MR690850

[Asc83c] Michael Aschbacher, *The uniqueness case for finite groups. II*, Ann. of Math. (2) **117** (1983), no. 3, 455–551, DOI 10.2307/2007034. MR701255

[Asc84] M. Aschbacher, *On the maximal subgroups of the finite classical groups*, Invent. Math. **76** (1984), no. 3, 469–514, DOI 10.1007/BF01388470. MR746539

[Asc94] Michael Aschbacher, *Sporadic groups*, Cambridge Tracts in Mathematics, vol. 104, Cambridge University Press, Cambridge, 1994. MR1269103

[Asc00] M. Aschbacher, *Finite group theory*, 2nd ed., Cambridge Studies in Advanced Mathematics, vol. 10, Cambridge University Press, Cambridge, 2000. MR1777008

[Asc15] Michael Aschbacher, *Classifying finite simple groups and 2-fusion systems*, ICCM Not. **3** (2015), no. 1, 35–42, DOI 10.4310/ICCM.2015.v3.n1.a5. MR3385504

[Bab16] László Babai, *Graph isomorphism in quasipolynomial time [extended abstract]*, STOC'16—Proceedings of the 48th Annual ACM SIGACT Symposium on Theory of Computing, ACM, New York, 2016, pp. 684–697. MR3536606

[BB04] Ayşe Berkman and Alexandre V. Borovik, *A generic identification theorem for groups of finite Morley rank*, J. London Math. Soc. (2) **69** (2004), no. 1, 14–26, DOI 10.1112/S0024610703004733. MR2025324

[BC13] Francis Buekenhout and Arjeh M. Cohen, *Diagram geometry*, Ergebnisse der Mathematik und ihrer Grenzgebiete. 3. Folge. A Series of Modern Surveys in Mathematics [Results in Mathematics and Related Areas. 3rd Series. A Series of Modern Surveys in Mathematics], vol. 57, Springer, Heidelberg, 2013. Related to classical groups and buildings. MR3014979

[BCN89] A. E. Brouwer, A. M. Cohen, and A. Neumaier, *Distance-regular graphs*, Ergebnisse der Mathematik und ihrer Grenzgebiete (3) [Results in Mathematics and Related Areas (3)], vol. 18, Springer-Verlag, Berlin, 1989. MR1002568

[BF55] Richard Brauer and K. A. Fowler, *On groups of even order*, Ann. of Math. (2) **62** (1955), 565–583, DOI 10.2307/1970080. MR0074414

[BFM15] Russell D. Blyth, Francesco Fumagalli, and Marta Morigi, *The solvability of groups with nilpotent minimal coverings*, J. Algebra **427** (2015), 375–386, DOI 10.1016/j.jalgebra.2014.12.033. MR3312310

[BGK14] Tatiana Bandman, Shelly Garion, and Boris Kunyavskiĭ, *Equations in simple matrix groups: algebra, geometry, arithmetic, dynamics*, Cent. Eur. J. Math. **12** (2014), no. 2, 175–211, DOI 10.2478/s11533-013-0335-4. MR3130678

[BGT11] Emmanuel Breuillard, Ben Green, and Terence Tao, *Suzuki groups as expanders*, Groups Geom. Dyn. **5** (2011), no. 2, 281–299, DOI 10.4171/GGD/128. MR2782174

[BGT12] Emmanuel Breuillard, Ben Green, and Terence Tao, *The structure of approximate groups*, Publ. Math. Inst. Hautes Études Sci. **116** (2012), 115–221, DOI 10.1007/s10240-012-0043-9. MR3090256

[BK88] T. R. Berger and R. Knörr, *On Brauer's height* 0 *conjecture*, Nagoya Math. J. **109** (1988), 109–116. MR931954

[BLO03] Carles Broto, Ran Levi, and Bob Oliver, *The homotopy theory of fusion systems*, J. Amer. Math. Soc. **16** (2003), no. 4, 779–856, DOI 10.1090/S0894-0347-03-00434-X. MR1992826

[BLS97] László Babai, Eugene M. Luks, and Ákos Seress, *Fast management of permutation groups. I*, SIAM J. Comput. **26** (1997), no. 5, 1310–1342, DOI 10.1137/S0097539794229417. MR1471984

[BP03] Robert W. Baddeley and Cheryl E. Praeger, *On primitive overgroups of quasiprimitive permutation groups*, J. Algebra **263** (2003), no. 2, 294–344, DOI 10.1016/S0021-8693(03)00113-3. MR1978653

[BPS07] Robert W. Baddeley, Cheryl E. Praeger, and Csaba Schneider, *Quasiprimitive groups and blow-up decompositions*, J. Algebra **311** (2007), no. 1, 337–351, DOI 10.1016/j.jalgebra.2006.11.006. MR2309892

[Bre14] Emmanuel Breuillard, *A brief introduction to approximate groups*, Thin groups and superstrong approximation, Math. Sci. Res. Inst. Publ., vol. 61, Cambridge Univ. Press, Cambridge, 2014, pp. 23–50. MR3220883

[Bro83] M. Broué, *La Z_p^*-conjecture de Glauberman*, Séminaire sur les Groupes Finis, vol. 1, Univ. Paris VII, 1983, (not in MathSciNet).

[Bro94] Kenneth S. Brown, *Cohomology of groups*, Graduate Texts in Mathematics, vol. 87, Springer-Verlag, New York, 1994. Corrected reprint of the 1982 original. MR1324339

[Bro00] Kenneth S. Brown, *The coset poset and probabilistic zeta function of a finite group*, J. Algebra **225** (2000), no. 2, 989–1012, DOI 10.1006/jabr.1999.8221. MR1741574

[BS08a] David J. Benson and Stephen D. Smith, *Classifying spaces of sporadic groups*, Mathematical Surveys and Monographs, vol. 147, American Mathematical Society, Providence, RI, 2008. MR2378355

[BS08b] R. A. Bryce and L. Serena, *Some remarks on groups with nilpotent minimal covers*, J. Aust. Math. Soc. **85** (2008), no. 3, 353–365, DOI 10.1017/S1446788708000670. MR2476446

[Bue79] Francis Buekenhout, *Diagrams for geometries and groups*, J. Combin. Theory Ser. A **27** (1979), no. 2, 121–151, DOI 10.1016/0097-3165(79)90041-4. MR542524

[BW04] Lowell W. Beineke and Robin J. Wilson (eds.), *Topics in algebraic graph theory*, Encyclopedia of Mathematics and its Applications, vol. 102, Cambridge University Press, Cambridge, 2004. MR2125091

[Cab88] Marc Cabanes, *Brauer morphism between modular Hecke algebras*, J. Algebra **115** (1988), no. 1, 1–31, DOI 10.1016/0021-8693(88)90280-3. MR937599

[Cab13] Marc Cabanes, *Two remarks on the reduction of Alperin's weight conjecture*, Bull. Lond. Math. Soc. **45** (2013), no. 5, 895–906, DOI 10.1112/blms/bdt012. MR3104980

[Cam81] Peter J. Cameron, *Finite permutation groups and finite simple groups*, Bull. London Math. Soc. **13** (1981), no. 1, 1–22, DOI 10.1112/blms/13.1.1. MR599634

[Cam99] Peter J. Cameron, *Permutation groups*, London Mathematical Society Student Texts, vol. 45, Cambridge University Press, Cambridge, 1999. MR1721031

[Car89] Roger W. Carter, *Simple groups of Lie type*, Wiley Classics Library, John Wiley & Sons, Inc., New York, 1989. Reprint of the 1972 original; A Wiley-Interscience Publication. MR1013112

[Car93] Roger W. Carter, *Finite groups of Lie type*, Wiley Classics Library, John Wiley & Sons, Ltd., Chichester, 1993. Conjugacy classes and complex characters; Reprint of the 1985 original; A Wiley-Interscience Publication. MR1266626

[CCN+85] J. H. Conway, R. T. Curtis, S. P. Norton, R. A. Parker, and R. A. Wilson, *Atlas of finite groups*, Oxford University Press, Eynsham, 1985. Maximal subgroups and ordinary characters for simple groups; With computational assistance from J. G. Thackray. MR827219

[Che13] Andrew Chermak, *Fusion systems and localities*, Acta Math. **211** (2013), no. 1, 47–139, DOI 10.1007/s11511-013-0099-5. MR3118305

[CNT82] Peter J. Cameron, Peter M. Neumann, and David N. Teague, *On the degrees of primitive permutation groups*, Math. Z. **180** (1982), no. 2, 141–149, DOI 10.1007/BF01318900. MR661693

[Con71] J. H. Conway, *Three lectures on exceptional groups*, Finite simple groups (Proc. Instructional Conf., Oxford, 1969), Academic Press, London, 1971, pp. 215–247. MR0338152

[Coo78] Bruce N. Cooperstein, *An enemies list for factorization theorems*, Comm. Algebra **6** (1978), no. 12, 1239–1288, DOI 10.1080/00927877808822291. MR499351

[Coo01] B. N. Cooperstein, *On the generation of some embeddable* GF(2) *geometries*, J. Algebraic Combin. **13** (2001), no. 1, 15–28, DOI 10.1023/A:1008719600293. MR1817701

[CPSS83] P. J. Cameron, C. E. Praeger, J. Saxl, and G. M. Seitz, *On the Sims conjecture and distance transitive graphs*, Bull. London Math. Soc. **15** (1983), no. 5, 499–506, DOI 10.1112/blms/15.5.499. MR705530

[CR90] Charles W. Curtis and Irving Reiner, *Methods of representation theory. Vol. I*, Wiley Classics Library, John Wiley & Sons, Inc., New York, 1990. With applications to finite groups and orders; Reprint of the 1981 original; A Wiley-Interscience Publication. MR1038525

[Cra11] David A. Craven, *The theory of fusion systems*, Cambridge Studies in Advanced Mathematics, vol. 131, Cambridge University Press, Cambridge, 2011. An algebraic approach. MR2808319

[CST89] Peter J. Cameron, Ron Solomon, and Alexandre Turull, *Chains of subgroups in symmetric groups*, J. Algebra **127** (1989), no. 2, 340–352, DOI 10.1016/0021-8693(89)90256-1. MR1028457

[Cur70] C. W. Curtis, *Modular representations of finite groups with split* (B, N)*-pairs*, Seminar on Algebraic Groups and Related Finite Groups (The Institute for Advanced Study, Princeton, N.J., 1968/69), Springer, Berlin, 1970, pp. 57–95. MR0262383

[Cur71] C. W. Curtis, *Chevalley groups and related topics*, Finite simple groups (Proc. Instructional Conf., Oxford, 1969), Academic Press, London, 1971, pp. 135–189. MR0340442

[Dav05] Brian Davies, *Whither mathematics?*, Notices Amer. Math. Soc. **52** (2005), no. 11, 1350–1356. MR2183872

[DGS85] A. Delgado, D. Goldschmidt, and B. Stellmacher, *Groups and graphs: new results and methods*, DMV Seminar, vol. 6, Birkhäuser Verlag, Basel, 1985. With a preface by the authors and Bernd Fischer. MR862622

[Dix69] John D. Dixon, *The probability of generating the symmetric group*, Math. Z. **110** (1969), 199–205, DOI 10.1007/BF01110210. MR0251758

[DM96] John D. Dixon and Brian Mortimer, *Permutation groups*, Graduate Texts in Mathematics, vol. 163, Springer-Verlag, New York, 1996. MR1409812

[DP13] Dieter Degrijse and Nansen Petrosyan, *$\mathfrak{F}$-structures and Bredon-Galois cohomology*, Appl. Categ. Structures **21** (2013), no. 6, 545–586, DOI 10.1007/s10485-012-9281-8. MR3127361

[DSC98] P. Diaconis and L. Saloff-Coste, *Walks on generating sets of groups*, Invent. Math. **134** (1998), no. 2, 251–299, DOI 10.1007/s002220050265. MR1650316

[Dyn52] E. B. Dynkin, *Semisimple subalgebras of semisimple Lie algebras* (Russian), Mat. Sbornik N.S. **30(72)** (1952), 349–462 (3 plates). MR0047629

[Fei80] Walter Feit, *Some consequences of the classification of finite simple groups*, The Santa Cruz Conference on Finite Groups (Univ. California, Santa Cruz, Calif., 1979), Proc. Sympos. Pure Math., vol. 37, Amer. Math. Soc., Providence, R.I., 1980, pp. 175–181. MR604576

[Fei82] Walter Feit, *The representation theory of finite groups*, North-Holland Mathematical Library, vol. 25, North-Holland Publishing Co., Amsterdam-New York, 1982. MR661045

[FKS81] Burton Fein, William M. Kantor, and Murray Schacher, *Relative Brauer groups. II*, J. Reine Angew. Math. **328** (1981), 39–57. MR636194

[FM00] Daniel Frohardt and Kay Magaard, *Grassmannian fixed point ratios*, Geom. Dedicata **82** (2000), no. 1-3, 21–104, DOI 10.1023/A:1005278605358. MR1789058

[FM01] Daniel Frohardt and Kay Magaard, *Composition factors of monodromy groups*, Ann. of Math. (2) **154** (2001), no. 2, 327–345, DOI 10.2307/3062099. MR1865973

[FMS08] Clara Franchi, Mario Mainardis, and Ronald Solomon, *A characterization of HN*, J. Group Theory **11** (2008), no. 3, 357–369, DOI 10.1515/JGT.2008.020. MR2419006

[FS73] Paul Fong and Gary M. Seitz, *Groups with a (B, N)-pair of rank 2. I, II*, Invent. Math. **21** (1973), 1–57; ibid. 24 (1974), 191–239, DOI 10.1007/BF01389689. MR0354858

[GG83] Robert H. Gilman and Robert L. Griess Jr., *Finite groups with standard components of Lie type over fields of characteristic two*, J. Algebra **80** (1983), no. 2, 383–516, DOI 10.1016/0021-8693(83)90007-8. MR691810

[GHM10] David J. Green, László Héthelyi, and Nadia Mazza, *On Oliver's p-group conjecture. II*, Math. Ann. **347** (2010), no. 1, 111–122, DOI 10.1007/s00208-009-0435-4. MR2593286

[GHM11] David J. Green, László Héthelyi, and Nadia Mazza, *On a strong form of Oliver's p-group conjecture*, J. Algebra **342** (2011), 1–15, DOI 10.1016/j.jalgebra.2011.05.025. MR2824526

[GL83] Daniel Gorenstein and Richard Lyons, *The local structure of finite groups of characteristic 2 type*, Mem. Amer. Math. Soc. **42** (1983), no. 276, vii+731, DOI 10.1090/memo/0276. MR690900

[GL13] David J. Green and Justin Lynd, *Weak closure and Oliver's p-group conjecture*, Israel J. Math. **197** (2013), no. 1, 497–507, DOI 10.1007/s11856-013-0010-2. MR3096626

[GL16] George Glauberman and Justin Lynd, *Control of fixed points and existence and uniqueness of centric linking systems*, Invent. Math. **206** (2016), no. 2, 441–484, DOI 10.1007/s00222-016-0657-5. MR3570297

[Gla66] George Glauberman, *Central elements in core-free groups*, J. Algebra **4** (1966), 403–420, DOI 10.1016/0021-8693(66)90030-5. MR0202822

[GLP+16] Michael Giudici, Martin W. Liebeck, Cheryl E. Praeger, Jan Saxl, and Pham Huu Tiep, *Arithmetic results on orbits of linear groups*, Trans. Amer. Math. Soc. **368** (2016), no. 4, 2415–2467, DOI 10.1090/tran/6373. MR3449244

[GLS94] Daniel Gorenstein, Richard Lyons, and Ronald Solomon, *The classification of the finite simple groups*, Mathematical Surveys and Monographs, vol. 40, American Mathematical Society, Providence, RI, 1994. MR1303592

[GLS96] Daniel Gorenstein, Richard Lyons, and Ronald Solomon, *The classification of the finite simple groups. Number 2. Part I. Chapter G*, Mathematical Surveys and Monographs, vol. 40, American Mathematical Society, Providence, RI, 1996. General group theory. MR1358135

[GLS98] Daniel Gorenstein, Richard Lyons, and Ronald Solomon, *The classification of the finite simple groups. Number 3. Part I. Chapter A*, Mathematical Surveys and Monographs, vol. 40, American Mathematical Society, Providence, RI, 1998. Almost simple K-groups. MR1490581

[GLS99] Daniel Gorenstein, Richard Lyons, and Ronald Solomon, *The classification of the finite simple groups. Number 4. Part II. Chapters 1–4*, Mathematical Surveys and Monographs, vol. 40, American Mathematical Society, Providence, RI, 1999. Uniqueness theorems; With errata: *The classification of the finite simple groups. Number 3. Part I. Chapter A* [Amer. Math. Soc., Providence, RI, 1998; MR1490581 (98j:20011)]. MR1675976

[GM02] Robert M. Guralnick and Gunter Malle, *Classification of $2F$-modules. I*, J. Algebra **257** (2002), no. 2, 348–372, DOI 10.1016/S0021-8693(02)00526-4. MR1947326

[Gor80] Daniel Gorenstein, *Finite groups*, 2nd ed., Chelsea Publishing Co., New York, 1980. MR569209

[Gor82] Daniel Gorenstein, *Finite simple groups*, University Series in Mathematics, Plenum Publishing Corp., New York, 1982. An introduction to their classification. MR698782

[Gor83] Daniel Gorenstein, *Finite simple groups*, University Series in Mathematics, Plenum Publishing Corp., New York, 1982. An introduction to their classification. MR698782

[GR93] Robert M. Guralnick and Geoffrey R. Robinson, *On extensions of the Baer-Suzuki theorem*, Israel J. Math. **82** (1993), no. 1-3, 281–297, DOI 10.1007/BF02808114. MR1239051

[Gre12] Ben Green, *What is… an approximate group?*, Notices Amer. Math. Soc. **59** (2012), no. 5, 655–656, DOI 10.1090/noti829. MR2954294

[Gri72] Robert L. Griess Jr., *Schur multipliers of the known finite simple groups*, Bull. Amer. Math. Soc. **78** (1972), 68–71, DOI 10.1090/S0002-9904-1972-12855-6. MR0289635

[Gri98] Robert L. Griess Jr., *Twelve sporadic groups*, Springer Monographs in Mathematics, Springer-Verlag, Berlin, 1998. MR1707296

[Gro82] Fletcher Gross, *Automorphisms which centralize a Sylow p-subgroup*, J. Algebra **77** (1982), no. 1, 202–233, DOI 10.1016/0021-8693(82)90287-3. MR665174

[Gro02] Jesper Grodal, *Higher limits via subgroup complexes*, Ann. of Math. (2) **155** (2002), no. 2, 405–457, DOI 10.2307/3062122. MR1906592

[GT90] Robert M. Guralnick and John G. Thompson, *Finite groups of genus zero*, J. Algebra **131** (1990), no. 1, 303–341, DOI 10.1016/0021-8693(90)90178-Q. MR1055011

[Gur83] Robert M. Guralnick, *Subgroups of prime power index in a simple group*, J. Algebra **81** (1983), no. 2, 304–311, DOI 10.1016/0021-8693(83)90190-4. MR700286

[Gur92] Robert M. Guralnick, *The genus of a permutation group*, Groups, combinatorics & geometry (Durham, 1990), London Math. Soc. Lecture Note Ser., vol. 165, Cambridge Univ. Press, Cambridge, 1992, pp. 351–363, DOI 10.1017/CBO9780511629259.031. MR1200273

[Gur17] ———, *Applications of the Classification of Finite Simple Groups*, Proceedings of the International Congress of Mathematicians. Seoul 2014, Seemingly these volumes will be published in 2017?, Viewable online, see pp165-177 at URL: http://www.icm2014.org/download/Proceedings_Volume_II.pdf.

[Hal76] Marshall Hall Jr., *The theory of groups*, Chelsea Publishing Co., New York, 1976. Reprinting of the 1968 edition. MR0414669

[Hal06] J. I. Hall, *Periodic simple groups of finitary linear transformations*, Ann. of Math. (2) **163** (2006), no. 2, 445–498, DOI 10.4007/annals.2006.163.445. MR2199223

[Hal07] J. I. Hall, *Locally finite simple Moufang loops*, Turkish J. Math. **31** (2007), no. suppl., 45–61. MR2368082

[Hel15] Harald A. Helfgott, *Growth in groups: ideas and perspectives*, Bull. Amer. Math. Soc. (N.S.) **52** (2015), no. 3, 357–413, DOI 10.1090/S0273-0979-2015-01475-8. MR3348442

[Her85] Christoph Hering, *Transitive linear groups and linear groups which contain irreducible subgroups of prime order. II*, J. Algebra **93** (1985), no. 1, 151–164, DOI 10.1016/0021-8693(85)90179-6. MR780488

[HH56] P. Hall and Graham Higman, *On the p-length of p-soluble groups and reduction theorems for Burnside's problem*, Proc. London Math. Soc. (3) **6** (1956), 1–42, DOI 10.1112/plms/s3-6.1.1. MR0072872

[Hol78] D. F. Holt, *Transitive permutation groups in which an involution central in a Sylow 2-subgroup fixes a unique point*, Proc. London Math. Soc. (3) **37** (1978), no. 1, 165–192, DOI 10.1112/plms/s3-37.1.165. MR0575516

[HR92] Derek F. Holt and Sarah Rees, *An implementation of the Neumann-Praeger algorithm for the recognition of special linear groups*, Experiment. Math. **1** (1992), no. 3, 237–242. MR1203878

[Hum78] James E. Humphreys, *Introduction to Lie algebras and representation theory*, Graduate Texts in Mathematics, vol. 9, Springer-Verlag, New York-Berlin, 1978. Second printing, revised. MR499562

[IKS95] I. M. Isaacs, W. M. Kantor, and N. Spaltenstein, *On the probability that a group element is p-singular*, J. Algebra **176** (1995), no. 1, 139–181, DOI 10.1006/jabr.1995.1238. MR1345299

[IMN07] I. M. Isaacs, Gunter Malle, and Gabriel Navarro, *A reduction theorem for the McKay conjecture*, Invent. Math. **170** (2007), no. 1, 33–101, DOI 10.1007/s00222-007-0057-y. MR2336079

[IS94] A. A. Ivanov and S. V. Shpectorov, *The flag-transitive tilde and Petersen-type geometries are all known*, Bull. Amer. Math. Soc. (N.S.) **31** (1994), no. 2, 173–184, DOI 10.1090/S0273-0979-1994-00511-7. MR1256977

[Isa06] I. Martin Isaacs, *Character theory of finite groups*, AMS Chelsea Publishing, Providence, RI, 2006. Corrected reprint of the 1976 original [Academic Press, New York; MR0460423]. MR2270898

[Iva94] A. A. Ivanov, *Distance-transitive graphs and their classification*, Investigations in algebraic theory of combinatorial objects, Math. Appl. (Soviet Ser.), vol. 84, Kluwer Acad. Publ., Dordrecht, 1994, pp. 283–378, DOI 10.1007/978-94-017-1972-8_10. MR1321634

[IY91] Nobuo Iiyori and Hiroyoshi Yamaki, *On a conjecture of Frobenius*, Bull. Amer. Math. Soc. (N.S.) **25** (1991), no. 2, 413–416, DOI 10.1090/S0273-0979-1991-16084-2. MR1102753

[Jam78] G. D. James, *The representation theory of the symmetric groups*, Lecture Notes in Mathematics, vol. 682, Springer, Berlin, 1978. MR513828

[Jan03] Jens Carsten Jantzen, *Representations of algebraic groups*, 2nd ed., Mathematical Surveys and Monographs, vol. 107, American Mathematical Society, Providence, RI, 2003. MR2015057

[JK81] Gordon James and Adalbert Kerber, *The representation theory of the symmetric group*, Encyclopedia of Mathematics and its Applications, vol. 16, Addison-Wesley Publishing Co., Reading, Mass., 1981. With a foreword by P. M. Cohn; With an introduction by Gilbert de B. Robinson. MR644144

[JLPW81] Christoph Jansen, Klaus Lux, Richard Parker, and Robert Wilson, *An atlas of Brauer characters*, London Mathematical Society Monographs. New Series, vol. 11, The Clarendon Press, Oxford University Press, New York, 1995. Appendix 2 by T. Breuer and S. Norton; Oxford Science Publications. MR1367961

[Kan79] William M. Kantor, *Subgroups of classical groups generated by long root elements*, Trans. Amer. Math. Soc. **248** (1979), no. 2, 347–379, DOI 10.2307/1998972. MR522265

[Kan85] William M. Kantor, *Some consequences of the classification of finite simple groups*, Finite groups—coming of age (Montreal, Que., 1982), Contemp. Math., vol. 45, Amer. Math. Soc., Providence, RI, 1985, pp. 159–173, DOI 10.1090/conm/045/822238. MR822238

[Kan87] William M. Kantor, *Primitive permutation groups of odd degree, and an application to finite projective planes*, J. Algebra **106** (1987), no. 1, 15–45, DOI 10.1016/0021-8693(87)90019-6. MR878466

[Kel05] Thomas Michael Keller, *The $k(GV)$-problem revisited*, J. Aust. Math. Soc. **79** (2005), no. 2, 257–276, DOI 10.1017/S144678870001048X. MR2176347

[Kem01] Gregor Kemper, *The depth of invariant rings and cohomology*, J. Algebra **245** (2001), no. 2, 463–531, DOI 10.1006/jabr.2001.8840. With an appendix by Kay Magaard. MR1863889

[KL88] Peter B. Kleidman and Martin W. Liebeck, *A survey of the maximal subgroups of the finite simple groups*, Geom. Dedicata **25** (1988), no. 1-3, 375–389, DOI 10.1007/BF00191933. Geometries and groups (Noordwijkerhout, 1986). MR925843

[KL90a] William M. Kantor and Alexander Lubotzky, *The probability of generating a finite classical group*, Geom. Dedicata **36** (1990), no. 1, 67–87, DOI 10.1007/BF00181465. MR1065213

[KL90b] Peter Kleidman and Martin Liebeck, *The subgroup structure of the finite classical groups*, London Mathematical Society Lecture Note Series, vol. 129, Cambridge University Press, Cambridge, 1990. MR1057341

[Kle97] Alexander Kleshchev, *On decomposition numbers and branching coefficients for symmetric and special linear groups*, Proc. London Math. Soc. (3) **75** (1997), no. 3, 497–558, DOI 10.1112/S0024611597000427. MR1466660

[KLN06] Martin Kassabov, Alexander Lubotzky, and Nikolay Nikolov, *Finite simple groups as expanders*, Proc. Natl. Acad. Sci. USA **103** (2006), no. 16, 6116–6119, DOI 10.1073/pnas.0510337103. MR2221038

[KLT87] W. M. Kantor, R. A. Liebler, and J. Tits, *On discrete chamber-transitive automorphism groups of affine buildings*, Bull. Amer. Math. Soc. (N.S.) **16** (1987), no. 1, 129–133, DOI 10.1090/S0273-0979-1987-15487-5. MR866031

[KM13] Radha Kessar and Gunter Malle, *Quasi-isolated blocks and Brauer's height zero conjecture*, Ann. of Math. (2) **178** (2013), no. 1, 321–384, DOI 10.4007/annals.2013.178.1.6. MR3043583

[Kov85] L. G. Kovács, *Two results on wreath products*, Arch. Math. (Basel) **45** (1985), no. 2, 111–115, DOI 10.1007/BF01270481. MR807442

[KR89] Reinhard Knörr and Geoffrey R. Robinson, *Some remarks on a conjecture of Alperin*, J. London Math. Soc. (2) **39** (1989), no. 1, 48–60, DOI 10.1112/jlms/s2-39.1.48. MR989918

[Lie87a] Martin W. Liebeck, *The affine permutation groups of rank three*, Proc. London Math. Soc. (3) **54** (1987), no. 3, 477–516, DOI 10.1112/plms/s3-54.3.477. MR879395

[Lie87b] Martin W. Liebeck, *The classification of finite simple Moufang loops*, Math. Proc. Cambridge Philos. Soc. **102** (1987), no. 1, 33–47, DOI 10.1017/S0305004100067025. MR886433

[Lie95] M. W. Liebeck, *Subgroups of simple algebraic groups and of related finite and locally finite groups of Lie type*, Finite and locally finite groups (Istanbul, 1994), NATO Adv. Sci. Inst. Ser. C Math. Phys. Sci., vol. 471, Kluwer Acad. Publ., Dordrecht, 1995, pp. 71–96. MR1362806

[Lie13] Martin W. Liebeck, *Probabilistic and asymptotic aspects of finite simple groups*, Probabilistic group theory, combinatorics, and computing, Lecture Notes in Math., vol. 2070, Springer, London, 2013, pp. 1–34, DOI 10.1007/978-1-4471-4814-2_1. MR3026185

[LM91] Alexander Lubotzky and Avinoam Mann, *On groups of polynomial subgroup growth*, Invent. Math. **104** (1991), no. 3, 521–533, DOI 10.1007/BF01245088. MR1106747

[LMS93] Alexander Lubotzky, Avinoam Mann, and Dan Segal, *Finitely generated groups of polynomial subgroup growth*, Israel J. Math. **82** (1993), no. 1-3, 363–371, DOI 10.1007/BF02808118. MR1239055

[LOST10] Martin W. Liebeck, E. A. O'Brien, Aner Shalev, and Pham Huu Tiep, *The Ore conjecture*, J. Eur. Math. Soc. (JEMS) **12** (2010), no. 4, 939–1008, DOI 10.4171/JEMS/220. MR2654085

[LPS87] Martin W. Liebeck, Cheryl E. Praeger, and Jan Saxl, *A classification of the maximal subgroups of the finite alternating and symmetric groups*, J. Algebra **111** (1987), no. 2, 365–383, DOI 10.1016/0021-8693(87)90223-7. MR916173

[LPS88] Martin W. Liebeck, Cheryl E. Praeger, and Jan Saxl, *On the O'Nan-Scott theorem for finite primitive permutation groups*, J. Austral. Math. Soc. Ser. A **44** (1988), no. 3, 389–396. MR929529

[LS74] Vicente Landazuri and Gary M. Seitz, *On the minimal degrees of projective representations of the finite Chevalley groups*, J. Algebra **32** (1974), 418–443, DOI 10.1016/0021-8693(74)90150-1. MR0360852

[LS85] Martin W. Liebeck and Jan Saxl, *The primitive permutation groups of odd degree*, J. London Math. Soc. (2) **31** (1985), no. 2, 250–264, DOI 10.1112/jlms/s2-31.2.250. MR809946

[LS90] Martin W. Liebeck and Gary M. Seitz, *Maximal subgroups of exceptional groups of Lie type, finite and algebraic*, Geom. Dedicata **35** (1990), no. 1-3, 353–387, DOI 10.1007/BF00147353. MR1066572

[LS91] Martin W. Liebeck and Jan Saxl, *Minimal degrees of primitive permutation groups, with an application to monodromy groups of covers of Riemann surfaces*, Proc.

London Math. Soc. (3) **63** (1991), no. 2, 266–314, DOI 10.1112/plms/s3-63.2.266. MR1114511

[LS92] Martin W. Liebeck and Jan Saxl, *Maximal subgroups of finite simple groups and their automorphism groups*, Proceedings of the International Conference on Algebra, Part 1 (Novosibirsk, 1989), Contemp. Math., vol. 131, Amer. Math. Soc., Providence, RI, 1992, pp. 243–259. MR1175777

[LS94] Martin W. Liebeck and Gary M. Seitz, *Subgroups generated by root elements in groups of Lie type*, Ann. of Math. (2) **139** (1994), no. 2, 293–361, DOI 10.2307/2946583. MR1274094

[LS95] Martin W. Liebeck and Aner Shalev, *The probability of generating a finite simple group*, Geom. Dedicata **56** (1995), no. 1, 103–113, DOI 10.1007/BF01263616. MR1338320

[LS98a] Martin W. Liebeck and Gary M. Seitz, *On the subgroup structure of classical groups*, Invent. Math. **134** (1998), no. 2, 427–453, DOI 10.1007/s002220050270. MR1650328

[LS98b] Martin W. Liebeck and Gary M. Seitz, *On the subgroup structure of exceptional groups of Lie type*, Trans. Amer. Math. Soc. **350** (1998), no. 9, 3409–3482, DOI 10.1090/S0002-9947-98-02121-7. MR1458329

[LS99a] Martin W. Liebeck and Gary M. Seitz, *On finite subgroups of exceptional algebraic groups*, J. Reine Angew. Math. **515** (1999), 25–72, DOI 10.1515/crll.1999.078. MR1717629

[LS99b] Martin W. Liebeck and Aner Shalev, *Simple groups, permutation groups, and probability*, J. Amer. Math. Soc. **12** (1999), no. 2, 497–520, DOI 10.1090/S0894-0347-99-00288-X. MR1639620

[LS03] Martin W. Liebeck and Gary M. Seitz, *A survey of maximal subgroups of exceptional groups of Lie type*, Groups, combinatorics & geometry (Durham, 2001), World Sci. Publ., River Edge, NJ, 2003, pp. 139–146, DOI 10.1142/9789812564481_0008. MR1994964

[LS04] Martin W. Liebeck and Gary M. Seitz, *The maximal subgroups of positive dimension in exceptional algebraic groups*, Mem. Amer. Math. Soc. **169** (2004), no. 802, vi+227, DOI 10.1090/memo/0802. MR2044850

[LSS87] Martin W. Liebeck, Jan Saxl, and Gary M. Seitz, *On the overgroups of irreducible subgroups of the finite classical groups*, Proc. London Math. Soc. (3) **55** (1987), no. 3, 507–537, DOI 10.1112/plms/s3-55.3.507. MR907231

[LST96] Martin W. Liebeck, Jan Saxl, and Donna M. Testerman, *Simple subgroups of large rank in groups of Lie type*, Proc. London Math. Soc. (3) **72** (1996), no. 2, 425–457, DOI 10.1112/plms/s3-72.2.425. MR1367085

[LST11] Michael Larsen, Aner Shalev, and Pham Huu Tiep, *The Waring problem for finite simple groups*, Ann. of Math. (2) **174** (2011), no. 3, 1885–1950, DOI 10.4007/annals.2011.174.3.10. MR2846493

[Lub94] Alexander Lubotzky, *Discrete groups, expanding graphs and invariant measures*, Progress in Mathematics, vol. 125, Birkhäuser Verlag, Basel, 1994. With an appendix by Jonathan D. Rogawski. MR1308046

[Luc99] Maria Silvia Lucido, *Prime graph components of finite almost simple groups*, Rend. Sem. Mat. Univ. Padova **102** (1999), 1–22. MR1739529

[Luk82] Eugene M. Luks, *Isomorphism of graphs of bounded valence can be tested in polynomial time*, J. Comput. System Sci. **25** (1982), no. 1, 42–65, DOI 10.1016/0022-0000(82)90009-5. MR685360

[Luk87] E. M. Luks, *Computing the composition factors of a permutation group in polynomial time*, Combinatorica **7** (1987), no. 1, 87–99, DOI 10.1007/BF02579204. MR905155

[LY79] Robert A. Liebler and Jay E. Yellen, *In search of nonsolvable groups of central type*, Pacific J. Math. **82** (1979), no. 2, 485–492. MR551705

[Lyn14] Justin Lynd, *The Thompson-Lyons transfer lemma for fusion systems*, Bull. Lond. Math. Soc. **46** (2014), no. 6, 1276–1282, DOI 10.1112/blms/bdu083. MR3291264

[Lyo72] Richard Lyons, *A characterization of the group* $U_3(4)$, Trans. Amer. Math. Soc. **164** (1972), 371–387, DOI 10.2307/1995982. MR0286881

[Mei95] U. Meierfrankenfeld, *Non-finitary locally finite simple groups*, Finite and locally finite groups (Istanbul, 1994), NATO Adv. Sci. Inst. Ser. C Math. Phys. Sci., vol. 471,

Kluwer Acad. Publ., Dordrecht, 1995, pp. 189–212, DOI 10.1007/978-94-011-0329-9_7. MR1362810

[Miy77] Masahiko Miyamoto, *On conjugation families*, Hokkaido Math. J. **6** (1977), no. 1, 46–51, DOI 10.14492/hokmj/1381758562. MR0442086

[MO09] John Maginnis and Silvia Onofrei, *On fixed point sets and Lefschetz modules for sporadic simple groups*, J. Pure Appl. Algebra **213** (2009), no. 6, 901–912, DOI 10.1016/j.jpaa.2008.09.011. MR2498784

[MS90] Avinoam Mann and Dan Segal, *Uniform finiteness conditions in residually finite groups*, Proc. London Math. Soc. (3) **61** (1990), no. 3, 529–545, DOI 10.1112/plms/s3-61.3.529. MR1069514

[MS16] Gunter Malle and Britta Späth, *Characters of odd degree*, Ann. of Math. (2) **184** (2016), no. 3, 869–908, DOI 10.4007/annals.2016.184.3.6. MR3549625

[MSS03] Ulrich Meierfrankenfeld, Bernd Stellmacher, and Gernot Stroth, *Finite groups of local characteristic p: an overview*, Groups, combinatorics & geometry (Durham, 2001), World Sci. Publ., River Edge, NJ, 2003, pp. 155–192, DOI 10.1142/9789812564481_0010. MR1994966

[MSW94] Gunter Malle, Jan Saxl, and Thomas Weigel, *Generation of classical groups*, Geom. Dedicata **49** (1994), no. 1, 85–116, DOI 10.1007/BF01263536. MR1261575

[MT11] Gunter Malle and Donna Testerman, *Linear algebraic groups and finite groups of Lie type*, Cambridge Studies in Advanced Mathematics, vol. 133, Cambridge University Press, Cambridge, 2011. MR2850737

[Mun84] James R. Munkres, *Elements of algebraic topology*, Addison-Wesley Publishing Company, Menlo Park, CA, 1984. MR755006

[MW15a] Kay Magaard and Rebecca Waldecker, *Transitive permutation groups where nontrivial elements have at most two fixed points*, J. Pure Appl. Algebra **219** (2015), no. 4, 729–759, DOI 10.1016/j.jpaa.2014.04.027. MR3282107

[MW15b] Kay Magaard and Rebecca Waldecker, *Transitive permutation groups with trivial four point stabilizers*, J. Group Theory **18** (2015), no. 5, 687–740, DOI 10.1515/jgth-2015-0016. MR3393412

[Net64] Eugen Netto, *The theory of substitutions and its applications to algebra*, Second edition. Revised by the author and translated with his permission by F. N. Cole, Chelsea Publishing Co., New York, 1964. MR0175908

[NP92] Peter M. Neumann and Cheryl E. Praeger, *A recognition algorithm for special linear groups*, Proc. London Math. Soc. (3) **65** (1992), no. 3, 555–603, DOI 10.1112/plms/s3-65.3.555. MR1182102

[NP10] Alice C. Niemeyer and Cheryl E. Praeger, *Estimating proportions of elements in finite groups of Lie type*, J. Algebra **324** (2010), no. 1, 122–145, DOI 10.1016/j.jalgebra.2010.03.018. MR2646035

[NPS13] Alice C. Niemeyer, Cheryl E. Praeger, and Ákos Seress, *Estimation problems and randomised group algorithms*, Probabilistic group theory, combinatorics, and computing, Lecture Notes in Math., vol. 2070, Springer, London, 2013, pp. 35–82, DOI 10.1007/978-1-4471-4814-2_2. MR3026186

[NS14] Gabriel Navarro and Britta Späth, *On Brauer's height zero conjecture*, J. Eur. Math. Soc. (JEMS) **16** (2014), no. 4, 695–747, DOI 10.4171/JEMS/444. MR3191974

[NT11] Gabriel Navarro and Pham Huu Tiep, *A reduction theorem for the Alperin weight conjecture*, Invent. Math. **184** (2011), no. 3, 529–565, DOI 10.1007/s00222-010-0295-2. MR2800694

[NT13] Gabriel Navarro and Pham Huu Tiep, *Characters of relative p'-degree over normal subgroups*, Ann. of Math. (2) **178** (2013), no. 3, 1135–1171, DOI 10.4007/annals.2013.178.3.7. MR3092477

[Oli04] Bob Oliver, *Equivalences of classifying spaces completed at odd primes*, Math. Proc. Cambridge Philos. Soc. **137** (2004), no. 2, 321–347, DOI 10.1017/S0305004104007728. MR2092063

[Oli06] Bob Oliver, *Equivalences of classifying spaces completed at the prime two*, Mem. Amer. Math. Soc. **180** (2006), no. 848, vi+102, DOI 10.1090/memo/0848. MR2203209

[Ono11] Silvia Onofrei, *Saturated fusion systems with parabolic families*, J. Algebra **348** (2011), 61–84, DOI 10.1016/j.jalgebra.2011.09.028. MR2852232

[OV09] Bob Oliver and Joana Ventura, *Saturated fusion systems over 2-groups*, Trans. Amer. Math. Soc. **361** (2009), no. 12, 6661–6728, DOI 10.1090/S0002-9947-09-04881-8. MR2538610

[PLN97] Cheryl E. Praeger, Cai Heng Li, and Alice C. Niemeyer, *Finite transitive permutation groups and finite vertex-transitive graphs*, Graph symmetry (Montreal, PQ, 1996), NATO Adv. Sci. Inst. Ser. C Math. Phys. Sci., vol. 497, Kluwer Acad. Publ., Dordrecht, 1997, pp. 277–318. MR1468792

[Pra83] Cheryl E. Praeger, *Finite simple groups and finite primitive permutation groups*, Bull. Austral. Math. Soc. **28** (1983), no. 3, 355–365, DOI 10.1017/S0004972700021080. MR729768

[Pra90] Cheryl E. Praeger, *The inclusion problem for finite primitive permutation groups*, Proc. London Math. Soc. (3) **60** (1990), no. 1, 68–88, DOI 10.1112/plms/s3-60.1.68. MR1023805

[Pra93] Cheryl E. Praeger, *An O'Nan-Scott theorem for finite quasiprimitive permutation groups and an application to 2-arc transitive graphs*, J. London Math. Soc. (2) **47** (1993), no. 2, 227–239, DOI 10.1112/jlms/s2-47.2.227. MR1207945

[Pra97] Cheryl E. Praeger, *Finite quasiprimitive graphs*, Surveys in combinatorics, 1997 (London), London Math. Soc. Lecture Note Ser., vol. 241, Cambridge Univ. Press, Cambridge, 1997, pp. 65–85, DOI 10.1017/CBO9780511662119.005. MR1477745

[PS] Cheryl E. Praeger and Csaba Schneider, *Permutation groups and Cartesian decompositions*, in preparation.

[PS11] Chris Parker and Gernot Stroth, *Strongly p-embedded subgroups*, Pure Appl. Math. Q. **7** (2011), no. 3, Special Issue: In honor of Jacques Tits, 797–858, DOI 10.4310/PAMQ.2011.v7.n3.a9. MR2848592

[PS14] Chris Parker and Gernot Stroth, $F_4(2)$ *and its automorphism group*, J. Pure Appl. Algebra **218** (2014), no. 5, 852–878, DOI 10.1016/j.jpaa.2013.10.005. MR3149638

[PSY87] Cheryl E. Praeger, Jan Saxl, and Kazuhiro Yokoyama, *Distance transitive graphs and finite simple groups*, Proc. London Math. Soc. (3) **55** (1987), no. 1, 1–21, DOI 10.1112/plms/s3-55.1.1. MR887281

[Pui06] Lluis Puig, *Frobenius categories*, J. Algebra **303** (2006), no. 1, 309–357, DOI 10.1016/j.jalgebra.2006.01.023. MR2253665

[Pui11] Lluis Puig, *On the reduction of Alperin's conjecture to the quasi-simple groups*, J. Algebra **328** (2011), 372–398, DOI 10.1016/j.jalgebra.2010.11.004. MR2745572

[Qui78] Daniel Quillen, *Homotopy properties of the poset of nontrivial p-subgroups of a group*, Adv. in Math. **28** (1978), no. 2, 101–128, DOI 10.1016/0001-8708(78)90058-0. MR493916

[Rob88] Geoffrey R. Robinson, *Some remarks on permutation modules*, J. Algebra **118** (1988), no. 1, 46–62, DOI 10.1016/0021-8693(88)90047-6. MR961325

[Rob90] Geoffrey R. Robinson, *The* Z_p^**-theorem and units in blocks*, J. Algebra **134** (1990), no. 2, 353–355, DOI 10.1016/0021-8693(90)90058-V. MR1074334

[Rob09] Geoffrey R. Robinson, *Central units in blocks and the odd* Z_p^**-theorem*, J. Algebra **321** (2009), no. 2, 384–393, DOI 10.1016/j.jalgebra.2008.10.017. MR2483272

[Rob11] Geoffrey R. Robinson, *On simple endotrivial modules*, Bull. Lond. Math. Soc. **43** (2011), no. 4, 712–716, DOI 10.1112/blms/bdr006. MR2820156

[Ron89] Mark Ronan, *Lectures on buildings*, Perspectives in Mathematics, vol. 7, Academic Press, Inc., Boston, MA, 1989. MR1005533

[Row81] Peter Rowley, *3-locally central elements in finite groups*, Proc. London Math. Soc. (3) **43** (1981), no. 2, 357–384, DOI 10.1112/plms/s3-43.2.357. MR628282

[RS80] M. A. Ronan and S. D. Smith, *2-local geometries for some sporadic groups*, The Santa Cruz Conference on Finite Groups (Univ. California, Santa Cruz, Calif., 1979), Proc. Sympos. Pure Math., vol. 37, Amer. Math. Soc., Providence, R.I., 1980, pp. 283–289. MR604595

[RS85] Mark A. Ronan and Stephen D. Smith, *Sheaves on buildings and modular representations of Chevalley groups*, J. Algebra **96** (1985), no. 2, 319–346, DOI 10.1016/0021-8693(85)90013-4. MR810532

[RS89] Mark A. Ronan and Stephen D. Smith, *Computation of 2-modular sheaves and representations for* $L_4(2), A_7, 3S_6,$ *and* M_{24}, Comm. Algebra **17** (1989), no. 5, 1199–1237. MR993399

[RSY90] A. J. E. Ryba, Stephen D. Smith, and Satoshi Yoshiara, *Some projective modules determined by sporadic geometries*, J. Algebra **129** (1990), no. 2, 279–311, DOI 10.1016/0021-8693(90)90222-A. MR1040940

[RT96] Geoffrey R. Robinson and John G. Thompson, *On Brauer's $k(B)$-problem*, J. Algebra **184** (1996), no. 3, 1143–1160, DOI 10.1006/jabr.1996.0304. MR1407890

[RV04] Albert Ruiz and Antonio Viruel, *The classification of p-local finite groups over the extraspecial group of order p^3 and exponent p*, Math. Z. **248** (2004), no. 1, 45–65, DOI 10.1007/s00209-004-0652-1. MR2092721

[Sag01] Bruce E. Sagan, *The symmetric group*, 2nd ed., Graduate Texts in Mathematics, vol. 203, Springer-Verlag, New York, 2001. Representations, combinatorial algorithms, and symmetric functions. MR1824028

[Sch07] Peter Schmid, *The solution of the $k(GV)$ problem*, ICP Advanced Texts in Mathematics, vol. 4, Imperial College Press, London, 2007. MR2369608

[Sco80] Leonard L. Scott, *Representations in characteristic p*, The Santa Cruz Conference on Finite Groups (Univ. California, Santa Cruz, Calif., 1979), Proc. Sympos. Pure Math., vol. 37, Amer. Math. Soc., Providence, R.I., 1980, pp. 319–331. MR604599

[Sei73] Gary M. Seitz, *Flag-transitive subgroups of Chevalley groups*, Ann. of Math. (2) **97** (1973), 27–56, DOI 10.2307/1970876. MR0340446

[Sei83] Gary M. Seitz, *The root subgroups for maximal tori in finite groups of Lie type*, Pacific J. Math. **106** (1983), no. 1, 153–244. MR694680

[Sei87] Gary M. Seitz, *The maximal subgroups of classical algebraic groups*, Mem. Amer. Math. Soc. **67** (1987), no. 365, iv+286, DOI 10.1090/memo/0365. MR888704

[Sei91] Gary M. Seitz, *Maximal subgroups of exceptional algebraic groups*, Mem. Amer. Math. Soc. **90** (1991), no. 441, iv+197, DOI 10.1090/memo/0441. MR1048074

[Smi80] Stephen D. Smith, *A characterization of orthogonal groups over* GF(2), J. Algebra **62** (1980), no. 1, 39–60, DOI 10.1016/0021-8693(80)90204-5. MR561116

[Smi11] Stephen D. Smith, *Subgroup complexes*, Mathematical Surveys and Monographs, vol. 179, American Mathematical Society, Providence, RI, 2011. MR2850680

[Ste68] Robert Steinberg, *Lectures on Chevalley groups*, Yale University, New Haven, Conn., 1968. Notes prepared by John Faulkner and Robert Wilson. MR0466335

[Tay92] Donald E. Taylor, *The geometry of the classical groups*, Sigma Series in Pure Mathematics, vol. 9, Heldermann Verlag, Berlin, 1992. MR1189139

[Tes88] Donna M. Testerman, *Irreducible subgroups of exceptional algebraic groups*, Mem. Amer. Math. Soc. **75** (1988), no. 390, iv+190, DOI 10.1090/memo/0390. MR961210

[Tho66] John G. Thompson, *Factorizations of p-solvable groups*, Pacific J. Math. **16** (1966), 371–372. MR0188296

[Tho68] John G. Thompson, *Nonsolvable finite groups all of whose local subgroups are solvable*, Bull. Amer. Math. Soc. **74** (1968), 383–437, DOI 10.1090/S0002-9904-1968-11953-6. MR0230809

[Tho69] John G. Thompson, *A replacement theorem for p-groups and a conjecture*, J. Algebra **13** (1969), 149–151, DOI 10.1016/0021-8693(69)90068-4. MR0245683

[Tie14] Pham Huu Tiep, *Representation of finite groups: conjectures, reductions, and applications*, Acta Math. Vietnam. **39** (2014), no. 1, 87–109, DOI 10.1007/s40306-013-0043-y. MR3176464

[Tim73] F. G. Timmesfeld, *A characterization of the Chevalley- and Steinberg-groups over F_2*, Geometriae Dedicata **1** (1973), no. 3, 269–321, DOI 10.1007/BF00147764. MR0330279

[Tim75] F. G. Timmesfeld, *Groups generated by root-involutions. I, II*, J. Algebra **33** (1975), 75–134; ibid. 35 (1975), 367–441, DOI 10.1016/0021-8693(75)90133-7. MR0372019

[Tim83] Franz G. Timmesfeld, *Tits geometries and parabolic systems in finitely generated groups. I*, Math. Z. **184** (1983), no. 3, 377–396, DOI 10.1007/BF01163511. MR716284

[Tit74] Jacques Tits, *Buildings of spherical type and finite BN-pairs*, Lecture Notes in Mathematics, Vol. 386, Springer-Verlag, Berlin-New York, 1974. MR0470099

[Tit81] J. Tits, *A local approach to buildings*, The geometric vein, Springer, New York-Berlin, 1981, pp. 519–547. MR661801

[Tob16] Imke Toborg, *Local arguments for a theorem of Z_3^*-type*, J. Algebra **452** (2016), 338–371, DOI 10.1016/j.jalgebra.2015.12.020. MR3461071

[TW02] Jacques Tits and Richard M. Weiss, *Moufang polygons*, Springer Monographs in Mathematics, Springer-Verlag, Berlin, 2002. MR1938841

[Var84] V. S. Varadarajan, *Lie groups, Lie algebras, and their representations*, Graduate Texts in Mathematics, vol. 102, Springer-Verlag, New York, 1984. Reprint of the 1974 edition. MR746308

[vB07] John van Bon, *Finite primitive distance-transitive graphs*, European J. Combin. **28** (2007), no. 2, 517–532, DOI 10.1016/j.ejc.2005.04.014. MR2287450

[Wal13] Rebecca Waldecker, *Isolated involutions in finite groups*, Mem. Amer. Math. Soc. **226** (2013), no. 1061, vi+150, DOI 10.1090/S0065-9266-2013-00684-3. MR3136255

[Web87] P. J. Webb, *Subgroup complexes*, The Arcata Conference on Representations of Finite Groups (Arcata, Calif., 1986), Proc. Sympos. Pure Math., vol. 47, Amer. Math. Soc., Providence, RI, 1987, pp. 349–365. MR933372

[Wei69] Edwin Weiss, *Cohomology of groups*, Pure and Applied Mathematics, Vol. 34, Academic Press, New York-London, 1969. MR0263900

[Whi00] Julius Whiston, *Maximal independent generating sets of the symmetric group*, J. Algebra **232** (2000), no. 1, 255–268, DOI 10.1006/jabr.2000.8399. MR1783924

[Wil86] R. A. Wilson, *Maximal subgroups of sporadic groups*, Proceedings of groups—St. Andrews 1985, London Math. Soc. Lecture Note Ser., vol. 121, Cambridge Univ. Press, Cambridge, 1986, pp. 352–358. MR896534

[Wil09] Robert A. Wilson, *The finite simple groups*, Graduate Texts in Mathematics, vol. 251, Springer-Verlag London, Ltd., London, 2009. MR2562037

[Wil17] R. A. Wilson, *Maximal subgroups of sporadic groups*, Proceedings of groups—St. Andrews 1985, London Math. Soc. Lecture Note Ser., vol. 121, Cambridge Univ. Press, Cambridge, 1986, pp. 352–358. MR896534

[Zel91] E. I. Zel′manov, *Solution of the restricted Burnside problem for 2-groups* (Russian), Mat. Sb. **182** (1991), no. 4, 568–592; English transl., Math. USSR-Sb. **72** (1991), no. 2, 543–565. MR1119009

[Zem54] Richard Anthony Zemlin, *ON A CONJECTURE ARISING FROM A THEOREM OF FROBENIUS*, ProQuest LLC, Ann Arbor, MI, 1954. Thesis (Ph.D.)–The Ohio State University. MR2612955

[Zha94] Ji Ping Zhang, *Studies on defect groups*, J. Algebra **166** (1994), no. 2, 310–316, DOI 10.1006/jabr.1994.1153. MR1279259

Index

Page locations for definitions, as well as for references which are particularly fundamental, are indicated in **boldface**.